MONOGRAPHS ON STATISTICS AND APPLIED PROBABILITY

General Editors

D.R. Cox, V. Isham, N. Keiding, N. Reid, H. Tong, and T. Louis

(Full details concerning this series are available from the Publishers).

Stochastic Geometry
Likelihood and Computation

Edited by

O. Barndorff-Nielsen

Professor of Theoretical Statistics
Institute of Mathematics
Aarhus
Denmark

W. Kendall

Professor of Statistics
University of Warwick
UK

M.N.M. Lieshout

Senior Researcher
Centre for Mathematics and Computer Science (CWI)
Amsterdam
The Netherlands

CHAPMAN & HALL/CRC

Boca Raton   London   New York   Washington, D.C.

Library of Congress Cataloging-in-Publication Data

Catalog record is available from the Library of Congress.

Contents

List of Contributors

Prof A J Baddeley
Department of Mathematics
University of Western Australia
Nedlands WA 6009, AUSTRALIA
Email: adrian@maths.uwa.edu.au

Prof C Geyer
Department Statistics
Univ Minnesota
Vincent Hall
Minneapolis MN 55455, USA
Email: charlie@stat.umn.edu

Prof J Møller
Department of Mathematics and Computer Science
Aalborg University
Fredrik Bajers Vej 7E
DK-9220 Aalborg, DENMARK
Email: jm@iesd.auc.dk

Dr L Vincent
1450 Oak Creek Drive Apt 206
Palo Alto, CA 94304, USA
Email: lucv@adoc.xerox.com

Dr I Molchanov
Department Statistics
University of Glasgow
Glasgow G12 8QW, Scotland, UK
Email: ilya@stats.gla.ac.uk

Dr I Dryden
Department of Statistics
University of Leeds
Leeds LS2 9JT, UK
Email: iand@amsta.leeds.ac.uk

Prof Laurent Saloff-Coste
Laboratoire de Stat et Prob
Université Paul Sabatier
118 rue de Narbonne
31062 Toulouse, FRANCE
Email: lsc@corail.cict.fr

Preface

The chapters in this volume represent the revised versions of the main papers given at the third Séminaire Européen de Statistique on 'Stochastic Geometry, Theory and Applications', held at Université Paul Sabatier, Toulouse, 13–18 May 1996. The aim of the Séminaire Européen de Statistique is to provide talented young researchers with an opportunity to get quickly to the forefront of knowledge and research in areas of statistical science which are of current major interest. As for the books based on the first two seminars in the series (*Networks and Chaos – Statistical and Probabilistic Aspects* and *Time Series Models*. In econometrics, finance and other fields), this volume is of tutorial character. In the present Séminaire about 40 young scientists from fifteen European countries participated. Nearly all participants gave short presentations about their recent work; these, while of high quality, are not reproduced here.

The first chapter of the book, written by A.J. Baddeley, is an introduction to the basic concepts which permeate stochastic geometry and thus the rest of the volume, particularly the vivid interplay between geometrical and probabilistic intuitions which is characteristic of the subject. In the second chapter Baddeley goes on to describe sampling issues relating to observations of spatial patterns; because the pattern is invariably observed through a window one must pay close attention to edge effects and resulting biases due to size dependence and censoring.

Inference for spatial statistics has been revolutionized by the advent of increased computer power and specifically by the possibility of doing likelihood inference using Markov Chain Monte Carlo (MCMC). In the third chapter C.J. Geyer introduces MCMC theory and practice with special reference to problems of inference for spatial point processes. J. Møller follows this up in chapter 4 with a discussion of more specific issues to do with MCMC for spatial point processes, including the application of very recent ideas of 'exact simulation' due to Propp and Wilson. Møller broadens the theme in chapter 5 by describing what stochastic geometry has to say about random sub-divisions of space ('tessellations') rather than point patterns.

The next three chapters move on from point patterns and sub-divisions to consider various kinds of random sets. In chapter 6, L. Vincent surveys the field of mathematical morphology, which studies the theory and algorithmic implementation of automatic manipulation and measurement of digital and continuous images, a fundamental concern for all kinds of image analysis. Chapter 7, by I.S. Molchanov, covers a related but more mathematical concern, surveying results and open problems in the theory of random closed sets. This in turn opens up the question of shape, and how to do statistics when the location, orientation and size of one's dataset are not of interest. I.L. Dryden shows in chapter 8 how the ideas of this rapidly advancing area can be viewed in a regression framework.

It is characteristic of stochastic geometry that it draws from a wide range of pure and applied mathematical science. The recent advances in inferential technique due to MCMC require us to take seriously the issues of Markov chain convergence to equilibrium, which has for a long time been the subject of attention of stochastic analysts. L. Saloff-Coste reports on this in chapter 9, which concludes this volume. Advances in this area place heavy mathematical demands on readers, and yet the great importance of MCMC means applied workers must still have a general acquaintance with the main results; Saloff-Coste comes to our rescue by providing simple finite-state-space examples which illustrate the fundamental issues.

It is inherent in a book of this nature that limitations of time and space mean that it can be only a partial snapshot of the discipline it covers. Omissions here include stereology (hugely important in practice), more general random processes modelling for example fibre and surface patterns, and exciting developments relating stochastic geometry to fractals. Nevertheless the coverage provided by this volume will enable readers to gain a perspective on contemporary concerns in current theoretical and applied research in stochastic geometry; it is the hope and expectation of the authors and editors that it will thus be of good service to the scientific community.

The third Séminaire Européen de Statistique was organized by O.E. Barndorff-Nielsen, Aarhus University; D. Bakry, Université Paul Sabatier, Toulouse; M. Casalis, Université Paul Sabatier, Toulouse; W.S. Kendall, University of Warwick; C. Klüppelberg, Technical University, Munich; G. Letac, Université Paul Sabatier, Toulouse; and M.N.M. van Lieshout, University of Warwick (now moved to CWI Amsterdam). The smooth running of the Séminaire was greatly aided by the enthusiastic administrative assistance provided by Catherine Price and Elke Thönnes, both studying for PhDs at Warwick at that time. The series of Séminaires Européen de Statistique has now been adopted by the European Regional Committee of the Bernoulli Society. Future Séminaires are currently being planned by the steering group of the series under its new chair Claudia Klüppelberg.

The third Séminaire Européen de Statistique was supported by the European Communities under an HCM Euroconferences contract (ERB CHEC CT93 0107) and a PECO East European participation contract (ERB CIPD CT94 0703), for which we express grateful thanks.

On behalf of the Organizers and Editors
W.S. Kendall
Warwick

CHAPTER 1

A crash course
in stochastic geometry

Adrian J. Baddeley
University of Western Australia

This chapter aims to give the reader a rapid introduction to the main ideas of stochastic geometry. It is not a literature review but rather a very selective presentation of some of the key points.

1.1 Introduction

Stochastic geometry is the study of random processes whose outcomes are geometrical objects or spatial patterns, that is, random subsets of $\mathbb{R}^d$ or some other given space. It has applications to digital image analysis, spatial statistics and stereology, and connections with many areas of probability and statistics. One of its most appealing, direct applications is to the analysis of spatial patterns, such as those shown in Figures 1.1 and 1.2.

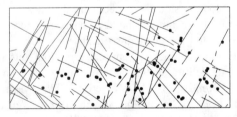

Figure 1.1 Left: *California redwood saplings (circles) in a 23 m square, extracted by Ripley (1981) from Strauss (1975).* Right: *copper deposits (dots) and geological faults (line segments) in a* 100×200 *km^2 region, from Berman (1986) Data reproduced by kind permission of Prof. B.D. Ripley and Dr. M. Berman.*

Figure 1.2 *Heather vegetation (black) and uncovered soil (white) in a 20 ×
10 m² sampling area, from a study by Diggle (1983). Data reproduced by kind
permission of Prof. P.J. Diggle.*

The earliest examples of random geometry were parlour games in
which a coin or stick is thrown haphazardly onto a flat surface, and the
gamble depends on the final location of the object thrown. From this
developed a theory of *geometrical probability*, concerned mainly with
problems in which rigid geometrical figures are randomly positioned in
the plane according to an appropriate uniform distribution or uniform
Poisson process. This is capable of modelling patterns such as those in
Figure 1.1. This classical theory, closely related to *integral geometry*,
reveals many fascinating connections between convex geometry and
probability. See Kendall and Moran (1963), Santaló (1976), Schneider
and Weil (1992), Solomon (1978).

Modern stochastic geometry handles random subsets of arbitrary
form, for example, the zero set of a random function, or a randomly-
generated fractal. It also deals with very general classes of probability models, such as stationary random sets in $\mathbb{R}^d$. These are capable
of modelling spatial patterns such as that in Figure 1.2. See Harding
and Kendall (1974), Matheron (1975), Mecke *et al.* (1990), Stoyan *et
al.* (1987), Stoyan and Stoyan (1992), Weil and Wieacker (1993) and
Cressie (1991, chap. 9), Serra (1982).

The fundamental idea of stochastic geometry is to find **connections
between geometry and probability**, i.e. between the geometrical and
probabilistic aspects of random spatial processes. This chapter identifies
some important instances of that idea.

1.2 Hitting probabilities and geometrical measures

The probability that a random set X intersects or 'hits' a fixed set B can often be related to a geometrical property of B, such as its length or area.

Example 1.1 (Uniform random point) Let X be a random point in the unit square $[0, 1]^2$ which we generate by assigning random values to the Cartesian coordinates X_1, X_2, independent and uniformly distributed on $[0, 1]$. The joint probability density of the coordinates is thus $f(x_1, x_2) = 1$ for $0 < x_1, x_2 < 1$ and zero otherwise. For any subset $B \subseteq [0, 1]^2$, the probability that X falls in B is

$$\mathbb{P}\{X \in B\} = \int_B f(x_1, x_2) \, dx_1 \, dx_2 = |B|$$

where $|B|$ denotes the area of B. In this case, probabilities are directly connected with measures of area.

More generally let $A \subset \mathbb{R}^2$ have finite positive area $|A|$. Let X be a random point uniformly distributed in A, i.e. the Cartesian coordinates of X have joint probability density

$$f(x_1, x_2) = \begin{cases} \frac{1}{|A|} & \text{if } (x_1, x_2) \in A \\ 0 & \text{otherwise} \end{cases}$$

Then for $B \subset A$

$$\mathbb{P}\{X \in B\} = \frac{|B|}{|A|}. \tag{1.1}$$

Example 1.2 (Uniform random lines) Any straight line in $\mathbb{R}^2$ is uniquely identified by its orientation θ and signed distance p from the origin, as

$$L_{\theta, p} = \{(x, y) : x \cos\theta + y \sin\theta = p\}$$

with $\theta \in [0, \pi)$ and $p \in \mathbb{R}$. See Figure 1.3. Let X be a random straight line through the unit disc $b(0, 1)$, generated by taking θ and p to be independent and uniformly distributed over $[0, \pi)$ and $[-1, 1]$.

If B is a circle of radius r contained in $b(0, 1)$, the probability that X intersects B is

$$\mathbb{P}\{X \cap B \neq \emptyset\} = \int_0^\pi \int_{-1}^1 \mathbf{1}\{L_{\theta, p} \cap B \neq \emptyset\} \frac{1}{2\pi} \, dp \, d\theta$$

$$= \int_0^\pi \frac{2r}{2\pi} \, d\theta$$

$$= r$$

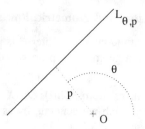

Figure 1.3 *The θ, p coordinates for a straight line in $\mathbb{R}^2$*

proportional to the length of the circumference of B, regardless of the location of B within $b(0, 1)$. Similarly if S is a line segment of length s and direction η contained in $b(0, 1)$, the probability that X hits S is

$$\mathbb{P}\{X \cap S \neq \emptyset\} = \int_0^\pi \int_{-1}^1 \mathbf{1}\{L_{\theta,p} \cap S \neq \emptyset\} \frac{1}{2\pi} \, dp \, d\theta$$

$$= \int_0^\pi s \, | \sin(\eta - \theta)| \frac{1}{2\pi} \, d\theta$$

$$= \frac{s}{\pi}$$

proportional to the length of S regardless of its orientation and location within $b(0, 1)$. That is, probabilities for X are related to measures of *length*.

More generally, for any convex, compact set $A \subset \mathbb{R}^2$ with nonempty interior, the measure of all lines intersecting A is

$$\int_0^\pi \int_{\mathbb{R}} \mathbf{1}\{L_{\theta,p} \cap A \neq \emptyset\} \, dp \, d\theta = \text{length}\,(\partial A)$$

equals the length of the boundary ∂A of A. Thus if A is compact and convex with nonempty interior, and X is a random line with (θ, p) coordinates jointly uniformly distributed, with density

$$f(\theta, p) = \begin{cases} \frac{1}{\text{length}(\partial A)} & \text{if } L_{\theta,p} \cap A \neq \emptyset \\ 0 & \text{otherwise} \end{cases}$$

then for $B \subseteq A$

$$\mathbb{P}\{X \cap B \neq \emptyset\} = \frac{\text{length}\,(\partial B)}{\text{length}\,(\partial A)} \qquad (1.2)$$

provided B is convex and has nonempty interior. For a line segment

$S \subset A$,

$$\mathbb{P}\{X \cap S \neq \emptyset\} = 2\,\frac{\text{length}(S)}{\text{length}(\partial A)} \tag{1.3}$$

which can be viewed as a special case of (1.2) if the boundary of S is taken to be a degenerate closed curve which traverses S twice.

These results are typical of stochastic geometry in that the hitting probabilities do not depend on the location of B within A, so that X can truly be said to sample A 'uniformly'; remarkably they do not depend on the shape of B; and the results hold only for one specific random set X which has a suitable 'uniform' distribution. The uniform distribution is a canonical one, characterised by the invariance property that hitting probabilities do not depend on location.

The calculation of hitting probabilities often calls for elegant geometrical arguments. For example, hitting probabilities exhibit geometrical duality in the sense that X intersects B if and only if B intersects X.

Example 1.3 (Uniform random disc) Write $b(x, r)$ for the disc of radius r centred at $x \in \mathbb{R}^2$. Let X be a random point uniformly distributed in $A \subset \mathbb{R}^2$, and consider the disc $Y = b(X, r)$ of fixed radius $r > 0$ centred at X.

Observe that Y covers a given point $y \in \mathbb{R}^2$ if and only if X belongs to $b(y, r)$. Hence the probability of covering y is

$$\mathbb{P}\{y \in Y\} = \frac{|b(y, r) \cap A|}{|A|}$$

More generally, consider the probability that the random disc Y hits a fixed set B. Let $B_{(+r)}$ be the set of all points lying at most r units distant from B,

$$B_{(+r)} = \{x \in \mathbb{R}^2 : d(x, B) \leq r\}$$

where

$$d(x, B) = \inf\{\|x - b\| : b \in B\}$$

is the shortest distance from a point x to B. Then Y intersects B if and only if $X \in B_{(+r)}$, see Figure 1.4. Thus

$$\mathbb{P}\{Y \cap B \neq \emptyset\} = \frac{|B_{(+r)} \cap A|}{|A|}$$

If $B_{(+r)} \subseteq A$ then this is

$$\mathbb{P}\{Y \cap B \neq \emptyset\} = \frac{|B_{(+r)}|}{|A|}$$

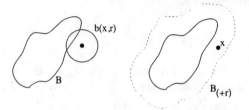

Figure 1.4 *The disc $b(x, r)$ intersects B iff $x \in B_{(+r)}$.*

and once again we have a hitting probability which does not depend on the location of B within some region. If additionally B is convex, then *Steiner's formula* states that

$$|B_{(+r)}| = |B| + r \, \text{length}\,(\partial B) + \pi \, r^2 \qquad (1.4)$$

so that the hitting probability depends on several geometrical characteristics of B and of Y.

There are many other results which express hitting probabilities as geometrical quantities, in the form

$$P\{X \cap B \neq \emptyset\} = \nu(B) \qquad (1.5)$$

where X is a random set with a specified distribution (usually 'uniform'), B is a fixed set, and ν is some geometrical functional. These results can be interpreted in several ways:

- as a simple, geometrical solution to a probability problem;
- as the basis of a statistical method for estimating $\nu(B)$ from random samples $X \cap B$;
- as a novel interpretation of the geometrical functional ν;
- as a characteristic property of the uniform distribution for X, for use in testing whether X is uniformly distributed.

These are broadly the approaches taken in *stochastic geometry, stereology, integral geometry* and *spatial statistics*, respectively.

1.3 Linearity of expectation

The results just mentioned can be extended greatly using the linearity of expectation, i.e. the property that $\mathbb{E}[X + Y] = \mathbb{E}[X] + \mathbb{E}[Y]$ for any random variables X, Y *whatever the dependence* between X and Y.

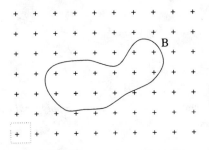

Figure 1.5 *A randomly translated square grid.*

Example 1.4 (Random grid of points) Consider a square grid of points in $\mathbb{R}^2$ with square side $a > 0$. A randomly-shifted copy of this grid is obtained by taking $X = \{(U + ma, V + na) : m, n \in \mathbb{Z}\}$ where the 'starting point' (U, V) is random and uniformly distributed in $A = [0, a) \times [0, a)$.

For any Borel set B in $\mathbb{R}^2$ with finite area, the number of points of X which fall in B has expectation

$$\mathbb{E}\left[\#(X \cap B)\right] = \frac{1}{a^2}|B|. \tag{1.6}$$

To prove this, note that $X_{m,n} = (U + ma, V + na)$ is uniformly distributed in the square $A_{m,n} = [ma, (m + 1)a) \times [na, (n + 1)a)$. Now

$$\#(X \cap B) = \sum_{m,n} \mathbf{1}\{X_{m,n} \in B\}$$

where $\mathbf{1}\{X_{m,n} \in B\}$ is the indicator variable which equals 1 if $X_{m,n} \in B$ and equals 0 otherwise. By the linearity of expectation,

$$\begin{aligned}
\mathbb{E}\left[\#(X \cap B)\right] &= \mathbb{E}\left[\sum_{m,n} \mathbf{1}\{X_{m,n} \in B\}\right] \\
&= \sum_{m,n} \mathbb{P}\{X_{m,n} \in B\} \\
&= \sum_{m,n} \frac{|B \cap A_{m,n}|}{|A_{m,n}|} \\
&= \frac{1}{a^2} \sum_{m,n} |B \cap A_{m,n}| \\
&= \frac{1}{a^2}|B|.
\end{aligned}$$

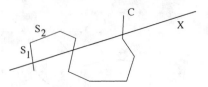

Figure 1.6 *Proof of* (1.7) *for a polygonal curve.*

This result is used in practice to estimate the area of a plane region, such as an area on a map or a shape in a photograph or microscope image. We superimpose over the region a transparent sheet marked with a regular grid of so-called *test points*, count the number of test points which fall in the region of interest, and multiply by the area of one grid square. By (1.6) this is an unbiased estimator of the area of the region. This is one of the basic techniques of *stereology*; see e.g. Baddeley (1991, 1993), Jensen (1997), Weibel (1979, 1980), Weil (1983), Stoyan *et al.* (1987, chap. 11) and Santaló (1976, chap. 16).

Again the result is typical of classical geometrical probability, in that it holds for essentially all B, but for only one specific random set X having a 'uniform' distribution. It gives only the expectation of $\#(X \cap B)$; this expectation does not depend on the shape of B because of the linearity of $\mathbb{E}$. Other properties of the distribution of $\#(X \cap B)$ are nigh intractable (however, some techniques have been developed; see Ambartzumian (1982)). For similar reasons, the majority of results in stochastic geometry concern first moments only.

Example 1.5 (Mean number of crossings) Let $A \subset \mathbb{R}^2$ be convex and compact, and X a uniformly distributed random line as in Example 1.2. Suppose $C \subset A$ is a rectifiable curve. Denote by $\#(X \cap C)$ the (random) number of intersection points between X and C. Then

$$\mathbb{E}\#(X \cap C) = 2 \, \frac{\text{length}\,(C)}{\text{length}\,(\partial A)}. \tag{1.7}$$

To prove this, first suppose C is a polygonal curve consisting of line segments $S_1, \ldots, S_n$. See Figure 1.6. Then (ignoring some special cases which have probability zero) the number of intersection points equals the number of segments S_i which are hit:

$$\#(X \cap C) = \sum_{i=1}^{n} \mathbf{1}\{X \cap S_i \neq \emptyset\}.$$

By the linearity of expectation,

$$\mathbb{E}\#(X \cap C) = \sum_{i=1}^{n} \mathbb{P}\{X \cap S_i \neq \emptyset\}$$

$$= 2\sum_{i=1}^{n} \frac{\text{length}\,(S_i)}{\text{length}\,(\partial A)}$$

$$= 2\frac{\text{length}\,(C)}{\text{length}\,(\partial A)}$$

from (1.3). This proves (1.7) for polygonal curves C, and by approximation, for arbitrary rectifiable curves.

This result (1.7) can also be used in practice to estimate the length of a plane curve; see e.g. Baddeley (1991), Jensen (1997), Weibel (1979, 1980), Weil (1983), Stoyan *et al.* (1987, chap. 11) and Santaló (1976, chap. 16).

The examples above give the mean number of intersection points between a random set and a fixed set. More generally one can find the mean size or 'content' of the intersection $X \cap B$ between a random set X and a fixed target set B:

$$\mathbb{E}\left[\mu(X \cap B)\right] = \nu(B) \tag{1.8}$$

where μ, ν are certain geometrical functionals. Such results can often be obtained using Fubini's theorem on exchanging the order of integration in a multiple integral. Fubini's theorem can be regarded as another form of the linearity of expectation.

Theorem 1.1 ('Robbins' theorem') Let X be a random set in $\mathbb{R}^d$, and μ a measure on $\mathbb{R}^d$. Then

$$\mathbb{E}\mu(X) = \int_{\mathbb{R}^d} \mathbb{P}\{x \in X\}\,d\mu(x) \tag{1.9}$$

provided X is measurable (as defined in the proof) and μ is σ-finite.

Proof. Consider the joint indicator function $\iota : \mathbb{R}^d \times \Omega \to \mathbb{R}$,

$$\iota(x, \omega) = \mathbf{1}\{x \in X(\omega)\}$$

where $(\Omega, \mathcal{F}, P)$ is the underlying probability space. If ι is jointly measurable in its arguments and μ is σ-finite then Fubini's theorem applies, so that

$$\int_{\Omega} \iota(x, \omega)\,dP(\omega) = \mathbb{P}\{x \in X\}$$

is a Borel measurable function of x,

$$\int_{\mathbb{R}^d} \iota(x, \omega) \, d\mu(x) = \mu(X(\omega))$$

is a measurable function of ω — i.e. $\mu(X)$ is a random variable — and (1.9) holds. $\square$

Equation (1.9) is due to Kolmogorov (1933, p. 41) and Robbins (1944 1945). Matheron (1975, Corollary 1, p. 47) established that any random closed set X satisfies the measurability condition. See also Kendall and Moran (1963), Bronowski and Neyman (1945), Solomon (1950, 1953), Baddeley and Molchanov (1997).

Example 1.6 (Random coverage) Let $Y = b(X, r)$ be a disc with fixed radius r and random centre X uniformly distributed in a plane region A. The probability that Y covers a point x is $\mathbb{P}\{x \in Y\} = |b(x, r) \cap A|/|A|$, as we saw in Example 1.3. Hence for any fixed $B \subseteq A$, applying Robbins' theorem to $Z = Y \cap B$,

$$\mathbb{E}|Y \cap B| = \int_{\mathbb{R}^2} \mathbb{P}\{x \in Y \cap B\} \, dx$$

$$= \int_B \mathbb{P}\{x \in Y\} \, dx$$

$$= \int_B \frac{|b(x, r) \cap A|}{|A|} \, dx.$$

In particular, if B is such that $B_{(+r)} \subseteq A$, then the integrand is equal to $\pi r^2/|A|$ for all $x \in B$ and

$$\mathbb{E}|Y \cap B| = \pi r^2 \frac{|B|}{|A|}.$$

Theorem 1.1 can also be used to calculate higher moments of the content of a random set. For example, if X is a random set in $\mathbb{R}^d$, applying Robbins' theorem to the product set $X \times X$ in $\mathbb{R}^{2d}$ yields

$$\mathbb{E}\left[|X \cap A|^2\right] = \int_A \int_A \mathbb{P}\{x \in X, y \in X\} \, dx \, dy$$

for $A \subset \mathbb{R}^d$.

A linearity result of a different type is Campbell's theorem for point processes (Daley and Vere-Jones 1988, p. 188, Reiss, 1993, p. 127 ff.). Recall that the *intensity measure* Λ of a point process X is defined by

$$\Lambda(B) = \mathbb{E}\left[\#(X \cap B)\right] \qquad (1.10)$$

for compact $B \subset \mathbb{R}^d$, i.e. $\Lambda(B)$ is the expected number of points of X falling in B. If X is stationary, for example if X is a homogeneous Poisson point process, then $\Lambda(B) = \lambda|B|$ where the constant λ is called the intensity.

Theorem 1.2 (Campbell's theorem) Let X be a point process in $\mathbb{R}^d$ with intensity measure Λ. Then for any measurable $f : \mathbb{R}^d \to \mathbb{R}$

$$\mathbb{E}\left[\sum_{x \in X} f(x)\right] = \int_{\mathbb{R}^d} f(x)\, d\Lambda(x) \qquad (1.11)$$

in the sense that if one side is finite then so is the other and they are equal.

In particular if X is stationary then (1.11) becomes

$$\mathbb{E}\left[\sum_{x \in X} f(x)\right] = \lambda \int_{\mathbb{R}^d} f(x)\, dx.$$

Proof. Equation (1.10) is the special case of (1.11) for functions $f(x) = 1\{x \in B\}$. By linearity of expectation, (1.11) holds for all linear combinations of such functions, i.e. for all step functions. Monotone convergence yields the result. $\square$

1.4 Integral geometry

'Mean content' formulae (1.8) are usually stochastic interpretations of identities of the type

$$\int_{\mathcal{T}} v(T \cap A)\, d\mu(T) = w(A) \qquad (1.12)$$

where v, w are geometrical functionals, A is a fixed set, $\mathcal{T}$ is a class of sets (such as the class of all lines in $\mathbb{R}^2$) and μ is a measure on $\mathcal{T}$. Typically μ is an appropriate analogue of Lebesgue measure, characterised by an invariance property, so that the left side of (1.12) represents 'uniform' integration over all T in $\mathcal{T}$. The intersection $T \cap A$ may be replaced by orthogonal projection or another geometrical operation.

Integral geometry (Santaló, 1976, Schneider and Weil, 1992) is the study of such formulae, particularly insofar as they provide information about the functional w. An important role is played by the criterion of invariance with respect to a group of transformations.

Let $F(n, r)$ be the set of all r-dimensional affine planes (i.e. planes not necessarily through the origin) in $\mathbb{R}^n$ for integers $0 \leq r < n$. It can be shown that, up to a constant factor, there is a unique measure

$\mu_{n,r}$ which is invariant under Euclidean rigid motions (i.e. under translations and rotations). For example $F(n, 0)$ is the set of points in $\mathbb{R}^n$ and Lebesgue measure is the unique measure on $\mathbb{R}^n$, up to a constant factor, which is invariant under translations and rotations. For the space $F(2, 1)$ of lines in $\mathbb{R}^2$, recall the (θ, p) parametrisation introduced in Example 1.2. While this parametrisation depends on the choice of origin in $\mathbb{R}^2$, it turns out that the measure $\mu_{2,1}$ with density $dp\, d\theta$ does not depend on the choice of origin, and is invariant under translations and rotations.

The most important functionals of integral geometry are the *quermass integrals* or *Minkowski functionals* defined for all compact convex sets $K \subset \mathbb{R}^n$ by

$$W_r^n(K) = \mu_{n,r}\{T \in F(n, r) : T \cap K \neq \emptyset\}$$

i.e. $W_r^n(K)$ is the invariant measure of the set of all r-planes intersecting K. The rth functional W_r^n is an $n - r$ dimensional measure of size: $W_0^n(K) = |K|$ is the Lebesgue volume of K and $W_{n-1}^n(K)$ is proportional to the $(n - 1)$ dimensional surface measure of ∂K. Conventionally $W_n^n(K) = 1$ for $K \neq \emptyset$. The Minkowski functionals W_r^n satisfy a '*section formula*'

$$\int_{F(n,s)} W_{s-(n-r)}^s(K \cap T)\, d\mu_{n,s}(T) = c_{n,r,s}\, W_r^n(K) \qquad (1.13)$$

for integers s and r with $0 < s < n$, $n - s \leq r \leq n$ and some constant $c_{n,r,s}$. This result, often called *Crofton's formula*, states that the rth quermassintegral of a set K in $\mathbb{R}^n$ can be obtained by taking the intersection of K with s-dimensional planes T, evaluating an appropriate quermassintegral of $K \cap T$, and integrating uniformly over all T. The integrand $W_{s-(n-r)}^s(K \cap T)$ is defined by identifying the s-dimensional plane T with $\mathbb{R}^s$ under any isometry.

The Minkowski functionals also satisfy a 'projection formula' (Cauchy's formula) which states for example that $W_r^n(K)$ for $0 < r < n$ is proportional to the average $(n - r)$-volume of the projection of K onto $(n - r)$-dimensional subspaces. They are also the coefficients in Steiner's formula:

$$|K_{(+t)}| = \sum_{r=0}^{n} t^r W_r^n(K), \qquad t \geq 0 \qquad (1.14)$$

(generalising (1.4)) for a compact convex set K, where $K_{(+t)}$ is the outer parallel set introduced in Example 1.3.

Integral geometric identities such as (1.13) are consequences of the coarea formula of geometric measure theory (Federer, 1969), or loosely speaking, of Fubini's theorem together with the Jacobian formula for a change of variables of integration. Classical integral geometry contains tools for computing the necessary Jacobians, or equivalently, for 'factorising' the 'densities' of invariant measures. See Santaló (1976), Jensen (1997).

Example 1.7 Lines in $\mathbb{R}^2$ can be parametrised by the coordinates (θ, p) defined in Example 1.2. Alternatively a line can be specified by the intercept x and angle α which it makes with the x-axis. The Jacobian of the map $(x, \alpha) \mapsto (\theta, p)$ is $|\sin \alpha|$, so we have the following equivalent 'factorisations' of the invariant density for lines:

$$dp \, d\theta = |\sin \alpha| \, dx \, d\alpha.$$

This result is useful in Example 1.20 below.

1.5 Independence, Poisson processes and Boolean models

Independent random variables are common in the study of one-dimensional random processes, but are scarcer in stochastic geometry. Random spatial processes generally do not exhibit complete independence, in the sense that two parts of the process $X \cap A$ and $X \cap B$ are usually dependent even if A and B are disjoint regions. Spatial structure induces stochastic dependence.

Independence arises most naturally in the context of a random process of geometrical figures X_i in which the X_i's can be assumed to be mutually independent.

The canonical 'completely random' spatial process is the uniform Poisson point process. See e.g. Daley and Vere-Jones (1988), Kingman (1993), Mecke *et al.* (1990), Miles (1970), Reiss (1993), Stoyan *et al.* (1987), Stoyan and Stoyan (1992). A *point process* in $\mathbb{R}^d$ is a random subset X for which the number $N(A) = \#(X \cap A)$ of points falling in a compact set $A \subset \mathbb{R}^d$ is almost surely finite. The *uniform Poisson point process* is characterised by

(PP1) $N(A)$ has a Poisson distribution with mean $\lambda|A|$, where the constant $\lambda > 0$ is the *intensity*;

(PP2) if $A_1, \ldots, A_m$ are disjoint subsets of $\mathbb{R}^d$ then $N(A_1,), \ldots,$ $N(A_m)$ are independent r.v.'s.

Important properties include

(conditional property) given $N(A) = n$, the n points falling in A are independent and uniformly distributed in A;

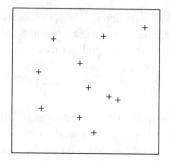
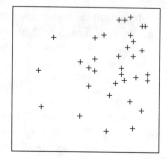

Figure 1.7 *Typical realisations of a Poisson point process. Left: uniform intensity; Right: non-uniform.*

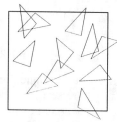
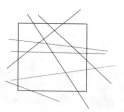

Figure 1.8 *Typical realisations of Poisson processes of d⁻cs, triangles and lines.*

(superposition property) the superposition (union) of independent Poisson point processes is again a Poisson point process, whose intensity is the sum of the intensities of the summands.

More generally the inhomogeneous Poisson point process with intensity measure Λ on a space S satisfies (PP2) and

(PP1′) $N(A)$ has a Poisson distribution with mean $\Lambda(A)$.

The superposition property continues to hold, and the conditional property becomes

Given $N(A) = n$, the n points falling in A are independent and identically distributed in A with probability distribution proportional to Λ, that is, $\mathbb{P}\{X_i \in B\} = \Lambda(B \cap A)/\Lambda(A)$.

Poisson processes of *geometrical objects* (such as lines and discs) are a popular object of study (see e.g. Kendall and Moran (1963), Kingman (1993), Matheron (1975), Miles (1964a, b, 1973, 1974), Serra (1982), Daley and Vere-Jones (1988, pp. 385–398), Reiss (1993, p. 193 ff.), Santaló (1976, pp. 56, 296). These may be constructed from Poisson point processes; calculations about them can be referred back to the underlying point process.

Example 1.8 (Poisson process of discs) Let X be a Poisson point process with intensity λ in $\mathbb{R}^2$. Replace each point x of X by a disc $b(x, r)$ of fixed radius $r > 0$ centred at that point. The result is a 'Poisson disc process' Y. From Example 1.3, for a fixed $A \subset \mathbb{R}^2$, a disc $b(x, r)$ intersects A iff $x \in A_{(+r)}$. Thus the number of discs of Y intersecting A equals the number $N(A_{(+r)})$ of points of X in $A_{(+r)}$; this number has a Poisson distribution with mean $\lambda |A_{(+r)}|$. For example the probability that A is not intersected by any disc is $\mathbb{P}\{N(A_{(+r)}) = 0\} = \exp(-\lambda |A_{(+r)}|)$.

Example 1.9 A Poisson process of lines can be regarded as a Poisson point process on the space $[0, \pi) \times \mathbb{R}$ of (θ, p) coordinate pairs introduced in Example 1.2. The number of lines intersecting any convex set $A \subset \mathbb{R}^2$ is a Poisson r.v. with mean $\lambda \text{length}\,(\partial A)$. If C is a rectifiable curve in $\mathbb{R}^2$ then, by Example 1.5 and Campbell's theorem 1.2, the expected total number of intersections between C and the lines of the Poisson line process is $2\lambda \text{length}\,(C)$.

A *Boolean model* is a random set formed by taking the set union of the individual objects in a Poisson process of geometrical objects. For example, taking the union of all the discs in a Poisson disc process (Example 1.8) would produce a Boolean model. Figure 1.9 shows a digital image which could plausibly be analysed as a realisation of such a random set. Boolean models have been extensively studied and applied to the analysis of real images, and still present a great many stimulating unsolved problems. See e.g. Hall (1988), Kendall and Moran (1963), Matheron (1975), Mecke *et al.* (1990), Molchanov (1994), Preteux and Schmitt (1988), Serra (1982), Stoyan *et al.* (1987), Cressie (1991, p. 753 ff.), Ripley (1988, chap. 6).

Frequently it is useful to augment a point process by attaching extra information (such as a 'colour') to each point. A *marked point process* on $\mathbb{R}^d$, with marks in a space S, is formally a point process on $\mathbb{R}^d \times S$ such that $N(A \times S) < \infty$ a.s. for compact $A \subset \mathbb{R}^d$. Thus a point $(x, s) \in \mathbb{R}^d \times S$ is viewed as a point $x \in \mathbb{R}^d$ labelled with a 'mark' $s \in S$.

Theorem 1.3 ('Marking Theorem', simplified version) *The following are equivalent*

- *a marked point process on $\mathbb{R}^d$ with marks in S, for which the underlying points are Poisson with intensity measure μ on $\mathbb{R}^d$, and*

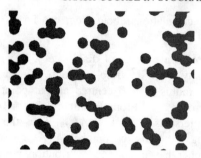

Figure 1.9 *Image from the impact plate of a nuclear accelerator reproduced from Kellerer (1985). A Boolean model would be a plausible starting point for the analysis of this image.*

conditional on the points, the marks are i.i.d. with probability distribution ν on S;

- a Poisson point process on $\mathbb{R}^d \times S$ with intensity measure of the form $\Lambda = \mu \otimes \nu$, i.e. $\Lambda(A \times B) = \mu(A)\,\nu(B)$ for $A \subset \mathbb{R}^d$, $B \subset S$, where μ, ν are measures on $\mathbb{R}^d$ and S respectively, and $\nu(S) = 1$.

See Kingman's (1993) excellent exposition of the Poisson process for further insights.

Example 1.10 (Random discs) A random process of discs with unequal radii can be formulated as a marked point process of centres x_i in $\mathbb{R}^2$ marked by the corresponding radii $r_i > 0$.

Assume the centres are uniform Poisson points with intensity λ and the radii are i.i.d. with distribution function F. Equivalently the pairs (x_i, r_i) form a Poisson process on $\mathbb{R}^2 \times (0, \infty)$ with intensity measure Λ satisfying $\Lambda(A \times [0, r]) = \lambda\,|A|\,F(r)$ for compact $A \subset \mathbb{R}^2$ and $r > 0$.

A disc $b(x, r)$ contains the origin 0 if and only if (x, r) belongs to

$$B = \{(x, r) : x \in b(0, r)\}.$$

By (PP1′) the number $N(B)$ of discs covering 0 is Poisson distributed, with mean

$$\begin{aligned}
\Lambda(B) &= \int_B \lambda\, dF(r)\, dx \\
&= \lambda \int_0^\infty |b(0, r)|\, dF(r) \\
&= \lambda \mathbb{E}\left[\pi R^2\right]
\end{aligned}$$

where R has distribution function F. Thus the expected number of discs covering a fixed point equals the intensity of centres times the expected area of a disc.

A corollary of Theorem 1.3 is the *thinning property* of the Poisson process, that if the points of (say) a uniform Poisson point process with intensity λ on $\mathbb{R}^d$ are randomly deleted or retained, independently for each point, with probability p of retention for each point, the resulting point process of retained points is Poisson with intensity $p\lambda$.

Some of the results presented here can be extended to general, non-Poisson processes; see e.g. Mecke *et al.* (1990), Stoyan *et al.* (1987), Stoyan and Stoyan (1992), Weil (1984).

1.6 Incidence and vacancy

If X is a random set and A a fixed set, we write

$$[A] = \{X \cap A \neq \emptyset\}$$

for the random event that X intersects A. The probabilities $\mathbb{P}([A])$ of such hitting events occupy a central place in stochastic geometry.

Incidence is 'one-sided' in the sense that

$$[A] \cup [B] = [A \cup B]$$

but

$$[A] \cap [B] \supseteq [A \cap B]$$

with *inequality in general*. This is reminiscent of the properties of minima and maxima of random variables.

Probabilities of complex events concerning a random set X can often be simplified into sums of hitting probabilities. For example

$$\mathbb{P}\{[A] \cup [B]\} = \mathbb{P}\{[A \cup B]\} \tag{1.15}$$

$$\mathbb{P}\{[A] \cap [B]\} = \mathbb{P}\{[A]\} + \mathbb{P}\{[B]\} - \mathbb{P}\{[A \cup B]\} \tag{1.16}$$

so that the probability of any combination of two hitting events can be expressed in terms of hitting probabilities. Special geometrical and combinatorial techniques apply to some types of random sets, including random lines (Ambartzumian, 1982) and pairs of random points (Sheng, 1985).

It is often natural to consider *vacancy events*

$$[A]^c = \{X \cap A = \emptyset\}$$

which satisfy

$$[A]^c \cap [B]^c = [A \cup B]^c.$$

In particular, stochastic independence is simpler to understand for vacancy events than for hitting events $[A]$.

Example 1.11 Let $X_1, \ldots, X_n$ be independent random points uniformly distributed in a plane region A, forming a random set $X = \{X_1, \ldots, X_n\}$. For $B \subset A$ the hitting probability $\mathbb{P}([B]) = \mathbb{P}\{X \cap B \neq \emptyset\}$ is the probability that some of the random points X_i fall in B. This is more easily computed using the complementary (vacancy) event $[B]^c$:

$$\mathbb{P}\{X \cap B \neq \emptyset\} = 1 - \mathbb{P}\{X \cap B = \emptyset\}$$

$$= 1 - \prod_{i=1}^{n} \mathbb{P}\{X_i \notin B\}$$

$$= 1 - \left[1 - \frac{|B|}{|A|}\right]^n.$$

Example 1.12 For the uniform Poisson point process with intensity λ,

$$\mathbb{P}\{[B]^c\} = \mathbb{P}\{N(B) = 0\} = \exp\{-\lambda |B|\}.$$

In fact, hitting probabilities (or avoidance probabilities) completely determine the entire distribution of a random set.

Theorem 1.4 (Choquet–D.G. Kendall-Matheron) *The distribution of a random closed set X in $\mathbb{R}^d$ is determined by its hitting probabilities*

$$T(K) = \mathbb{P}\{X \cap K \neq \emptyset\} \tag{1.17}$$

for all compact sets $K \subset \mathbb{R}^d$.

See Matheron (1975, Theorem 2-2-1, p. 30) for a precise statement. This result is plausible if we recall (1.15)–(1.16) and note that the avoidance probabilities

$$1 - T(K) = \mathbb{P}\{X \cap K = \emptyset\} = \mathbb{P}\{X \subseteq K^c\}$$

are analogous to the cumulative distribution function $F(x) = \mathbb{P}\{X \leq x\}$ which characterises the distribution of a real random variable X.

Theorem 1.4 is extremely useful, both in proofs and in applications. For example, it implies a random set X may also be characterised by the functional

$$\Psi(K) = -\log(1 - T(K)) = -\log \mathbb{P}\{X \cap K = \emptyset\} \tag{1.18}$$

which is useful because independence properties of X are translated into additivity properties of Ψ. For example if X is a Poisson point process of intensity λ then $T(K) = \mathbb{P}\{N(K) > 0\} = 1 - \exp(-\lambda |K|)$ so $\Psi(K) = \lambda |K|$. If X is a Poisson line process in $\mathbb{R}^2$ then $\Psi(K) = \lambda \text{length}(\partial K)$ for convex K. If X is a Boolean model formed from a Poisson process of compact convex sets, then there is an expression for $\Psi(K)$ generalising Steiner's formula (1.4), (1.14). This expression can be used to fit a Boolean model to image data such as Figure 1.9; see Hall (1988), Serra (1982), Stoyan *et al.* (1987), Stoyan and Stoyan (1992), Cressie (1991, p. 759 ff.), and Molchanov (1994), Preteux and Schmitt (1988) for related alternatives.

1.7 Distances and waiting times

If X is a random set in $\mathbb{R}^d$ and $y \in \mathbb{R}^d$ a fixed point, consider

$$R = d(y, X) = \inf\{\|x - y\| : x \in X\},$$

the shortest distance from y to X. The random variable R is like a 'waiting time' and can indeed be visualised as the time taken for an inflating balloon to first touch X when it starts at time zero as a sphere of zero diameter at the location y and inflates spherically at a uniform rate.

It is a fundamental fact, albeit trivial, that distance distributions are connected with hitting or vacancy probabilities. For a random closed set X, we have $R > r$ if and only if X does not intersect the ball $b(y, r)$ of radius r around y. Thus the distribution of R is given by

$$\mathbb{P}\{R \leq r\} = \mathbb{P}\{X \cap b(y, r) \neq \emptyset\} = T(b(y, r)).$$

First contact distances are typically exponential or gamma distributed, similar to lifetime distributions in one-dimensional stochastic processes (Miles, 1970, Møller Zuyev, 1996) For example for the Poisson point process in $\mathbb{R}^2$ we have $\mathbb{P}\{R > r\} = \exp(-\lambda \pi r^2)$ so that R^2 has an exponential distribution with parameter $(= 1/\text{mean}) \lambda \pi$. For the Poisson line process in $\mathbb{R}^2$, we have $\mathbb{P}\{R > r\} = \exp(-\lambda 2\pi r)$ so that R is exponential $(2\lambda \pi)$ distributed. For Boolean models composed of compact convex sets in $\mathbb{R}^2$, an appropriate quadratic function of R is exponentially distributed given $R > 0$.

1.8 Weighted distributions

Weighted distributions are ubiquitous in stochastic geometry. They arise from sampling bias effects, and as conditional or marginal distributions.

If (p_i) is a discrete probability distribution and (w_i) is a set of nonnegative weights, the corresponding 'w_i-weighted distribution' is (q_i) where $q_i = w_i p_i/(\sum_j w_j p_j)$. For a probability density $f(x)$ on $[0, \infty)$ and weight function $w(x) \geq 0$ the w-weighted density is $g(x) = w(x)f(x)/m$ provided $m = \int_0^\infty w(x)f(x)\,dx$ is finite and nonzero. In general if P is a probability measure and W a nonnegative r.v. on the same probability space with $0 < \mathbb{E}\,W < \infty$, the W-weighted distribution is the probability measure

$$P_W(A) = \frac{\mathbb{E}\,[\mathbf{1}_A\,W]}{\mathbb{E}\,W}$$

where $\mathbb{E}$ is expectation with respect to P, and $\mathbf{1}_A$ is the indicator r.v. of the event A. It is thus possible to speak of the area-weighted distribution of a random set X, and so forth.

1.8.1 Sampling bias

The act of selecting a geometrical object from amongst a population of objects usually introduces a sampling bias related to size.

Example 1.13 (Point-sampling) Let $B_1, \ldots, B_m$ be disjoint compact subsets of $\mathbb{R}^2$. Suppose we choose one of the B_i at random by generating a random point X uniformly distributed in $A = \bigcup_{i=1}^m B_i$ and identifying which object contains X. Let the random variable I be the index of the set hit by X, that is, $I = i$ where $X \in B_i$. Then

$$\mathbb{P}\{I = i\} = \mathbb{P}\{X \in B_i\} = \frac{|B_i|}{|A|} = \frac{|B_i|}{\sum_j |B_j|},$$

that is, the probability of selecting an object is proportional to its area, rather than each object having equal probability.

Example 1.14 (Point-sampling a Poisson disc process) Construct a Poisson process of discs of *random* radius by taking a uniform Poisson point process with intensity λ in $\mathbb{R}^2$ and replacing each point x_i by a disc $b(x_i, R_i)$ where the R_i are i.i.d. positive real r.v.'s with cumulative distribution function F.

Let us sample the process by selecting only the discs which contain the origin 0. A disc $b(x, r)$ contains 0 if and only if $x \in b(0, r)$. Referring to Example 1.10, the number of discs sampled is Poisson with mean $\lambda\mathbb{E}\left[\pi\,R^2\right]$ where R has distribution function F, and the radii of the sampled discs are independent and identically distributed, with distribution function

$$F_1(r) = \frac{\int_0^r \pi r^2 \, dF(r)}{\mathbb{E}\left[\pi R^2\right]},$$

i.e. the point-sampled discs are area-weighted with respect to the original disc distribution.

The 'bus paradox' is a familiar example from classical probability.

Example 1.15 (Bus paradox) Buses arrive at a bus stop at random times which form a Poisson point process on $\mathbb{R}$ with intensity λ. The bus company can truthfully report that the time interval between two successive buses has an exponential (λ) distribution with density

$$f(x) = \lambda e^{-\lambda x}$$

with mean $1/\lambda$. However, to a passenger coming to the bus stop at time $t = 0$ say, the observed interval between buses is the sum of the elapsed time since the last bus and the time until the next bus; these are independent and exponential (λ) distributed, so their sum has a gamma $(2, \lambda)$ distribution with density

$$g(x) = \lambda^2 x e^{-\lambda x}$$

with mean $2/\lambda$, twice the value reported by the bus company. This paradox arises because the time interval observed by the passenger is effectively 'point sampled' (it is the interval containing 0) and is length-weighted with respect to the 'typical' time interval observed by the bus company.

This raises the question of how to define coherently the distribution of the 'typical' interval, or in general the 'typical' geometrical object in a random process of geometrical objects. Let $\mathcal{K}$ be the class of all compact sets in $\mathbb{R}^d$; a 'random compact set process' is a point process on $\mathcal{K}$.

One approach, pioneered by Miles (1964a b, 1973) (see Weil and Wieacker (1984, 1993)) is to average over those sets X_i in the process which are contained in a sphere of radius $R \to \infty$. Consider the limiting proportion of such sets which belong to some class $C \subset \mathcal{K}$:

$$P^*(C) = \lim_{R \to \infty} \frac{\sum_{X_i \subset b(0,R)} \mathbf{1}\{X_i \in C\}}{\#\{i : X_i \subset b(0, R)\}}. \tag{1.19}$$

Here C should be translation-invariant (e.g. C might be the class of all compact sets with volume or diameter less than t). For suitable random processes, this limit exists in the sense of almost sure convergence to a constant. Then P^* is a probability distribution on $\mathcal{K}$ which can be interpreted as the distribution of the typical object. Note that P^* only assigns values to the sub-σ-field of translation-invariant events.

The introduction of techniques from marked point process theory has aided and extended the abovementioned analysis (see Stoyan *et al.* (1987), Stoyan and Stoyan (1992)). A random spatial process $\{X_i\}$ of compact sets X_i may be formulated as a 'germ-grain model' $X_i = K_i + x_i$ where the 'germs' x_i are the points of a point process, and the 'grains' K_i are random compact sets (Matheron, 1975). Here

$$K + x = \{x + y : y \in K\}$$

denotes the translation of a compact set K by a vector $x \in \mathbb{R}^d$. Thus we visualise a two-stage process where the 'locations' x_i are formed by a point process and the 'shapes' K_i, typically centred at the origin in some fashion, are translated to those locations. Example 1.14 could be formulated as a germ-grain model where the germ process $\{x_i\}$ is Poisson and the grains $K_i b(0, R_i)$ are i.i.d. random sets.

View the germ-grain model as a marked point process on $\mathbb{R}^d$ with marks in $\mathcal{K}$. It is called *first-order stationary* if, for all measurable $B \subset \mathbb{R}^d$ and $C \subset \mathcal{K}$, the expected number $\mathbb{E}N(B \times C)$ of marked points (x_i, K_i) with $x_i \in B$ and $K_i \in C$ is invariant under translations of B. Since Lebesgue measure is the only translation-invariant measure on $\mathbb{R}^d$ up to a constant factor, we must have

$$\mathbb{E}N(B \times C) = \lambda |B| Q(C) \qquad (1.20)$$

where $\lambda > 0$ is the intensity of the germ process $\{x_i\}$, and Q is a probability distribution on $\mathcal{K}$ which can be dubbed the *distribution of the typical grain*.

This definition (1.20) of the typical grain applies to processes which are merely stationary, and not necessarily ergodic as required by (1.19). The two definitions agree for stationary ergodic processes.

Example 1.16 (Typical interval for the Poisson process)   Return to the bus paradox. If t_i are the arrival times of buses in the Poisson point process, let $x_i = t_i$ and let X_i be the interval $[t_i, t_{i+1}]$ so that $K_i = [0, t_{i+1} - t_i]$. The K_i are in fact completely determined by the underlying point process $\{x_i\}$, but this is not problematic. Let C be the class of all closed intervals of length $\leq u$. Then it can be shown that

$$\frac{\#\{i : x_i \in (-R, R), K_i \in C\}}{2R} \to \lambda(1 - e^{-\lambda u})$$

a.s. and in L^1. Hence $P^*(C) = Q(C) = 1 - e^{-\lambda u}$, that is, the typical interval has exponentially distributed length.

The sampling bias inherent in point-sampling and other operations can be determined from the following version of Campbell's formula (1.11) (see also Theorem 1.9).

Theorem 1.5 (Campbell–Mecke formula) *Consider a germ-grain process $\{(x_i, K_i)\}$ with $x_i \in \mathbb{R}^d$ and $K_i \in \mathcal{K}$ which is first-order stationary. For any measurable function $f : \mathcal{K} \to \mathbb{R}_+$,*

$$\mathbb{E} \sum_i f(X_i) = \lambda \int_{\mathbb{R}^d} \int_{\mathcal{K}} f(K + x) \, dQ(K) \, dx, \qquad (1.21)$$

where λ is the intensity of $\{x_i\}$, and Q is the distribution of the typical grain. More succinctly

$$\mathbb{E} \sum_i f(X_i) = \lambda \int_{\mathbb{R}^d} \mathbb{E}^0 f(K_0 + x) \, dx = \lambda \mathbb{E}^0 \int_{\mathbb{R}^d} f(K_0 + x) \, dx$$

$$(1.22)$$

where K_0 is 'the typical grain' (a random compact set with distribution Q) and $\mathbb{E}^0$ denotes expectation with respect to Q.

Example 1.17 (Point-sampling, general) Let $\{(x_i, K_i)\}$ be a stationary germ-grain process on $\mathbb{R}^d$. Suppose we sample objects X_i iff they cover the origin, i.e. iff $0 \in X_i$. Given a translation-invariant, measurable subclass $C \subseteq \mathcal{K}$, apply (1.22) to $f(X) = \mathbf{1}\{0 \in X\} \, \mathbf{1}\{X \in C\}$ to obtain

$$\mathbb{E}\#\{i : 0 \in X_i, X_i \in C\} = \mathbb{E}\left[\sum_i f(X_i) \right]$$

$$= \lambda \mathbb{E}^0 \int_{\mathbb{R}^d} \mathbf{1}\{(K_0 + x) \ni 0\} \, \mathbf{1}\{(K_0 + x) \in C\} \, dx$$

$$= \lambda \mathbb{E}^0 \left[\mathbf{1}\{K_0 \in C\} \int_{\mathbb{R}^d} \mathbf{1}\{(K_0 + x) \ni 0\} \, dx \right]$$

$$= \lambda \mathbb{E}^0 \left[\mathbf{1}\{K_0 \in C\} \, |K_0| \right]$$

For the last line, we used the fact that $0 \in (K + x)$ if and only if $x \in \check{K}$ where $\check{K} = \{-a : a \in K\}$, and we have $|\check{K}| = |K|$. It follows that

$$\frac{\mathbb{E}\#\{i : X_i \ni 0, X_i \in C\}}{\mathbb{E}\#\{i : X_i \ni 0\}} = \frac{\mathbb{E}^0 \{\mathbf{1}\{K_0 \in C\} \, |K_0|\}}{\mathbb{E}^0 |K_0|}, \qquad (1.23)$$

i.e. point-sampled objects are volume-weighted relative to the distribution of the typical object.

Generalisations of the bus paradox to higher dimensions were first studied by Miles (1964a, b, 1970); see also Cowan (1978), Matheron (1975), Møller (1994), Zuyev (1996), Serra (1982), Santaló (1976), Solomon (1978).

Example 1.18 Consider a Poisson line process in $\mathbb{R}^2$ with intensity λ. The lines partition $\mathbb{R}^2$ into polygonal cells. The distribution of the 'typical' cell is obtained from (1.19) or (1.20). The cell containing the origin is area-weighted with respect to this distribution. See e.g. Miles (1964a), Santaló (1976, p. 55 ff.).

1.8.2 Geometrical coordinates

When a random geometrical object is parametrised by coordinates, the marginal distributions of each coordinate are usually non-uniform even if their joint distribution is uniform.

In the nineteenth century, the following apparent paradox was adduced as purported evidence of the logical inconsistency of probability theory (Bertrand, 1907).

Example 1.19 (Bertrand's paradox) Suppose we wish to 'draw a line at random' through the unit disc $b(0, 1)$. Three possible mechanisms are:

1. choose two points at random on the circumference (independently and uniformly distributed on the circle) and draw the line joining them;

2. let the orientation θ of the line and its distance p from the centre 0 be independent and uniformly distributed over $[0, 2\pi)$ and $[0, 1]$ respectively;

3. let the point X at the foot of the perpendicular from 0 to the line, be uniformly distributed over the disc.

Bertrand showed that these mechanisms produce different probability distributions for the length of a 'random chord' of the unit circle. See Bertrand (1907, p. 9).

Indeed, in terms of the (θ, p) coordinates introduced in Example 1.2, the probability densities of the random lines are respectively

$$f_1(\theta, p) = \frac{1}{2\pi \sqrt{1 - p^2}}$$

$$f_2(\theta, p) = \frac{1}{2\pi}$$

$$f_3(\theta, p) = \frac{1}{\pi} p.$$

The paradox arises because the word 'random' is misinterpreted as implying that coordinates are independent and uniformly distributed. This would indeed be inconsistent, since

(a) if random variables are jointly uniformly distributed, it does not follow that they are marginally uniform nor that they are independent. For example if X is a random point uniformly distributed in $A \subset \mathbb{R}^2$, the marginal distribution of the first coordinate has density

$$f_1(x_1) = \int f(x_1, x_2) \, dx_2 = \frac{\text{length} \left(A \cap L_{0, x_1} \right)}{|A|}$$

where $L_{0, p}$ is the line $x_1 = p$.

(b) independence and uniformity of coordinates are not preserved under a change of coordinates, because of the intervention of the Jacobian. For example the *polar* coordinates (R, Θ) of a random point X uniformly distributed in the unit disc are independent, and R^2, Θ are uniformly distributed on $[0, 1]$ and $[0, 2\pi)$ respectively.

Consequently there are many instances in stochastic geometry when a geometrical operation results, unexpectedly, in a non-uniform distribution.

Example 1.20 Take a uniform Poisson line process in $\mathbb{R}^2$ with intensity λ and observe the locations x_i and angles α_i at which the lines of the process intersect the x-axis. By Examples 1.7 and 1.9 and the Marking Theorem 1.3, the points x_i form a Poisson point process on $\mathbb{R}$ with intensity 2λ and the α_i are i.i.d. with probability density

$$f(\alpha) = \frac{1}{2} |\sin \alpha|$$

on $[0, \pi)$. Thus, although the lines of the line process have 'uniformly distributed orientations' in some sense, the angles of incidence with any fixed axis are not uniformly distributed. Acute angles are less likely, and this may be regarded as a sampling bias effect: the probability of 'catching' a random line in a given sampling interval of the x-axis depends on the orientation of the line.

Example 1.21 Let $A \subset \mathbb{R}^2$ be convex and compact, and X a random point uniformly distributed in A. Draw the line Y through X in a random direction Θ which is uniformly distributed in $[0, \pi)$ independently of X. Then Y is **not** a uniformly distributed random line in the sense of Example 1.2; instead its (θ, p) coordinates have density

$$f(\theta, p) = \frac{\text{length}\left(L_{\theta,p} \cap A\right)}{\pi \, |A|} \, dp \, d\theta \qquad (1.24)$$

which is length-weighted with respect to the uniform distribution. This may also be regarded as a sampling bias effect.

An elegant way to derive (1.24), see e.g. Kendall and Moran (1963), Santaló (1976), Jensen (1997), is to consider the pair (x, L) consisting of a point $x \in A$ and a line $L = L_{\theta,p}$ constrained by $x \in L$. The pair is completely specified by the coordinates (x_1, x_2) of the point together with the orientation θ of the line; or dually by the (θ, p) coordinates of the line together with a one-dimensional coordinate t giving the position of the point on this line. Simple calculation gives for the Jacobian

$$dx_1 \, dx_2 \, d\theta = dp \, d\theta \, dt.$$

In our example the joint density of (x_1, x_2, θ) is jointly uniform over $A \times [0, \pi)$. Changing coordinates as above and integrating out the t variable yields (1.24).

1.9 Conditioning and variational methods

Conditional probabilities and expectations are used frequently in stochastic geometry.

1.9.1 Simple conditioning

Let $X_1, \ldots, X_n$ be random sets and $Y = f(X_1, \ldots, X_n)$ a quantity whose expectation we want to compute. This may be done via the conditional expectations $\mathbb{E}[Y \mid X_1]$ or $\mathbb{E}[Y \mid X_2, \ldots, X_n]$, etc., which often coincide with simple geometrical quantities, in the manner of Section 1.2.

Example 1.22 Let X_1, X_2 be independent random lines uniformly distributed through a convex compact set $A \subset \mathbb{R}^2$ as in Example 1.2.

Consider the probability that the intersection point $X_1 \cap X_2$ lies inside A. This can be obtained by first conditioning on X_1:

$$\mathbb{P}\{X_1 \cap X_2 \in A\} = \mathbb{E}\left[\mathbb{P}\{X_2 \cap (X_1 \cap A) \neq \emptyset \mid X_1\}\right]$$

For fixed X_1, $\mathbb{P}\{X_2 \cap (X_1 \cap A) \neq \emptyset \mid X_1\}$ is the probability that X_2 intersects the line segment $(X_1 \cap A)$. Thus

$$\mathbb{P}\{X_1 \cap X_2 \in A\} = \mathbb{E}\left[\frac{2\mathrm{length}\,(X_1 \cap A)}{\mathrm{length}\,(\partial A)}\right]$$
$$= \frac{2\pi\,|A|}{\mathrm{length}\,(\partial A)^2}$$

Example 1.23 (Efron (1965)) Let $X_1, \ldots, X_n$ be independent random points uniformly distributed in a convex compact set $A \subset \mathbb{R}^2$. We want to determine the expected number of vertices of the convex hull of these points. Write $C_n = \mathrm{conv}\,(X_1, \ldots, X_n)$ for the convex hull, and N_n for the number of vertices of C_n. We have

$$\mathbb{E}\,N_n = \sum_{i=1}^{n} \mathbb{P}\{X_i \text{ is a vertex of } C_n\}$$
$$= n\mathbb{P}\{X_1 \text{ is a vertex of } C_n\}$$

by symmetry. Now X_1 is a vertex of C_n iff $X_1 \notin \mathrm{conv}\,(X_2, \ldots, X_n)$. Since X_1 is independent of the other points and uniformly distributed we have

$$\mathbb{P}\{X_1 \text{ is a vertex of } C_n\} = \mathbb{P}\{X_1 \notin \mathrm{conv}\,(X_2, \ldots, X_n)\}$$
$$= \mathbb{E}\left[\mathbb{P}\{X_1 \notin \mathrm{conv}\,(X_2, \ldots, X_n) \mid X_2, \ldots, X_n\}\right]$$
$$= \mathbb{E}\left[1 - \frac{|\mathrm{conv}\,(X_2, \ldots, X_n)|}{|A|}\right]$$

so that

$$\mathbb{E}N_n = n\left[1 - \frac{\mathbb{E}|C_{n-1}|}{|A|}\right]$$

The case $n = 4$ is related to *Sylvester's problem* (Kendall and Moran, 1963, p. 42 ff.) of finding the probability p that four independent uniform random points in a convex set A form a convex quadrilateral. Since N_4 takes the value 4 with probability p and 3 with probability $(1 - p)$ we have $\mathbb{E}N_4 = 3 + p$. By the argument above, $\mathbb{E}N_4 = 4[1 - \mathbb{E}|C_3|/|A|]$. Hence $p = 1 - 4\mathbb{E}|C_3|/|A|$. This latter result could also be derived directly by considering the complementary event that the quadrilateral is not convex.

1.9.2 Crofton's perturbation method

Many elementary problems in stochastic geometry concern the expectation $\mathbb{E}Y$ of a real-valued functional $Y = f(X)$ of a random set X,
where the distribution of X depends on a domain $A \subset \mathbb{R}^d$. Effectively
$\mathbb{E}Y$ is a function of the set argument A, say $F(A)$, and we may use
various analytic tools to study F.

Crofton (1885) introduced a technique which involves applying a
small perturbation to the domain A. In effect, we consider a family of
sets A_t indexed by a parameter $t > 0$, and differentiate $F(A_t)$ with
respect to t. This produces a differential equation which often yields F
or at least simplifies the problem. The underlying tool is the following
(see Kendall and Moran (1963, p. 26), Baddeley (1977), Ruben and
Reed (1973), Solomon (1978), Sheng (1985)).

Theorem 1.6 *Let*

$$I(t) = \int_{A_t} f(x)\, dx$$

*where $f : \mathbb{R}^d \to \mathbb{R}$ is Lebesgue integrable and $(A_t, t \in [0, T])$ is a
family of compact subsets of $\mathbb{R}^d$. Suppose that the sets A_t are smoothly
changing in the sense that the 'graph' $\Gamma = \{(x, t) : x \in A_t\}$ is a twice
continuously differentiable, $d+1$ dimensional, embedded manifold-with-
boundary in $\mathbb{R}^d \times \mathbb{R}$. Then I is differentiable almost everywhere, and*

$$\frac{d}{dt}I(t) = \int_{\partial A_t} v(x, t)\, f(x)\, d'x$$

*where $v(x, t)$ is a certain quantity which can be interpreted as the
(signed) velocity of expansion of A_t normal to the boundary ∂A_t, at
the location $x \in \partial A_t$. On the right hand side, integration $d'x$ is with
respect to $d - 1$ dimensional surface area measure.*

See Baddeley (1977) for a version of this result. The differentiability
condition on A_t can be relaxed substantially.

Example 1.24 If $A_t = b(0, t)$ is the disc of radius t in $\mathbb{R}^2$, the expansion velocity is $v(x, t) = 1$ for all $x \in \partial b(0, t)$. Taking $f(x) \equiv 1$
yields $I'(t) = \frac{d}{dt}|A_t| = \frac{d}{dt}\pi t^2 = 2\pi t = \text{length}\,(\partial A_t)$.

Theorem 1.7 (Crofton's mean value theorem) *Let $(A_t, t \geq 0)$ be a
differentiable family of sets as in the previous Theorem, and consider*

$$M(t) = \mathbb{E}f(X_1, \ldots, X_m)$$

where $X_1, \ldots, X_m$ are i.i.d. random points uniformly distributed in A_t, *and* $f : (\mathbb{R}^d)^m \to \mathbb{R}$ *is symmetric in its arguments. Then M has derivative*

$$M'(t) = m \frac{\frac{d}{dt}|A_t|}{|A_t|}(M_1(t) - M(t))$$

(almost everywhere), where

$$M_1(t) = \mathbb{E}f(Y, X_2, \ldots, X_m)$$

is the expectation of the same functional in which $Y, X_2, \ldots, X_m$ *are independent random points,* X_i *is uniformly distributed in* A_t, *and* Y *is a random point on* ∂A_t *with density proportional to* $v(x, t)$ *relative to* $d - 1$ *dimensional surface area measure.*

The proof is a matter of applying Theorem 1.6 to the functions $f, g : (\mathbb{R}^d)^m \to \mathbb{R}$ on the Cartesian product $B_t = (A_t)^m$, where $g \equiv 1$. The symmetry of f is purely a convenience, and more general versions of this formula exist. See Baddeley (1977), Ruben and Reed (1973).

Example 1.25 Consider the mean distance $M(t) = \mathbb{E}\|X - Y\|$ between two independent random points X, Y uniformly distributed in the disc $A_t = b(0, t)$ in $\mathbb{R}^2$. From the previous example we have $|A_t| = \pi t^2$, $\frac{d}{dt}|A_t| = 2\pi t$,

$$M'(t) = \frac{4}{t}(M_1(t) - M(t)).$$

Here $M_1(t) = \mathbb{E}\|X - Z\|$ with X uniformly distributed in the disc A_t and Z uniformly distributed on the circumference ∂A_t since $v(\cdot, t) \equiv 1$. By rotational symmetry we can assume Z is a fixed point on ∂A_t which reduces $M_1(t)$ to a two-dimensional integral. Note also that scaling properties imply $M(t), M_1(t)$ must be proportional to t. By direct integration or other means we obtain $M_1(t) = \frac{32}{9\pi}t$ so that

$$M(t) = \frac{128}{45\pi}t.$$

1.9.3 Palm distributions

Let X be a point process on $\mathbb{R}^d$. The *Palm distribution* P^x with respect to X at a location $x \in \mathbb{R}^d$ is, intuitively, the conditional probability distribution given that X has a point at x:

$$P^x(E) = \mathbb{P}\{E \mid x \in X\}$$

for events E. Since the conditioning event $\{x \in X\}$ typically has probability zero, we cannot use the elementary definition of conditional probability of events, but must regard the Palm probability as a Radon–Nikodym density.

Palm distributions are formally constructed along the following lines (Reiss, 1993, p. 176 ff., Daley and Vere-Jones, 1988, chap. 12). Let $(\Omega, \mathcal{F}, P)$ be the underlying probability space. Writing $N(A) = \#(X \cap A)$ for the number of points of the process in $A \subset \mathbb{R}^d$, assume that the intensity measure $\Lambda(A) = \mathbb{E}\,N(A)$ exists and is finite for all compact $A \subset \mathbb{R}^d$. Define the *Campbell measure* $\mathbf{C}$ on $\mathbb{R}^d \times \Omega$ by

$$\mathbf{C}(A \times D) = \mathbb{E}\,[N(A)\,1_D]$$

for all compact $A \subset \mathbb{R}^d$ and all events $D \in \mathcal{F}$. For each fixed event $D \in \mathcal{F}$, the measure μ_D on $\mathbb{R}^d$ defined by

$$\mu_D(A) = \mathbf{C}(A \times D)$$

for all bounded Borel sets $A \subset \mathbb{R}^d$, has

$$\mu_D(A) = \mathbf{C}(A \times D) = \mathbb{E}\,[N(A)\,1_D] \leq \mathbb{E}N(A) = \Lambda(A)$$

so that $\mu_D \ll \Lambda$, and by the Radon-Nikodym theorem there is a function $f_D : \mathbb{R}^d \to \mathbb{R}_+$ such that

$$\mu_D(A) = \int_A f_D(x)\,d\Lambda(x)$$

for all measurable $A \subset \mathbb{R}^d$. Then $f_D(x)$ can be interpreted the probability $P^x(D)$ of the event D under the Palm distribution at x. We require that $(x, D) \mapsto P^x(D)$ be a **conditional probability kernel**, i.e.

- for all $D \in \mathcal{F}$ the function $x \mapsto P^x(D)$ is measurable;

- $P^x(\cdot)$ is a probability measure on Ω for Λ-almost all $x \in \mathbb{R}^d$.

Theorem 1.8 *Let X be a point process on $\mathbb{R}^d$ with locally finite intensity measure Λ. Assume the underlying probability space $(\Omega, \mathcal{F}, P)$ has a countably generated σ-algebra. Then there exists a conditional probability kernel $(x, D) \mapsto P^x(D)$ which is characterised up to a.e. equivalence by*

$$\mathbb{E}\,[N(A)\,1_D] = \int_A P^x(D)\,d\Lambda(x) \qquad (1.25)$$

for all compact $A \subset \mathbb{R}^d$ and all events $D \in \mathcal{F}$.

Example 1.26 For a Poisson point process, the Palm distribution is equivalent to the original distribution of the process modified by adding a fixed point at x. Namely, *Slivnyak's theorem* states that

$$P^x = \delta_x * P$$

where P is the original distribution, $*$ denotes the superposition of independent point processes, and δ_x is the distribution of the degenerate point process consisting of a single fixed point at x. (This is in fact a characterisation of the Poisson process.)

To partially verify this we can calculate the Palm probability of a specific event such as $D = \{N(B) = k\}$ for the uniform Poisson process, where $B \subset \mathbb{R}^d$ is compact and $k \in \mathbb{N}$. First consider subsets $A \subseteq B^c$. By the independence property of the Poisson process we have

$$\mathbb{E}\left[N(A)\,1_D\right] = \mathbb{E}N(A)\,\mathbb{P}(D) = \lambda\,|A|\,P(D)$$

and comparing this with (1.25) we find $P^x(D) \equiv P(D)$ for $x \notin B$.

For subsets $A \subseteq B$ write

$$\mathbb{E}\left[N(A)\,1_D\right] = P(D)\mathbb{E}\left[N(A) \mid D\right]$$

and observe that by the conditional property of the Poisson process, the conditional distribution of $N(A)$ given D is Binomial k, p) where $p = |A|/|B|$. Hence

$$\mathbb{E}\left[N(A)\,1_D\right] = P(D)\,k\frac{|A|}{|B|}$$

$$= e^{-\lambda|B|}\frac{(\lambda|B|)^k}{k!}\,k\frac{|A|}{|B|}$$

$$= e^{-\lambda|B|}\frac{(\lambda|B|)^{k-1}}{(k-1)!}\lambda|A|$$

so that $P^x(D) = \exp\{-\lambda|B|\}(\lambda|B|)^{k-1}/(k-1)! = P\{N(B) = k - 1\}$ for $x \in B$. That is, for $x \in B$, the distribution of $N(B) - 1$ under P^x is Poisson($\lambda|B|$).

Note that the Palm distribution P^x assigns probabilities to all events in the probability space, and not merely to events concerning the point process X.

Example 1.27 Let U be a positive r.v. with $0 < \mathbb{E}U < \infty$. Conditional on $U = u$, generate a uniform Poisson point process on $\mathbb{R}^d$ with intensity u. The result is a 'mixed Poisson process' X.

The intensity measure of X is $\Lambda(A) = \mathbb{E}\,N(A) = \mathbb{E}\left[\mathbb{E}[N(A) \mid U]\right] = (\mathbb{E}U)\,|A|$. For the event $D = \{U \leq u\}$ we obtain

$$\mathbb{E}\left[N(A)\,1_D\right] = \mathbb{E}\left[\mathbb{E}[N(A) \mid U] \mid D\right]\,P(D) = |A|\,\mathbb{E}\left[U \mid D\right]\,P(D)$$

hence

$$P^x(D) = \frac{\mathbb{E}\,[U\,1_D]}{\mathbb{E}\,U}$$

i.e. the distribution of U under P^x is the U-weighted counterpart of its distribution under P.

By linearity of expectation and monotone convergence, equation (1.25) extends to the following more general version of Theorem 1.5.

Theorem 1.9 (Campbell–Mecke) *Let X and $(\Omega, \mathcal{F}, P)$ be as in the previous theorem. Then for any jointly measurable $f : \mathbb{R}^d \times \Omega \to \mathbb{R}$*

$$\mathbb{E}\left[\sum_{x \in X} f(x, \omega)\right] = \int_A \mathbb{E}^x f(x, \omega)\, d\Lambda(x) \qquad (1.26)$$

in the sense that if one side is finite then both sides are finite and equal, where $\mathbb{E}^x$ denotes expectation with respect to the Palm distribution P^x.

This is a generalisation of Theorem 1.5 in the sense that the distribution of the typical grain in a stationary germ-grain process equals the Palm distribution of the grain attached to any given point, i.e. the distribution under P^x of the grain K_x attached to x.

Example 1.28 (Bus paradox) This was discussed in Examples 1.15 and 1.16. If t_i are the arrival times of buses in the Poisson point process, we need

$$\mathbb{E}\left[\sum_{t_i \in [0,T]} 1\{t_{i+1} - t_i \leq u\}\right]$$

By (1.25) this equals

$$\lambda\,T\,P^0\{t_1 \leq u\}$$

where P^0 is the Palm distribution at 0 and t_1 is the first arrival after time 0. That is, the distribution of the typical interval is the Palm distribution of the interval starting at 0. Applying Slivnyak's theorem we have

$$P^0\{t_1 \leq u\} = P\{t_1 \leq u\} = 1 - e^{-\lambda u}$$

i.e. the typical interval has exponential (λ) distributed length.

Acknowledgements

I thank many colleagues and especially the editors for their encouragement and helpful comments on the draft.

References

R.V. Ambartzumian. *Combinatorial Integral Geometry, with applications to mathematical stereology*. John Wiley and Sons, Chichester, 1982.

A.J. Baddeley. Integrals on a moving manifold and geometrical probability. *Advances in Applied Probability*, 9:588–603, 1977.

A.J. Baddeley. Stereology. In *Spatial Statistics and Digital Image Analysis*, chapter 10, pages 181–216. National Research Council USA, Washington DC, 1991.

A.J. Baddeley. Stereology and survey sampling theory. *Bulletin of the International Statistical Institute*, 50, book 2:435–449, 1993.

A.J. Baddeley and I.S. Molchanov. On the expected measure of a random set. In D. Jeulin, editor, *Advances in Theory and Applications of Random Sets*, pages 3–20, Singapore, 1997. World Scientific Publishing.

M. Berman. Testing for spatial association between a point process and another stochastic process. *Applied Statistics*, 35:54–62, 1992.

J. Bertrand. *Calcul des probabilités*. Paris, 1907.

J. Bronowski and J. Neyman. The variance of the measure of a two-dimensional random set. *Annals of Mathematical Statistics*, 16:330–341, 1945.

R.J. Cowan. The use of the ergodic theorems in random geometry. *Advances in Applied Probability*, 10 (Supplement):47–57, 1978.

N.A.C. Cressie. *Statistics for spatial data*. John Wiley and Sons, New York, 1991.

M.W. Crofton. Probability. In *Encyclopaedia Britannica*. 9th edition, 1885.

D.J. Daley and D. Vere-Jones. *An introduction to the theory of point processes*. Springer Verlag, New York, 1988.

P.J. Diggle. Binary mosaics and the spatial pattern of heather. *Biometrics*, 37:531–539, 1983.

B. Efron. The convex hull of a random set of points. *Biometrika*, 52:331–344, 1965.

H. Federer. *Geometric Measure Theory*. Springer Verlag, Heidelberg, 1969.

P. Hall. *An introduction to the theory of coverage processes*. John Wiley and Sons, New York, 1988.

E.F. Harding and D.G. Kendall, editors. *Stochastic geometry: a tribute to the memory of Rollo Davidson*. John Wiley and Sons, London-New York-Sydney-Toronto, 1974.

E.B.V. Jensen. *Local Stereology*. Singapore. World Scientific Publishing, 1997.

A.M. Kellerer. Counting figures in planar random configurations. *jap*, 22:68–81, 1985.

M.G. Kendall and P.A.P. Moran. *Geometrical Probability*. Charles Griffin, London, 1963. Griffin's Statistical Monographs and Courses no. 10.

J.F.C. Kingman. *Poisson processes*. Oxford University Press, 1993.

A. Kolmogoroff. *Grundbegriffe der Wahrscheinlichkeitsrechnung*. Ergebnisse der Mathematik und ihrer Grenzgebiete. Schriftleitung Zentralblatt für Mathematik, Berlin, 1933. Reprinted by Chelsea, New York, 1946.

G. Matheron. *Random sets and integral geometry*. John Wiley and Sons, New York, 1975.

J. Mecke, R.G. Schneider, D. Stoyan, and W.R.R. Weil. *Stochastische Geometrie*. DMV Seminar Band 16. Birkhäuser, Basel, 1990.

R.E. Miles. Random polygons determined by random lines in a plane, I. *Proceedings of the National Academy of Sciences USA*, 52:901–907, 1964.

R.E. Miles. Random polygons determined by random lines in a plane, II. *Proceedings of the National Academy of Sciences USA*, 52:1157–1160, 1964.

R.E. Miles. On the homogeneous planar Poisson point process. *Mathematical Biosciences*, 6:85–127, 1970.

R.E. Miles. The various aggregates of random polygons determined by random lines in a plane. *Advances in Math.*, 10:256–290, 1973.

R.E. Miles. A synopsis of 'Poisson flats in Euclidean spaces'. In E F Harding and D G Kendall, editors, *Stochastic Geometry*, pages 202–227. John Wiley and Sons, 1974.

I.S. Molchanov. On statistical analysis of Boolean models with nonrandom grains. *Scand. J. Statist.*, 21(1):73–82, 1994.

J. Møller. *Lectures on random Voronoi tessellations*. Number 87 in Lecture Notes in Statistics. Springer-Verlag, 1994.

J. Møller and S. Zuyev. Gamma-type results and other related properties of Poisson processes. *Adv. in Appl. Probab.*, 28(3):662–673, 1996.

F. Preteux and M. Schmitt. Boolean texture analysis and synthesis. In J Serra, editor, *Image analysis and mathematical morphology. Volume II: Theoretical Advances*, chapter 18, pages 379–400. John Wiley and Sons, 1988.

R.-D. Reiss. *A course on point processes*. Springer, 1993.

B.D. Ripley. *Spatial statistics*. John Wiley and Sons, New York, 1981.

B.D. Ripley. *Statistical inference for spatial processes*. Cambridge University Press, 1988.

H.E. Robbins. On the measure of a random set. *Annals of Mathematical Statistics*, 15:70–74, 1944.

H.E. Robbins. On the measure of a random set II. *Annals of Mathematical Statistics*, 16:342–347, 1945.

H. Ruben and W. J. Reed. A more general form of a theorem of Crofton. *J. Appl. Probability*, 10:479–482, 1973.

L.A. Santaló. *Integral Geometry and Geometric Probability*. Encyclopedia of Mathematics and Its Applications, vol. 1. Addison-Wesley, 1976.

R. Schneider and W. Weil. *Integralgeometrie*. Teubner Skripten zur Mathematischen Stochastik. [Teubner Texts on Mathematical Stochastics]. B. G. Teubner, Stuttgart, 1992.

J. Serra. *Image analysis and mathematical morphology*. Academic Press, London, 1982.

T.K. Sheng. The distance between two random points in plane regions. *Advances in Applied Probability*, 17(4):748–773, 1985.

H. Solomon. A coverage distribution. *Annals of Mathematical Statistics*,

21:139–140, 1950.

H. Solomon. Distribution of the measure of a random two-dimensional set. *Annals of Mathematical Statistics*, 24:650–656, 1953.

H. Solomon. *Geometric Probability*. Number 28 in CBMS-NSF Regional Conference Series in Applied Mathematics. Society for Industrial and Applied Mathematics, Philadelphia, Pennsylvania, 1978.

D. Stoyan, W.S. Kendall, and J. Mecke. *Stochastic Geometry and its Applications*. John Wiley and Sons, Chichester, 1987.

D. Stoyan and H. Stoyan. *Fraktale — Formen — Punktfelder*. Akademie Verlag, Berlin, 1992.

D.J. Strauss. A model for clustering. *Biometrika*, 63:467–475, 1975.

E.R. Weibel. *Stereological Methods, 1. Practical Methods for Biological Morphometry*. Academic Press, London, 1979.

E.R. Weibel. *Stereological Methods, 2. Theoretical Foundations*. Academic Press, London, 1980.

W. Weil. Stereology: a survey for geometers. In P M Gruber and J M Wills, editors, *Convexity and its applications*, pages 360–412. Birkhauser, Basel, Boston, Stuttgart, 1983.

W. Weil. Point processes of cylinders, particles and flats. *Acta Applicandae Mathematicae*, 9:103–136, 1984.

W. Weil and J.A. Wieacker. Densities for stationary random sets and point processes. *Adv. in Appl. Probab.*, 16(2):324–346, 1984.

W. Weil and J.A. Wieacker. Stochastic geometry. In *Handbook of convex geometry, Vol. A, B*, pages 1391–1438. North-Holland, Amsterdam, 1993.

CHAPTER 2

Spatial sampling and censoring

Adrian J. Baddeley

University of Western Australia

When a spatial pattern is observed through a bounded window, inference about the pattern is hampered by sampling effects known as 'edge effects'. This chapter identifies two main types of edge effects: *size-dependent sampling bias* and *censoring effects*. Sampling bias can be eliminated by changing the sampling technique, or 'corrected' by weighting the observations. Censoring effects can be tackled using the methods of survival analysis.

Introduction

We shall consider spatial patterns like those sketched in Figure 2.1. The pattern may consist of distinct features or objects, such as points (which might represent the locations of trees, meteorite impacts or bird nests), line segments (geological faults, microscopic fibres or cracks) or other shapes (biological cells or mineral grains). Alternatively the pattern may be simply a two-colour image, for example representing the presence or absence of vegetation in a region.

The exploratory data analysis of such patterns begins by computing summary statistics analogous to the sample moments of numerical data. Examples are the average number of objects per unit area, and the

Figure 2.1 *Four spatial patterns, each observed within a rectangular sampling window. Left to right: (a) points; (b) line segments; (c) irregular shapes; (d) mosaic.*

empirical distributions of object sizes, of object orientations, and of distances between pairs of objects.

Each pattern in Figure 2.1 has been observed within a rectangular sampling 'window' W while the pattern itself extends beyond the window, with potentially infinite extent. Normally we assume the spatial pattern is a realisation of a stationary random spatial process X. Since information is available only within a window, sampling effects known as '**edge effects**' arise, which affect statistical inference.

Suppose for example that we wish to estimate the average number of fibres per unit area in paper, from a microscope image like Figure 2.1(b). In counting fibres, should we include those which cross the boundary of the window and apparently extend outside? Ignoring such fibres would clearly underestimate the fibre density, while including them would yield an overestimate.

In general, when observation of a spatial pattern is restricted to a bounded window, two types of edge effects arise:

sampling bias is present when the probability of observing a geometrical object depends on its size or shape;

censoring effects occur when we are prevented from observing the full extent of a geometrical object that lies partially within the window.

Spatial sampling bias was encountered in the 1940s by statisticians working on textile applications (Cox, 1949, 1969, Daniels, 1942) but the first general treatment was presented by Miles (1974). General techniques for eliminating sampling bias were developed by Miles (1978), Lantuéjoul (1978a, b), Gundersen (1977, 1978) and Jensen and Sundberg (1986). For the special case of point processes, edge effects have been extensively discussed, and corrections introduced by Ripley, Lantuéjoul, Hanisch, Stoyan, Ohser and others; see Diggle (1983), Hanisch (1984), Ohser (1983), Ripley (1977, 1981), Serra (1982, p. 246). For surveys see Ripley (1988, chap. 3), Stoyan *et al.* (1987, pp. 122–131), Cressie (1991, chap. 8), Baddeley (1993).

Much more recently, it was noticed that there is an analogy between edge effects for spatial processes and the random censoring of lifetimes in survival analysis. Laslett (1982a, b) drew attention to this analogy for spatial patterns of line segments. The observed lengths of line segments, after they have been 'clipped' within a sampling window, can be compared to censored survival times. Wijers (1995) found the optimal estimator of the segment length distribution. Baddeley, Gill and Hansen (Baddeley and Gill, 1993, 1997, Hansen *et al.*, 1996, to appear) noted a similar analogy for distance distributions in point patterns and random sets. Zimmerman (1991) used artificial censoring to control edge effects.

Edge effects are severe in dimensions $d > 2$ and when the window is small or complex in shape. The most effective strategies which have been adopted to deal with them are as follows.

unbiased sampling rules: Edge effects depend on the sampling rule which we use to decide which objects should be counted or sampled as lying 'in' the window of observation. It may be possible to use an alternative sampling rule which has no sampling bias.

additive statistics: Certain summary statistics are not susceptible to edge effects, namely those which are additive functionals of the pattern. It may be possible to modify the statistic of interest to have this property.

data-dependent weighting: Sampling bias may be corrected by weighting the contribution from each sampled object by the reciprocal of its sampling probability. This is closely related to the Horvitz–Thompson device in survey sampling.

survival analysis: Censoring effects may be countered using the methods of survival analysis.

This chapter describes the four strategies in detail. Section 2.1 discusses the general issue of sampling bias. Sections 2.2, 2.3 and 2.4 outline the three strategies of unbiased sampling rules, additive statistics, and data-dependent weighting. In section 2.5 we apply these techniques to the special case of point patterns.

Sections 2.6 and 2.8 develop the analogy between edge effects and random censoring. In section 2.6 we recall some general concepts of random censoring, study censoring effects for point processes, and construct Kaplan–Meier style estimators of the standard point process functions F, G and K. Section 2.8 treats censoring effects for general random sets.

2.1 Spatial sampling bias

In this section, we discuss spatial sampling bias in the general case where the spatial pattern X consists of distinct objects X_i in $\mathbb{R}^d$ (which might be points, lines or other compact sets). The observation window W is a fixed, known compact set in $\mathbb{R}^d$. Visualise the sampling situation as in Figure 2.2.

It is useful to distinguish between a **clipping window** and a **sampling frame**. A clipping window supplies information within W only;

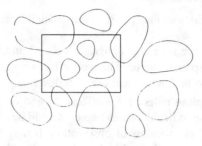

Figure 2.2 *A spatial pattern of distinct objects X_i observed through a rectangular sampling window W.*

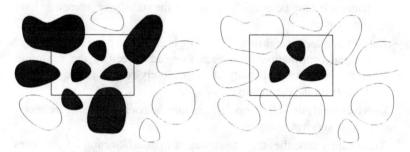

Figure 2.3 *Plus-sampling (left) and minus-sampling (right) in a rectangular sampling frame or window W. The shaded objects are sampled.*

the data consist of the 'clipped' pattern $X \cap W$. For example, the images produced by cameras and satellite sensors are clipped within a rectangular boundary. On the other hand, a sampling frame is a region W of known size and shape, outside which we may still be able to observe the pattern. In forestry and ecology, one can sample a field of vegetation by throwing a rectangular wooden frame at random into the field. In optical microscopy, it is common to delineate a rectangular sampling frame of known size within the field of view. The (vaguely defined) visible region outside W is sometimes called the 'guard area'.

The pioneering paper of Miles (1974) discussed spatial sampling bias in general and treated two basic sampling operations:

plus-sampling, where we sample any object X_i that intersects the frame, $X_i \cap W \neq \emptyset$;

minus-sampling, where we sample only those objects that lie within the frame, $X_i \subseteq W$.

These are illustrated in Figure 2.3. Plus-sampling uses information from outside the frame W, while minus-sampling requires only a clipping window.

It is intuitively clear that plus-sampling introduces a sampling bias in favour of larger objects X_i, while minus-sampling favours smaller objects. For example, under minus-sampling it is impossible to sample an object larger than the window.

These and other sampling rules can be analysed using the methods of marked point processes described in section 1.7.1 of Chapter 1. Regard the objects X_i as arising from a germ-grain process, that is, represent each X_i as the translation $X_i = K_i + x_i$ of a compact set K_i to a location x_i, where $\{(x_i, K_i)\}$ is a stationary marked point process in $\mathbb{R}^d$ with marks in the space $\mathcal{K}$ of compact sets in $\mathbb{R}^d$. The point process of germs $\{x_i\}$ has intensity α and the grains K_i have common distribution Q. Note that the representation $X_i = K_i + x_i$ is a mathematical convenience; it is always possible to represent a random process of compact sets in this way (Weil and Wieacker, 1987), but the germs x_i are not necessarily observable in practice.

For plus-sampling, the Campbell–Mecke formula (equation (1.26) in Theorem 9 of Chapter 1, section 1.8.3) yields, for the expected number of objects sampled,

$$\mathbb{E}\#\{i \ : \ X_i \cap W \neq \emptyset\} = \mathbb{E} \sum_i \mathbf{1}\{X_i \cap W \neq \emptyset\}$$

$$= \alpha \, \mathbb{E}^0 \left[\int_{\mathbb{R}^d} \mathbf{1}\{(x + K_0) \cap W \neq \emptyset\} \, dx \right]$$

$$= \alpha \, \mathbb{E}^0 \left| W \oplus \check{K}_0 \right|$$

Here K_0 is 'the typical grain', a random compact set with distribution Q, and $\mathbb{E}^0$ may be interpreted in this context as the expectation with respect to Q. The symbol $|\cdot|$ denotes Lebesgue area, and $W \oplus \check{K}$ is the dilation of W by K (Serra, 1982)

$$W \oplus \check{K} = \{x \in \mathbb{R}^d \ : \ (x + K) \cap W \neq \emptyset\}.$$

The operation of dilation is illustrated in Figure 2.4.

More generally, for some functional $f : \mathcal{K} \to \mathbb{R}$, consider the expectation of the sample total of the values of $f(X_i)$ for all plus-sampled

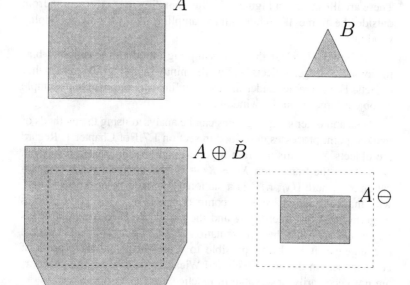

Figure 2.4 *Top row: two plane sets A and B; Bottom left: dilation $A \oplus \check{B}$; Bottom right: erosion $A \ominus B$.*

objects X_i. The Campbell–Mecke formula gives

$$\mathbb{E} \sum_{X_i \cap W \neq \emptyset} f(X_i) = \mathbb{E} \sum_i f(X_i) \, \mathbf{1}\{X_i \cap W \neq \emptyset\}$$

$$= \alpha \, \mathbb{E}^0 \left[\int_{\mathbb{R}^d} f(K_0 + x) \, \mathbf{1}\{(K_0 + x) \cap W \neq \emptyset\} \, dx \right].$$

Assume f is translation-invariant,

$$f(K + x) = f(K) \quad \text{for all } x \in \mathbb{R}^d, \ K \in \mathcal{K}$$

so that $f(X_i) = f(K_i)$ and we get

$$\mathbb{E} \sum_{X_i \cap W \neq \emptyset} f(X_i) = \alpha \, \mathbb{E}^0 \left[f(K_0) \int_{\mathbb{R}^d} \mathbf{1}\{(K_0 + x) \cap W \neq \emptyset\} \, dx \right]$$

$$= \alpha \, \mathbb{E}^0 \left[f(K_0) \, \left| W \oplus \check{K}_0 \right| \right].$$

Hence

$$\frac{\mathbb{E}\sum_{X_i \cap W \neq \emptyset} f(X_i)}{\mathbb{E}\#\{i \ : \ X_i \cap W \neq \emptyset\}} = \frac{\mathbb{E}^0\left[f(K_0)\left|W \oplus \check{K}_0\right|\right]}{\mathbb{E}^0\left|W \oplus \check{K}_0\right|} \qquad (2.1)$$

The right-hand side of (2.1) is the expectation of f under the size-biased distribution $Q_\oplus$ which has density

$$\frac{dQ_\oplus}{dQ}(K) \propto \left|W \oplus \check{K}\right|$$

with respect to Q. That is, plus-sampling introduces a sampling bias proportional to the area of the dilation $W \oplus \check{X}_i$ of each object X_i. This is a bias in favour of larger objects.

Clearly a similar analysis can be applied to minus-sampling, and indeed to any sampling rule which decides whether to include or exclude each object X_i based only on information about X_i.

Theorem 2.1 *Let $\{X_i\}$ be a stationary germ-grain model in $\mathbb{R}^d$ with germ intensity α and compact grains with distribution Q. Consider any sampling rule such that X_i is included in the sample if and only if $I(X_i) = 1$ where $I : \mathcal{K} \to \{0, 1\}$ is measurable.*

Then for any translation-invariant, measurable $f : \mathcal{K} \to \mathbb{R}$

$$\mathbb{E}\sum_{\text{sample}} f(X_i) = \mathbb{E}\sum I(X_i)f(X_i) = \alpha\,\mathbb{E}^0\left[f(K_0)\pi(K_0)\right] \qquad (2.2)$$

where for $K \in \mathcal{K}$

$$\pi(K) = \int_{\mathbb{R}^d} I(K + x)\,dx \qquad (2.3)$$

is the volume of the set of all translation vectors x such that $K + x$ would be included in the sample. The objects sampled by this rule are π-weighted in the sense that

$$\frac{\mathbb{E}\sum_{\text{sample}} f(X_i)}{\mathbb{E}(\text{number in sample})} = \frac{\mathbb{E}\sum_i I(X_i)\,f(X_i)}{\mathbb{E}\sum_i I(X_i)} = \frac{\mathbb{E}^0\left[f(K_0)\pi(K_0)\right]}{\mathbb{E}^0\left[\pi(K_0)\right]} \qquad (2.4)$$

is the expectation of f under the distribution which is π-weighted with respect to Q.

The theorem applies to minus-sampling, which turns out to have sampling bias factor

$$\pi(K) = |W \ominus K|$$

where $W \ominus K$ is the erosion

$$W \ominus K = \{x \in W \; : \; K + x \subseteq W\}.$$

The operation of erosion is also illustrated in Figure 2.4.

2.2 Unbiased sampling rules

For applications in optical microscopy, forestry, ecology and other fields which favor manual counting methods, the most appropriate solution to the problem of spatial sampling bias is to avoid it altogether by adopting an alternative sampling rule. An *unbiased sampling rule* is one for which the sampling bias factor $\pi(K)$ in (2.3) is constant. This produces 'unbiased samples' in the sense that the right side of (2.4) is then simply the expectation of f under Q, the unweighted distribution of the typical grain.

This situation is typical of estimators in spatial statistics, most of which are not unbiased but instead are ratios of two unbiased consistent estimators

$$\widehat{\theta} = \frac{Y}{X} \quad \text{where } \theta = \frac{\mathbb{E}\, Y}{\mathbb{E}\, X}$$

with $X, Y \geq 0$, $\mathbb{P}\{X > 0\} > 0$ and $X = 0 \Rightarrow Y = 0$, typically arising as the mean of a weighted empirical distribution where the weights are random variables (Baddeley *et al.*, 1993, Ripley, 1988). We call such estimators 'ratio-unbiased' (Baddeley, 1993) and accept this property as a substitute for the generally unobtainable unbiasedness.

The *associated point rule* introduced by Miles (1978) is an unbiased sampling rule. Associate with each object $K \in \mathcal{K}$ a unique point $c(K) \in \mathbb{R}^d$, such as the centroid, lowest point, or circumcentre of K. The choice of this 'associated point' is arbitrary provided it is equivariant under translations, $c(K + x) = c(K) + x$ for all $K \in \mathcal{K}$ and $x \in \mathbb{R}^d$. Then we sample an object X_i if and only if its associated point $c(X_i)$ falls in the window W. See Figure 2.5.

The associated point $c(X_i)$ should not be confused with the germ point x_i featuring in the germ-grain process construction. The germ point is a mathematical convenience which need not be observable in practice. The associated point $c(X_i)$ is observable since it can be determined from X_i alone.

For any $K \in \mathcal{K}$ we have that $K + x$ is sampled if and only if $c(K + x) \in W$, or equivalently if $c(K) + x \in W$, that is, if $x \in (W - c(K))$. The latter set is congruent to W. Thus the bias factor (2.3)

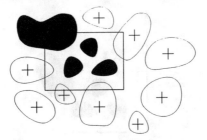

Figure 2.5 *Miles' associated point rule. An object X_i is sampled if and only if its associated point $c(X_i)$ (marked $+$) falls in the sampling frame W. The shaded objects are sampled.*

is

$$\pi(K) = \int_{\mathbb{R}^d} \mathbf{1}\{c(K + x) \in W\}\, dx$$
$$= |W - c(K)|$$
$$= |W|.$$

Hence the associated point rule is unbiased.

Additionally, since the value of $\pi(K)$ is known, the 'sample total' formula (2.2) can be used in practice:

$$\mathbb{E} \sum_{\text{sample}} f(X_i) = \alpha\, |W|\, \mathbb{E}^0 \left[f(K_0) \right] \tag{2.5}$$

so that

$$\widehat{\alpha} = \frac{\text{number in sample}}{|W|} \tag{2.6}$$

is an unbiased estimator of α.

An alternative method, Gundersen's (1977, 1978) *tiling rule*, has become very popular in microscopy Gundersen *et al.* (1988a, b). Its practical implementation is sketched in Figure 2.6. Any object X_i intersecting the rectangular sampling frame W will be sampled, provided it does not intersect any of the 'forbidden' lines marked in bold (namely one side of the rectangle and two half-infinite lines extending from it).

An equivalent description of the tiling rule is as follows. Tessellate $\mathbb{R}^2$ with copies of W, say $W_{m,n} = W + (ma, nb)$ where a, b are the side lengths of W. Order the tiles $W_{m,n}$ according to an arbitrary total order. Then we sample X_i in tile W if it intersects W and does not intersect any tile which is 'earlier' in this ordering. Here we have

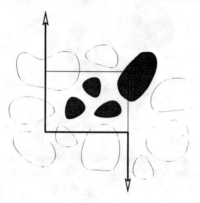

Figure 2.6 *Gundersen's tiling rule in a rectangular sampling window W. The shaded objects are sampled.*

used the ordering in which the tiles to the left of W or below W are 'earlier' than W,

$$W_{m,n} \prec W_{m',n'} \quad \text{iff} \quad m < m' \text{ or } (m = m' \text{ and } n < n').$$

It is easy to check that this ordering implies the 'forbidden line' rule enunciated above.

The key fact is that any object X_i will be sampled by *exactly one* of the tiles $W_{m,n}$ (namely the 'earliest' tile which intersects X_i). Fix $K \in \mathcal{K}$ and let $S(K, m, n)$ be the set of all translations $x \in \mathbb{R}^2$ such that $K + x$ will be sampled in $W_{m,n}$. The sets $S(K, m, n), m, n \in \mathbb{Z}$ are congruent, disjoint and cover the whole of $\mathbb{R}^2$. Hence $|S(K, m, n)| = |W|$ for all m, n and

$$\pi(K) = |S(K, 0, 0)| = |W|.$$

The tiling rule is therefore unbiased, and again we have (2.5)–(2.6).

Both the associated point rule and the tiling rule generally require a guard area around the sampling frame, in order to determine which objects X_i are sampled. This can be dispensed with in some cases, for example, when it is known beforehand that all X_i are convex. The tiling rule is easier to perform manually, but applies only when the window shape tessellates the plane. The associated point rule is easy for computers to carry out and applies to arbitrary window shapes.

2.3 Additive functionals

An extension of the foregoing strategy is to allow objects X_i to be counted potentially more than once, but to compensate by assigning each object a fractional weight (DeHoff, 1977, Hall, 1985, Jensen and Sundberg, 1986, Miles, 1978).

For example, in a spatial pattern of line segments, Hall (1985, 1988, pp. 216–217) observed that since every line segment has two end-points, we could estimate the expected number of segments per unit area by counting the number of segment endpoints in W and dividing by $2|W|$. The sample of line segments intersecting W (i.e. selected by plus-sampling) can be converted into an unbiased sample by assigning a weight of $\frac{1}{2}$ to segments with only one endpoint in W, and a weight of 1 to segments with both endpoints in W. This is a variant of Miles' associated point method in which each segment has two associated points, namely its endpoints. Independently Jensen and Sundberg (1986) proposed the use of $m > 1$ associated points for each object.

In general, let $u(X_i, W)$ be the weight which we attach to an object X_i when it is sampled in window W. If we can arrange that the identity

$$\int_{\mathbb{R}^d} I(K + x) u(K + x, W) \, dx = 1 \qquad (2.7)$$

holds for all fixed K and W, where I is the sampling rule as in Theorem 2.1, then the bias is corrected, in the sense that (2.2) is replaced by

$$\mathbb{E} \sum_{\text{sample}} u(X_i, W) f(X_i) = \alpha \, \mathbb{E}^0 \left[f(K_0) \right]. \qquad (2.8)$$

Following are some mechanisms which guarantee (2.7).

- Equip each object X_i with $m > 1$ different associated points $c_1(X_i)$, $\ldots, c_m(X_i)$, and weight X_i by the number of associated points of X_i which fall in the sampling window Jensen and Sundberg (1986). Thus

$$u(K, W) = \frac{1}{m} \sum_{j=1}^{m} \mathbf{1}\{c_j(K) \in W\};$$

it is easy to verify (2.7).

- Weight each object K proportional to its area of intersection with the window,

$$u(K, W) = \frac{|K \cap W|}{|K|};$$

standard integral geometric results give (2.7).

- Tessellate the plane with copies of the sampling window, and weight X_i by the reciprocal of the number of tiles which intersect X_i. We can verify (2.7) using arguments like those for Gundersen's tiling rule.

- Other weight functions in $\mathbb{R}^2$ include the integral of curvature of $W \cap \partial X_i$, if the boundary ∂X_i of each X_i is a simple closed curve; and 4 minus the number of boundary intersections, $1 - \#(\partial W \cap \partial X_i)/4$, if both X_i and W are convex.

2.4 Horvitz–Thompson estimators

Miles (1974) showed that spatial sampling bias can be corrected by weighting each sampled object X_i by a quantity analogous to the reciprocal of sampling probability. Lantuéjoul (1978a, b) further explored and extended the results.

Weighting by the reciprocal of sampling probability is a well-known technique which goes back to the Horvitz–Thompson estimator of survey sampling theory.

2.4.1 Horvitz–Thompson estimator in a finite population

We shall first review this estimator (Horvitz and Thompson, 1952, Cochran, 1977, pp. 259–261, Krishnaiah and Rao, 1988, pp. 291, 313, 428, Brewer and Hanif, 1983).

Suppose we observe a (non-independent, non-uniform) random sample from a finite population. The sample size may be random; the only condition is that $\pi_i = \mathbb{P}\{i \in \text{sample}\}$ be known and positive for all i. Then we can estimate any population total by weighting each element i in the sample by $1/\pi_i$. Let

$$Y = \sum_{i \in \text{population}} y_i$$

be the population total of some variable y. The **Horvitz–Thompson estimator**

$$\widehat{Y}_{HT} = \sum_{i \in \text{sample}} \frac{y_i}{\pi_i} \tag{2.9}$$

is unbiased for Y since

$$\mathbb{E}\widehat{Y}_{HT} = \mathbb{E}\left[\sum_{i \in \text{population}} \mathbb{1}\{i \in \text{sample}\}\frac{y_i}{\pi_i}\right]$$

$$= \sum_{i \in \text{population}} \pi_i \frac{y_i}{\pi_i} = Y. \tag{2.10}$$

2.4.2 Spatial Horvitz–Thompson estimators

Returning to the context of Theorem 2.1, consider a stationary germ-grain model with germ intensity α and compact grains K_i with common distribution Q. The sample consists of all objects X_i such that $I(X_i) = 1$. Adapting the Horvitz–Thompson approach to this spatial random sample is more complicated than (2.10) in that we require averages over an infinite population, and we cannot simply exchange the expectation and summation. The solution is again to use the Campbell–Mecke formula (Chapter 1, section 1.8.3). Suppose we weight each sampled object X_i by $1/\pi(X_i)$, the reciprocal of the sampling bias factor defined at (2.3). This requires that $\pi(X_i)$ be known and almost surely positive for the typical grain. Then

$$\mathbb{E}\left[\sum_{i \in \text{sample}} \frac{1}{\pi(X_i)}\right] = \alpha$$

$$\mathbb{E}\left[\sum_{i \in \text{sample}} \frac{1}{\pi(X_i)}f(X_i)\right] = \alpha\,\mathbb{E}^0 f(K_0);$$

Hence

$$\frac{\mathbb{E}\left[\sum_{i \in \text{sample}} f(X_i)/\pi(X_i)\right]}{\mathbb{E}\left[\sum_{i \in \text{sample}} 1/\pi(X_i)\right]} = \mathbb{E}^0 f(K_0) \tag{2.11}$$

so that the Horvitz–Thompson style estimator

$$\frac{\sum_{i \in \text{sample}} f(X_i)/\pi(X_i)}{\sum_{i \in \text{sample}} 1/\pi(X_i)} \tag{2.12}$$

is ratio-unbiased, and (under suitable regularity conditions) consistent and approximately unbiased for $\mathbb{E}^0 f(K_0)$.

For example, the bias introduced by plus-sampling can be corrected by weighting each object X_i in the sample by $1/|W \oplus \check{X}_i|$. Similarly the bias of minus-sampling can be corrected using the weights

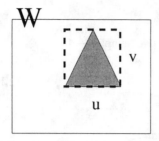

Figure 2.7 *Calculation of the minus sampling correction* $|W \ominus K|$ *in a rectangular sampling window* W.

$1/|W \ominus X_i|$, but here we require that $|W \ominus K_0| > 0$ almost surely, i.e. every grain must be small enough to fit in the sampling window.

Lantuéjoul (1978a, b) noted that the minus-sampling correction is easy to compute when W is an $a \times b$ rectangle, for in that case $|W \ominus X_i| = (a - h)(b - v)$ where h and v are the dimensions of the smallest rectangle (aligned with W) containing K. See Figure 2.7.

Unfortunately the analogy with classical Horvitz–Thompson estimators is not strong enough to enable the estimation of variances in a simple way. The Sen–Yates–Grundy and similar variance estimators (Krishnaiah and Rao, 1988)) depend on specific model assumptions in the finite population case, and in our context the required information about joint sampling probabilities would also have to be estimated from the data.

2.5 Sampling bias for point processes

We now focus on the special case of spatial point processes. Sampling bias effects for point processes have been extensively discussed, and corrections introduced by Ripley, Lantuéjoul, Hanisch, Stoyan, Ohser and others; see Diggle (1983), Hanisch (1984), Ohser (1983), Ripley (1977, 1981), Serra (1982, p. 246). For surveys see Ripley (1988, chap. 3), Stoyan *et al.* (1987, pp. 122–131), Cressie (1991, chap. 8), Baddeley (1993).

In the exploratory data analysis of a point pattern, one starts typically by estimating certain distance distributions: $F(t)$, the distribution of the distance from an arbitrary point in space to the nearest point of the process; $G(t)$, the distribution of the distance from a typical point of the process to the nearest other point of the process; and $K(t)$, the expected number of other points within distance t of a typical point of the process, divided by the intensity. For a homogeneous Poisson process F, G and K take known functional forms, and deviations of estimates of F, G, K from these forms are taken as indications of 'clustered' or 'inhibited' alternatives (Diggle, 1983, Ripley, 1981, 1988). We will first define these functions and then consider the sampling bias effects.

2.5.1 Definitions

For $x \in \mathbb{R}^d$ and any subset $A \subset \mathbb{R}^d$ let

$$\rho(x, A) = \inf\{\|x - a\| : a \in A\} \tag{2.13}$$

be the shortest Euclidean distance from x to A, and

$$A_{(+r)} = \{x \in \mathbb{R}^d : \rho(x, A) \leq r\}$$
$$A_{(-r)} = \{x \in A : \rho(x, A^c) > r\}$$

where ^{c} denotes complement. For closed sets A, these are respectively the dilation and erosion of A by a ball of radius r. Write $b(x, r)$ for the closed ball of radius r and centre x in $\mathbb{R}^d$.

Let Φ be a simple point process in $\mathbb{R}^d$ which is a.s. stationary under translations, with finite positive intensity α. For $r \geq 0$ define

$$F(r) = \mathbb{P}\{\rho_B(0, X) \leq r\} \tag{2.14}$$

By stationarity the point 0 here may be replaced by any arbitrary point x. Thus F is the cumulative distribution function of the random distance $\rho_B(0, X)$ from an arbitrary point 0 to the nearest random point. Also define

$$G(r) = \mathbb{P}^0\{\rho(0, \Phi \setminus \{0\}) \leq r\} \tag{2.15}$$

where $\mathbb{P}^0$ denotes the Palm distribution of Φ at 0. Thus G is the c.d.f. of the distance *from a typical random point of the process* to the nearest other random point. Alternatively, treating Φ as a random measure, we could write

$$F(r) = \mathbb{P}\{\Phi(b(0, r)) > 0\} \tag{2.16}$$
$$G(r) = \mathbb{P}^0\{\Phi(b(0, r) \setminus \{0\}) > 0\} \tag{2.17}$$

where $\Phi(A)$ denotes the random number of points of Φ falling in $A \subset \mathbb{R}^d$.

Further define Ripley's K-function

$$K(r) = \alpha^{-1}\mathbb{E}^0\left[\Phi(b(0, r) \setminus \{0\})\right] \tag{2.18}$$

where $\mathbb{E}^0$ denotes the expectation with respect to the Palm distribution $\mathbb{P}^0$. Thus $\alpha K(r)$ is the expected number of further points within a distance r of a typical random point of the process.

It turns out (see Theorem 2.2 of section 2.6.5) that F is always differentiable, while G and K need not have any special continuity properties. In fact G may be degenerate and K purely discrete, as in the case of a randomly translated lattice.

The following results are useful for estimation of F, G and K. Trivially

$$F(r) = \mathbb{P}\{\rho_B(0, X) \leq r\} = \mathbb{P}\{0 \in \Phi_{(+r)}\}$$

so that by Robbins' Theorem (Theorem 1 of Chapter 1, section 1.2)

$$F(r) = \frac{\mathbb{E}\left|\Phi_{(+r)} \cap A\right|_d}{|A|_d} \qquad (2.19)$$

for arbitrary Borel sets A with $0 < |A|_d < \infty$, where $|\cdot|_d$ denotes Lebesgue volume in $\mathbb{R}^d$. Using the Campbell–Mecke formula (equation (1.26) in Theorem 9 of Chapter 1, section 1.8.3)

$$G(r) = \frac{\mathbb{E}\sum_{x \in \Phi \cap A} \mathbf{1}\{\rho(x, \Phi \setminus \{x\}) \leq r\}}{\mathbb{E}\Phi(A)} \qquad (2.20)$$

and

$$\alpha K(r) = \frac{\mathbb{E}\sum_{x \in \Phi \cap A} \Phi(b(x, r) \setminus \{x\})}{\mathbb{E}\Phi(A)} \qquad (2.21)$$

The latter applies to any second-order stationary process (Stoyan *et al.*, 1987).

2.5.2 Estimation problem

The point process Φ is observed through a window $W \subset \mathbb{R}^d$. We assume W is compact and inner regular (it is the closure of its interior), and denote its boundary by ∂W.

Estimators of $F(r)$, $G(r)$ and $K(r)$ can be formed by taking the appropriate sample averages. From (2.19)–(2.21) we have that for each fixed r and for any $A \subset \mathbb{R}^d$ with $0 < |A|_d < \infty$,

$$\widehat{F}(r) = \frac{\left|\Phi_{(+r)} \cap A\right|_d}{|A|_d} \qquad (2.22)$$

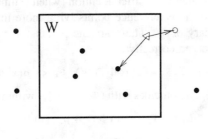

Figure 2.8 *Edge effect for distances in point patterns. Filled dots: point pattern* Φ. *Open circle: reference point* x, *either a fixed point in* W *or a point of the pattern* Φ. *The edge effect arises because the nearest point of* Φ *to* x *could lie outside the window of observation* W.

is an unbiased estimator of $F(r)$, while

$$\widehat{G}(r) = \frac{\sum_{x \in \Phi \cap A} 1\{\rho(x, \Phi \setminus \{x\}) \leq r\}}{\Phi(A)} \tag{2.23}$$

is ratio-unbiased for $G(r)$ and

$$\widehat{\alpha K(r)} = \frac{\sum_{x \in \Phi \cap A} \Phi(b(x, r) \setminus \{x\})}{\Phi(A)} \tag{2.24}$$

is ratio-unbiased for $\alpha K(r)$.

However, *these estimators are not feasible* in general, since they require information from outside the window W. The essential problem is an edge effect: $\rho_B(x, X)$ is not determined from information inside W alone, since for a given point $x \in W$ the closest point of Φ may lie outside W. The available data $\Phi \cap W$ yield $\rho(x, \Phi \cap W)$ rather than $\rho_B(x, X)$ for all $x \in W$. See Figure 2.8.

2.5.3 Border method

Edge-corrected estimators for F, G and K based on observation of Φ in W are reviewed in Ripley (1988, chap. 3), Stoyan *et al.* (1987), pp. 122–131, Cressie (1991, chap. 8). See Baddeley *et al.* (1993), Barendregt and Rottschäfer (1991), Doguwa (1989, 1990, 1992), Doguwa and Choji (1991), Doguwa and Upton (1989, 1990), Fiksel (1988), Stein (1991).

The simplest approach is the 'border method' (Diggle, 1979, Ripley, 1981, 1988) in which we restrict attention (when estimating F, G or K at distance r) to those reference points lying more than r units away from the boundary of W. These are the points x for which distances up to r are observed correctly:

$$\rho_B(x, \partial W) > r \quad \Rightarrow \quad (\rho(x, \Phi \cap W) \leq r \Leftrightarrow \rho(x, \Phi) \leq r). \quad (2.25)$$

Equivalently, $\Phi_{(+r)}$ coincides with $(\Phi \cap W)_{(+r)}$ within the mask $W_{(-r)}$:

$$\Phi_{(+r)} \cap W_{(-r)} = (\Phi \cap W)_{(+r)} \cap W_{(-r)}. \quad (2.26)$$

This is an instance of the 'local knowledge principle' of mathematical morphology (Serra, 1982, pp. 49, 233, Baddeley and Heijmans, 1995).

The **border method** estimators of F, G and K are the sample averages within the eroded window:

$$\widehat{F}^b(r) = \frac{\left|W_{(-r)} \cap \Phi_{(+r)}\right|_d}{\left|W_{(-r)}\right|_d} \quad (2.27)$$

$$\widehat{G}^b(r) = \frac{\sum_{x \in \Phi \cap W_{(-r)}} \mathbf{1}\{\rho(x, \Phi \setminus \{x\}) \leq r\}}{\Phi(W_{(-r)})} \quad (2.28)$$

$$\widehat{K}^b(r) = \frac{\sum_{x \in \Phi \cap W_{(-r)}} \Phi(b(x, r) \setminus \{x\})}{\widehat{\alpha}\, \Phi(W_{(-r)})} \quad (2.29)$$

where

$$\widehat{\alpha} = \frac{\Phi(W)}{|W|_d}.$$

These are the estimators (2.22)–(2.24) using $A = W_{(-r)}$. By (2.26) the numerator in each case is observable in the sense that it is determined by the data $\Phi \cap W$. This approach was introduced by Diggle (1979) and dubbed the 'border method' by Ripley (1988).

The border method estimator of F is pointwise unbiased, $\mathbb{E}\widehat{F}^b(r) = F(r)$ for all r satisfying $\left|W_{(-r)}\right|_d > 0$. The estimator of G is ratio-unbiased, and $\widehat{\alpha}\, \widehat{K}^b(r)$ is ratio-unbiased for $\alpha K(r)$. However, $\widehat{F}^b$ and $\widehat{G}^b$ may not be distribution functions. The three estimators may fail to be monotone functions of r, and $\widehat{F}^b$ and $\widehat{G}^b$ may have maximum values either greater or less than unity. The border method also discards much of the data; in three dimensions (Baddeley et al., 1993) it seems to be unacceptably wasteful, especially when estimating G.

The variances of all three estimators increase with r. One possibility for reducing variance in the estimators of G and K is to replace the denominator $\Phi(W_{(-r)})$ by a less variable estimate of its expectation,

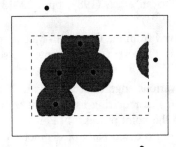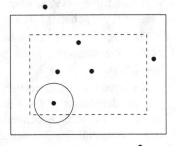

Figure 2.9 *Geometry of the border method estimators. Spatial process* Φ *indicated by filled dots. The eroded window* $W_{(-r)}$ *is the dashed rectangle. Left: The estimator of F is the fraction of area shaded in the eroded window. Right: The estimator of G or K is a sum of contributions from points* x_i *in the eroded window.*

$\widehat{\alpha} \, |W_{(-r)}|_d$. For example

$$G_2(r) = \frac{|W|_d}{\Phi(W)} \frac{\sum_{x \in \Phi \cap W_{(-r)}} \mathbf{1}\{\rho(x, \Phi \setminus \{x\}) \le r\}}{|W_{(-r)}|_d}, \qquad (2.30)$$

see Stoyan *et al.* (1987, p. 178). However, this may not reduce the variance, since it is plausible that the numerators and denominators of (2.28) and (2.29) may be positively correlated. Such estimators also fail to be monotone.

2.5.4 Edge corrections

'Edge-corrected' estimators are generally an improvement on the border method for G and K; these are weighted empirical distributions of the distances between points. The weight $c(x, y)$ attached to the observed distance $\|x - y\|$ between two points x, y is the reciprocal of the 'probability' of observing this distance, under invariance assumptions (stationarity under translation and/or rotation). In other words, these are Horvitz–Thompson style estimators. Corrections of this type were first suggested by Miles (1974) and developed by Ripley, Lantuéjoul, Hanisch, Stoyan, Ohser and others (Diggle, 1983, Hanisch, 1984, Ohser, 1983, Ripley, 1977, 1981, Serra, 1982, p. 246). For surveys see Ripley (1988, chap. 3), Stoyan *et al.* (1987, pp. 122–131), Cressie (1991, chap. 8), Baddeley (1993).

The edge corrections can be derived from the Campbell–Mecke formula for a stationary point process (Theorem 9 of Chapter 1, section 1.8.3).

A simple way to appreciate the various edge corrections is to interpret the definitions of $G(r)$ and $K(r)$ as requiring us to count certain geometrical objects associated with the point process Φ. For $K(r)$ we should count all line segments joining ordered pairs of distinct points x, y in Φ such that $\|x - y\| \leq r$. For $G(r)$ we count all discs $b(x_i, r)$ centred on points of Φ which contain no other points of Φ. Then the edge corrections are instances of the corrections already described for spatial processes of geometrical objects.

First consider the estimation of G. The notation for estimators will be simplified if we write for each point x_i of $\Phi \cap W$

$$s_i = \rho(x_i, \Phi \setminus \{x_i\} \cap W) \tag{2.31}$$
$$c_i = \rho_B(x_i, \partial W) \tag{2.32}$$

for the distances from x_i to its nearest neighbour and to the edge of the window, respectively.

Suppose we construct from Φ the random process of all discs or spheres $b(x_i, r)$ of radius r centred on points x_i of Φ which do not contain any further points of Φ. Minus-sampling for these discs is equivalent to sampling the points $x_i \in W_{(-r)}$. The minus-sampling bias factor for a disc $b(x, r)$ is

$$\pi(b(x, r)) = \left| W_{(-r)} \right|_d$$

so that the Horvitz–Thompson style estimator for G based on minus sampling is just the border method estimator G_2 of (2.28),

$$\widehat{G}_2(r) = \frac{\sum_i 1\{s_i \leq r\} \, 1\{c_i > r\}}{\left| W_{(-r)} \right|_d}.$$

Alternatively, consider the process of all discs or spheres $b(x_i, s_i)$ where s_i is the nearest neighbour distance as above. Minus-sampling for these discs is equivalent to sampling those $x_i \in \Phi \cap W$ for which $s_i \leq c_i$ or equivalently $x_i \in W_{(-s_i)}$. The corresponding Horvitz–Thompson style estimator is

$$\widehat{\alpha}\,\widehat{G}_4(r) = \sum_{x \in \Phi \cap W} \frac{1\{\rho(x, \Phi \setminus \{x\}) \le \rho_B(x, \partial W)\}\,1\{\rho(x, \Phi \setminus \{x\}) \le r\}}{|W_{(-\rho(x, \Phi \setminus \{x\}))}|_d}$$

$$= \sum_i \frac{1\{s_i \le c_i\}\,1\{s_i \le r\}}{|W_{(-s_i)}|_d}. \tag{2.33}$$

This estimator was introduced *ad hoc* by Hanisch (1984), along with the corresponding ratio-of-counts estimator $G_3(r)$ in which the denominator of (2.33) is replaced by

$$\sum_i 1\{s_i \le c_i\}.$$

Other estimators of G are described in Stoyan *et al.* (1987, p. 128), Cressie (1991, p. 614, 637–638), Doguwa (1989), Doguwa and Upton (1990) and Floresroux and Stein (1996), and in sections 2.5.5, 2.6.3 and 2.6.7 below.

Next consider the estimation of K. Construct the process of all line segments joining distinct points x and y of Φ with length $\|x - y\| \le r$. Each segment is counted once for each endpoint. Minus-sampling of these line segments is equivalent to simply observing all ordered pairs of distinct points in the window which are at most r units apart. The minus-sampling bias factor for a line segment joining the points x and y is

$$\pi(\{x, y\}) = |W \ominus \{x, y\}|_d$$
$$= |(W + x) \cap (W + y)|_d = |W \cap (W + y - x)|_d$$

so that

$$\widehat{\alpha}^2\,\widehat{K}_2(r) = \sum_{x, y \in \Phi \cap W} \frac{1\{0 < \|x - y\| \le r\}}{|(W + x) \cap (W + y)|_d} \tag{2.34}$$

is unbiased for $\alpha^2 K(r)$, provided $|W \cap (W + z)|_d > 0$ for all $z \in b(0, r)$. This is dubbed the *translation correction*. Recall that the sampling bias factor is easy to compute for rectangular windows, see Figure 2.7.

If the point process Φ is also isotropic (invariant under rotations) we may invoke a variant of Theorem 2.1 in which the bias factor $\pi(K)$ is replaced by a rotational average. This leads to Ripley's (1976, 1977) estimator (slightly corrected by Ohser (1983)) in two dimensions,

$$\widehat{\alpha^2} \, \widehat{K}_3(r) = \sum_{x,y \in \Phi \cap W} \frac{1\{0 < \|x - y\| \leq r\}}{w(x, r) \, v(\|x - y\|)} \tag{2.35}$$

which again is unbiased for $\alpha^2 K(r)$, where

$$w(x, r) = \frac{1}{2\pi r} \, \text{length}(W \cap \partial b(x, r))$$

i.e. $w(x, \|x - y\|)$ is the fraction of the circumference of the circle centred at x and passing through y which lies in W, and

$$v(r) = |\{x \in W \ : \ W \cap \partial b(x, r) \neq \emptyset\}| \, .$$

See also Diggle (1983), Ohser (1983), Ripley (1977, 1981, 1988, chap. 3), Stoyan *et al.* (1987, pp. 122–131), Cressie (1991, pp. 616–619, 639–644) and recent investigations in Doguwa (1990), Doguwa and Upton (1989), Stein (1991).

Small-sample variances of these estimators are intractable. They have been studied by Ripley (1979, 1988, p. 40) and Ohser (1983) for the Poisson and binomial processes. Asymptotic limiting distributions have been obtained by Heinrich (1988a, b) for Poisson cluster processes; see also Heinrich (1991). Various limiting regimes have been studied by Stein (1991, 1993, 1995) and in Baddeley and Gill (1997).

2.5.5 Stein's variance reduction techniques

A general problem with the Horvitz–Thompson style estimators is that objects X_i with very small values of $\pi(X_i)$ will be given very large weights if they are encountered in the sample. This is a substantial source of variability. In the edge corrections described above, contributions from points close to the edge of the window will be given large weights, inflating the variance of the estimator, despite the paucity of data near the edge.

One strategy for variance reduction is to downweight the more variable contributions. Stein (1991) pointed out that any modification of Ripley's estimator (2.35) of the form

$$\widehat{\alpha^2} \, \widehat{K}_4(r) = \sum_{x,y \in \Phi \cap W} u(x, r) \frac{1\{0 < \|x - y\| \leq r\}}{w(x, r) \, v(\|x - y\|)} \tag{2.36}$$

where $u(\cdot, \cdot)$ is a weight function, is still unbiased for $\alpha^2 K(r)$ provided

$$\int_{W[r]} u(x, r) \, dx = 1$$

where $W[r] = \{x \in W : \partial b(x, r) \cap W \neq \emptyset\}$. This estimator may have smaller variance than (2.35) if the values of u decrease near the boundary of W. There is one natural choice of u which has superior asymptotic properties when the underlying process is Poisson.

A second strategy for variance reduction is *projection* (Stein, 1993). Take a statistic of the form

$$T = \sum_{x_i, x_j \in \Phi \cap W} \phi(x_i, x_j)$$

such as (2.34)–(2.35). Consider the modified statistic

$$T' = \sum_{x_i, x_j \in \Phi \cap W} \phi(x_i, x_j) - \sum_{x_i \in \Phi \cap W} \left[g(x_i) - \frac{1}{|W|_d} \int_W g(x) \, dx \right]$$

$$(2.37)$$

where $g : W \to \mathbb{R}$ is any function. If T is unbiased then so is T', because the expectation of the second sum on the right of (2.37) is zero, by the Campbell–Mecke formula. The Hájek Projection Lemma (Hájek, 1968, Lemma 4.1) identifies the choice of g which minimises the conditional variance of (2.37) when the process is Poisson. Stein (1993) applies this modification to the rigid motion correction estimator for $\alpha^2 K(r)$ and demonstrates that T' has substantially lower variance than the Ripley isotropic correction.

These variance reduction techniques clearly should apply to any of the Horvitz–Thompson style estimators. A practical problem with projection is that the optimal g is often difficult to compute.

2.6 Censoring effects for point processes

Edge effects can also be interpreted as a type of censoring. In this section we deal with censoring effects for point processes.

The estimation problem for F, G and K from a point pattern in a bounded window W has a clear analogy to the estimation of a survival function based on a sample of randomly censored survival times. Essentially the distance from a given reference point x to Φ is right-censored by its distance to the boundary of W. We shall first recall some basic theory of random censoring.

2.6.1 Survival data

Following is a brief account of random censoring. Suppose $T_1, \ldots, T_n$ are i.i.d. positive r.v.'s with distribution function F and survival function $S = 1 - F$. Let $C_1, \ldots, C_n$ be independent of the T_i's and i.i.d. with d.f. H. Let $\tilde{T}_i = T_i \wedge C_i$, $D_i = 1\{T_i \leq C_i\}$ where $a \wedge b$ denotes $\min\{a, b\}$. Then $(\tilde{T}_1, D_1), \ldots, (\tilde{T}_n, D_n)$ is a sample of censored survival times $\tilde{T}_i$ with censoring indicators D_i (really, *non*-censoring indicators).

The *reduced-sample estimator* of F is

$$\widehat{F}^{\text{rs}}(t) = \frac{\#\{i : \tilde{T}_i \leq t \leq C_i\}}{\#\{i : C_i \geq t\}} \tag{2.38}$$

This requires that we can observe the censoring times C_i themselves, or at least the event $\{C_i \geq t\}$ for all t for which $F(t)$ must be estimated. This estimator is clearly pointwise unbiased for F and has values in $[0, 1]$ but may not be a monotone function of t.

The optimal estimator of F is the *Kaplan–Meier estimator* (Kaplan and Meier, 1958),

$$\widehat{F}(t) = 1 - \prod_{s \leq t} \left(1 - \frac{\#\{i : \tilde{T}_i = s, D_i = 1\}}{\#\{i : \tilde{T}_i \geq s\}} \right) \tag{2.39}$$

Note that the product in (2.39) is effectively only a finite product in which s ranges over the observed failure times $\tilde{T}_i$. Thus $\widehat{F}(t)$ jumps only at these values of t. The validity of the Kaplan–Meier estimator can be understood intuitively by considering $\#\{i : \tilde{T}_i \in \mathbf{d}s, D_i = 1\}0/\#\{i : \tilde{T}_i \geq s\}$, for a small interval $\mathbf{d}s = [s, s + \mathbf{d}s)$, as an estimator of $\mathbb{P}\{T_i \in \mathbf{d}s \mid T_i \geq s\}$. The complement of this probability is therefore $\mathbb{P}\{T_i \geq s + \mathbf{d}s \mid T_i \geq s\}$. Multiplying over small intervals $[s, s + \mathbf{d}s)$ partitioning $[0, t + dt)$ produces $\mathbb{P}\{T_i > t\} = 1 - F(t)$.

More formally, introduce

$$N_n(t) = \frac{1}{n}\#\{i : \tilde{T}_i \leq t, D_i = 1\} \tag{2.40}$$

$$Y_n(t) = \frac{1}{n}\#\{i : \tilde{T}_i \geq t\} \tag{2.41}$$

$$\widehat{\Lambda}_n(t) = \int_0^t \frac{dN_n(s)}{Y_n(s)} \tag{2.42}$$

$$\Lambda(t) = \int_0^t \frac{dF(s)}{1 - F(s-)}. \tag{2.43}$$

Then Λ is the *cumulative hazard* associated with F, and $\widehat{\Lambda}_n$ is the

Nelson-Aalen estimator of Λ. One can write

$$1 - F(t) = \prod_0^t (1 - d\Lambda(s)),$$

$$1 - \widehat{F}_n(t) = \prod_0^t (1 - d\widehat{\Lambda}_n(s)) \tag{2.44}$$

where $\prod$ denotes **product integration**:

$$\prod_0^t (1 + dA(s)) = \lim_{\max |t_i - t_{i-1}| \to 0} \prod_{i=1}^m (1 + A(t_i) - A(t_{i-1})),$$

the limit of the product over increasingly fine partitions $0 = t_0 < \cdots < t_m = t$ of the interval $(0, t]$. See Gill (1994), Gill and Johansen (1990) for further information on the product integral.

If F is absolutely continuous with density f then defining the *hazard rate*

$$\lambda(t) = f(t)/(1 - F(t))$$

one has $\Lambda(t) = \int_0^t \lambda(s)\,ds$ and

$$1 - F(t) = \prod_0^t (1 - d\Lambda(s)) = \exp(-\Lambda(t)).$$

However if F has a discrete component the relation $\Lambda = -\log(1 - F)$ no longer holds.

Under random censorship the empirical processes N_n, Y_n satisfy a mean value relation

$$\mathbb{E}N_n(t) = \int_0^t \mathbb{E}Y_n(s)\,d\Lambda(s) \tag{2.45}$$

which may be interpreted loosely as saying that $dN_n(t)/Y_n(t)$ is ratio-unbiased for $d\Lambda(t)$.

2.6.2 Analogy between censoring and edge effects

Returning to the spatial pattern context, let Φ be an a.s. stationary point process in $\mathbb{R}^d$, and W a fixed compact window W with nonempty interior. We observe Φ only within W.

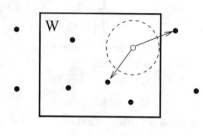

Figure 2.10 *Censoring occurs when the reference point x (open circle) is closer to the boundary of the observation window W than to the nearest point of the spatial pattern (filled dots).*

Consider the estimation of the empty space function F. Every point x in the window W contributes one possibly censored observation of the distance from an arbitrary point in space to the nearest point of Φ. The analogy with survival times is to regard the distance

$$T = \rho_B(x, X)$$

as the 'distance (time) to failure', and

$$C = \rho_B(x, \partial W)$$

as the censoring distance. The observation is censored if $\rho_B(x, \partial W) < \rho_B(x, X)$. See Figure 2.10.

Similar remarks apply to the nearest neighbour distance distribution G. Let x_i be a point of the pattern $\Phi \cap W$. In this case we regard the nearest neighbour distance

$$T = \rho(x_i, \Phi \setminus \{x_i\})$$

as the distance (time) to failure, and the observation is censored if $\rho_B(x_i, \partial W) < \rho(x_i, \Phi \setminus \{x_i\})$.

Under this analogy, *the reduced sample estimator (2.38) for G is precisely the border method estimator (2.28)*. Both are obtained using only those observations for which the censoring time or distance is at least r when estimating the probability of survival to time or distance r. The border method estimator (2.27) of F is also a reduced sample estimator, as we explain in section 2.6.5. Since the reduced sample estimator is known to be inefficient in the case of i.i.d. random censoring, it is of interest to explore analogues of the Kaplan–Meier estimator for point processes.

First we note that Kaplan–Meier style estimators are feasible. For the empty space function, from the data $\Phi \cap W$ we can compute $T^* = \rho(x, \Phi \cap W)$ and $C = \rho_B(x, \partial W)$ for each x. Note that

$$\rho_B(x, X) \wedge \rho_B(x, \partial W) = \rho(x, \Phi \cap W) \wedge \rho_B(x, \partial W) \qquad (2.46)$$

(another application of the local knowledge principle (Serra, 1982, pp. 49, 233)), that is, $T \wedge C = T^* \wedge C$. Thus we can indeed observe $\tilde{T} = T \wedge C$ and $D = \mathbf{1}\{T \leq C\}$ for each x, as required for the Kaplan–Meier estimator.

Similarly for the nearest neighbour distance, replacing Φ by $\Phi \setminus \{x_i\}$, the analogue of (2.46) holds and the other statements above continue to hold.

In the next three sections we discuss Kaplan–Meier style estimators of G, K and F respectively, developed in Baddeley and Gill (1997).

2.6.3 Kaplan–Meier estimator of G for point patterns

It is simplest to begin with the estimation of the nearest neighbour distance distribution function G defined in (2.15), (2.20).

Let $\Phi \cap W = \{x_1, \ldots, x_m\}$ be the observed point pattern and $s_i = \rho(x_i, \Phi \setminus \{x_i\} \cap W)$, $c_i = \rho_B(x_i, \partial W)$. By the analogy sketched above, the set

$$\{x_i \ : \ s_i \wedge c_i \geq r\}$$

can be thought of as the set of points '*at risk* of failure at distance r', and

$$\{x_i \ : \ s_i = r, \, s_i \leq c_i\}$$

are the '*observed* failures at distance r'. These two sets are analogous to the points counted in the empirical functions $Y_n(s)$, $dN_n(s)$ respectively in the definition of the Kaplan–Meier estimator in section 2.6.1. Counting them as for censored data, let

$$Y^G(r) = \#\{x \in \Phi \cap W : r \leq \rho(x, \Phi \setminus \{x\}) \wedge \rho_B(x, \partial W)\}$$
$$= \#\{i : s_i \wedge c_i \geq r\}$$

and

$$N^G(r) = \#\{x \in \Phi \cap W : \rho(x, \Phi \setminus \{x\}) \leq \rho_B(x, \partial W) \wedge r\}$$
$$= \#\{i : s_i \leq c_i \wedge r\}.$$

Continuing the analogy, define the Nelson-Aalen estimator

$$\widehat{\Lambda}^G(r) = \int_0^r \frac{dN^G(s)}{Y^G(s)} \tag{2.47}$$

and the Kaplan–Meier style estimator of G

$$\widehat{G}(r) = 1 - \prod_0^r (1 - d\widehat{\Lambda}^G(s))$$

$$= 1 - \prod_{s \leq r} \left(1 - \frac{\#\{i : s_i = s, \, s_i \leq c_i\}}{\#\{i : s_i \geq s, \, c_i \geq s\}} \right) \tag{2.48}$$

where s in the product ranges over the finite set $\{s_i\}$.

It follows from the Campbell–Mecke formula (see (2.20)) that the numerator and denominator of (2.47) satisfy the same mean-value relation (2.45) as for i.i.d. randomly censored data,

$$\mathbb{E}N^G(r) = \int_0^r \mathbb{E}Y^G(s) \, d\Lambda^G(s), \tag{2.49}$$

where Λ^G is the cumulative hazard associated with G as in (2.43), $d\Lambda^G(s) = dG(s)/(1 - G(s-))$.

Compare this to the reduced-sample (border method) estimator

$$\widehat{G}^b(r) = \frac{\#\{i : s_i \leq r, \, c_i \geq r\}}{\#\{i : c_i \geq r\}}. \tag{2.50}$$

Since the observed, censored distances are highly interdependent, classical theory from survival analysis has little to say about statistical properties of the Kaplan–Meier estimator here. Modest simulation experiments (Baddeley and Gill, 1997) show that the Kaplan–Meier estimator $\widehat{G}$ is generally more efficient than the reduced sample estimator $\widehat{G}^b$, as expected. However, $\widehat{K}$ is not uniformly better than $\widehat{G}^b$, and appears to be less efficient than some of the edge corrected estimators of section 2.5.4. It seems that the spatial dependence has destroyed the uniform optimality enjoyed by the Kaplan–Meier estimator under i.i.d. random censoring. Further investigation is needed.

2.6.4 Kaplan–Meier estimator of K

The function $K(r)$ was defined in (2.18). One may also write

$$\alpha K(r) = \sum_{n=0}^{\infty} G_n(r) \tag{2.51}$$

where $G_n(r) = \mathbb{P}^0 \{\Phi(b(0, r)) > n\}$ is the distribution function of the distance from a typical point of Φ to the nth nearest point. For each of the distance distributions G_n one can form a Kaplan–Meier estimator $\widehat{G}_n$, since the distance from a point $x \in \Phi$ to its nth nearest neighbour is also censored just as before by its distance to the boundary. The sequence of Kaplan–Meier estimators always satisfies the natural stochastic ordering of the distance distributions.

The pointwise sum of these estimators $\widehat{G}_n$ yields an estimator $\widehat{K}$ of K which is always a nondecreasing, right-continuous function, with jumps at the observed interpoint distances (between all pairs of points of the pattern). This estimator was studied briefly in Baddeley and Gill (1997) but needs further investigation.

2.6.5 Kaplan–Meier estimator of F

The estimation of F poses a new problem, since one has a *continuum* of observations: for each point in the sampling window, a censored distance to the nearest point of the process.

The set

$$\{x \in W \ : \ \rho_B(x, X) \wedge \rho_B(x, \partial W) \geq r\}$$

can be thought of as the set of points '*at risk* of failure at distance r', and

$$\{x \in W : \rho_B(x, X) = r, \quad \rho_B(x, X) \leq \rho_B(x, \partial W)\}$$

are the '*observed* failures at distance r'. Geometrically these sets are the closures of $W_{(-r)} \setminus \Phi_{(+r)}$ and $\partial (\Phi_{(+r)}) \cap W_{(-r)}$ respectively. See Figure 2.11.

Define the Kaplan–Meier style estimator $\widehat{F}$ of F, based on data $\Phi \cap W$, to be

$$\widehat{F}(r) = 1 - \exp\left\{ -\int_0^r \frac{\left| \partial (\Phi_{(+s)}) \cap W_{(-s)} \right|_{d-1}}{\left| W_{(-s)} \setminus \Phi_{(+s)} \right|_d} \, ds \right\} \tag{2.52}$$

where $|\cdot|_{d-1}$ denotes $d - 1$ dimensional Hausdorff measure ('surface area' or 'length'). Note that the estimator is a proper distribution function and is even absolutely continuous, with hazard rate

$$\widehat{\lambda}(r) = \frac{\left| \partial (\Phi_{(+r)}) \cap W_{(-r)} \right|_{d-1}}{\left| W_{(-r)} \setminus \Phi_{(+r)} \right|_d} \tag{2.53}$$

for almost all r.

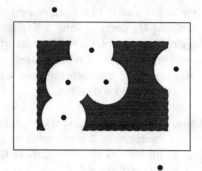

Figure 2.11 *Geometry of the Kaplan–Meier estimator of F. Spatial process* Φ
*indicated by filled dots. Points x at risk are shaded, and observed failures con-
stitute the curved boundary of the shaded region.*

Here $\widehat{F}$ is the Kaplan–Meier estimator based on the continuum of
observations generated by all $x \in W$.

An alternative representation, showing the contribution from each
point x, is

$$\widehat{F}(r) = 1 - \exp\left\{-\int_W \frac{\mathbf{1}\{t(x) \le c(x)\}\,\mathbf{1}\{t(x) \le r\}}{\left|W_{(-t(x))} \setminus \Phi_{(+t(x))}\right|_d}\,dx\right\} \qquad (2.54)$$

where $t(x) = \rho_B(x, X)$ and $c(x) = \rho_B(x, \partial W)$.

In practice, one would compute the estimator by discretizing W,
superimposing a regular lattice L of points, calculating for each $x_i \in
W \cap L$ the censored distance $\rho_B(x_i, X) \wedge \rho_B(x_i, \partial W)$ and the indicator
$\mathbf{1}\{\rho_B(x_i, X) \le \rho_B(x_i, \partial W)\}$. Then one would calculate the ordinary
Kaplan–Meier estimator (2.39) based on this finite dataset. As the lattice
becomes finer, the discrete Kaplan–Meier estimates converge to the
continuous estimator $\widehat{F}$, uniformly on any compact interval in $[0, R)$.
See Baddeley and Gill (1997).

The censoring approach leads to the following insights about F
which are of general interest.

Theorem 2.2 (Baddeley and Gill (1997)) *Let Φ be any stationary point
process with intensity $0 < \alpha < \infty$. Then*

(a) the empty space function F is absolutely continuous;

(b) the hazard rate of F equals

$$\lambda(r) = \frac{\mathbb{E}\left|W \cap \partial\left(\Phi_{(+r)}\right)\right|_{d-1}}{\mathbb{E}\left|W \setminus \Phi_{(+r)}\right|_{d}}$$

for almost all r, for any compact window W such that the denominator is positive.

It follows that the Kaplan–Meier estimator (2.53) of $\lambda(r)$ is ratio-unbiased for almost all r. The estimator $\widehat{F}(r)$ respects the smoothness of the true empty space function F. The border method estimator (2.27) is not even necessarily monotone.

The key to Theorem 2.2 and the representations (2.53)–(2.54) is the identity

$$\left|Z \cap A_{(+r)}\right|_{d} = \left|Z \cap A\right|_{d} + \int_{0}^{r}\left|Z \cap \partial\left(A_{(+s)}\right)\right|_{d-1} ds$$

holding for compact $Z, A \subset \mathbb{R}^{d}$ where A is sufficiently regular. This is related to Crofton's perturbation method (Baddeley, 1977, Crofton, 1869), see section 1.8.2 of Chapter 1). Geometrical techniques are also enlisted (Baddeley and Gill, 1997, Hansen *et al.*, 1996, to appear) to show that $\left|Z \cap \partial\left(\Phi_{(+r)}\right)\right|_{d-1}$ is uniformly bounded over possible realisations of Φ, so that dominated convergence justifies interchanges of expectation and integration or differentiation.

The numerator and denominator of (2.53) satisfy the same mean-value relation as for ordinary randomly censored data,

$$\mathbb{E}N(r) = \int_{0}^{r} \mathbb{E}Y(s)\, d\Lambda(s).$$

It does not seem to be widely known in spatial statistics (cf. Cressie (1991, p. 764), Diggle and Matérn (1981), Doguwa (1992)) that computation of the distances $\rho(x, \Phi \cap W)$, $\rho_B(x, \partial W)$ for all points x in a fine rectangular lattice can be performed very efficiently using the *distance transform* algorithm of image processing (Borgefors, 1984, 1986, Rosenfeld and Pfalz, 1966, 1968). Thus the reduced-sample and Kaplan–Meier estimators are equivalent in computational cost when a fine grid is used.

Again, classical statistical theory for the Kaplan–Meier estimator is not applicable here because of the strong dependence between observations at different points. Simulations (Baddeley and Gill, 1997) suggest that $\widehat{F}$ is substantially more efficient than $\widehat{F}^{b}$ in most situations.

2.6.6 Hanisch-type estimator

The empty space function F could also be estimated by a continuous analogue of Hanisch's (1984) estimator (2.33) for G:

$$\widehat{F}_{\text{cs}}(r) = \int_W \frac{1\{t(x) \le c(x)\} \, 1\{t(x) \le r\}}{\left|W_{(-t(x))}\right|_d} dx \qquad (2.55)$$

where $t(x) = \rho_B(x, X)$ and $c(x) = \rho_B(x, \partial W)$. Chiu and Stoyan (1995) attribute this estimator to earlier work of Hanisch (see also Stoyan et al. (1995, pp. 138, 215)) although it clearly originates in Chiu and Stoyan (1995).

This estimator can be rewritten in a form comparable to (2.52):

$$\widehat{F}_{\text{cs}}(r) = \int_0^r \frac{\left|\partial(\Phi_{(+s)}) \cap (W_{(-s)})\right|_{d-1}}{\left|W_{(-s)}\right|_d} \, ds. \qquad (2.56)$$

It follows from Theorem 2.2 above, and associated continuity results in Baddeley and Gill (1997), that $\widehat{F}_{\text{cs}}(r)$ is unbiased for $F(r)$.

2.6.7 Imputation estimators

The foregoing estimators do not use all 'information' available from a point pattern, in the following sense. Write $C(x)$ for the censoring distance $\rho_B(x, \partial W)$ at a point x, and $T(x)$ for the *true* failure distance $\rho(x, \Phi)$ or $\rho(x, \Phi \setminus \{x\})$ as appropriate. Also write $T^*(x)$ for the *observed* failure distance $\rho(x, \Phi \cap W)$ or $\rho(x, \Phi \cap W \setminus \{x\})$. Then the border method estimate at distance r depends only on those points x where $C(x) \ge r$. The Kaplan–Meier and Hanisch/Chiu-Stoyan estimates use these points, but also use cases where $T^*(x) \le C(x) < r$. However, neither estimator makes use of censored cases where $C(x) < T^*(x)$ and it seems plausible that these may still contain usable information. Indeed the weighted edge-correction estimators for G and K use information from cases where $C(x) < T^*(x) \le r$.

Doguwa (1992) argued that information should be used from all six possible orderings of $C(x), T^*(x), r$. For five of these orderings, it is known with certainty whether the true failure distance satisfies $T(x) \le r$. The sixth ordering, $C(x) < r < T^*(x)$, is the 'maybe' case where the ball of radius r centred on x does not include any observed points of the pattern but also extends outside W.

Consider the estimation of the nearest neighbour distance distribution function G. Let

$$
\begin{aligned}
H(r, x) &= \mathbb{P}^x \left\{ T(x) \leq r \mid T^*(x) > r \right\} \\
&= \mathbb{P}^x \left\{ \rho(x, \Phi) \leq r \mid \rho(x, \Phi \cap W) > r \right\} \\
&= \mathbb{P}^x \left\{ \Phi(b(x, r)) > 2 \mid \Phi(b(x, r) \cap W) = 1 \right\}
\end{aligned}
$$

be the conditional probability that there is another random point within a distance r of x but outside W, given that there is no such point inside W (and given also that x is a point of the process). For the uniform Poisson process of intensity α, $H(r, x) = 1 - \exp\{-\alpha \, |b(x, r) \setminus W|_d\}$. Doguwa (1992) suggested estimating G by

$$
\widehat{G}_6(t) = \frac{1}{n} \sum_{i=1}^{n} [\mathbf{1}\{s_i \leq r\} + \mathbf{1}\{s_i > r\} \mathbf{1}\{c_i < r\} \widehat{H}(r, x_i)] \qquad (2.57)
$$

where $s_i = \rho(x_i, \Phi \setminus \{x_i\} \cap W)$, $c_i = \rho_B(x_i, \partial W)$ and $\widehat{H}(r, x_i) = 1 - \exp\{-\widehat{\alpha} \, |b(x_i, r) \setminus W|_d\}$. This would effectively impute a fraction of the 'maybe' cases to the favorable cases. Doguwa also proposed analogous kernel estimators.

Floresroux and Stein (1996) noted that $\widehat{G}_6$ could have substantial bias if the process Φ is not Poisson. They proposed instead that $H(r, x_i)$ be estimated nonparametrically from the data, by averaging over spatial configurations in the dataset that are analogous to the neighbourhood of x_i. That is, since $H(r, x_i)$ is itself a Palm probability, it should first be estimated from the data using the Campbell–Mecke formula. The improved Floresroux–Stein estimator performs well on a range of simulated patterns.

2.7 Line segment processes

Laslett (1982a, b) first noted the analogy between edge effects and censoring for the case of line segments. Figure 2.1(b) shows a line segment process observed within a rectangular window. The segments may be uncensored (both endpoints visible within the window), censored at one end (one endpoint visible) or doubly censored (neither endpoint visible). Laslett (1982b) proposed estimating the line segment length distribution essentially using the Kaplan–Meier estimator based upon the uncensored and singly-censored lengths. However Wijers (1995) showed that the optimal (nonparametric maximum likelihood) estimator of the length distribution, in a Poisson line segment process, is another, complicated estimator determined implicitly as the solution of an integral equation involving data from all segments.

2.8 Censoring of random sets

Finally we switch attention to the case where the spatial pattern X is a random closed set in $\mathbb{R}^d$, assumed to be stationary. A typical application is to the study of vegetation patterns where X could represent that part of the surveyed region which is covered by a particular vegetation type.

Summary statistics for random closed sets are described in Serra (1982) and Stoyan *et al.* (1987), §6.2–6.3). For several such statistics, notably the spatial covariance function, edge effects can be handled in a straightforward fashion using the local knowledge principle (Serra (1982, pp. 49, 233, Daley and Vere-Jones, 1988, p. 374). In this section we discuss the more complex question of censoring effects for random sets.

2.8.1 Empty space function F

Of particular interest here is the empty space function, defined analogously to (2.14) as the distribution function of the distance from an arbitrary point in space to the nearest point of X:

$$F(r) = \mathbb{P}\{\rho(x, X) \le r\}, \quad r \ge 0$$

where the point $x \in \mathbb{R}^d$ is arbitrary and may be taken to be the origin 0. The empty space function is a useful summary of the 'size' of voids between the components of X. For stationary Poisson processes of points, lines or other figures, F takes known functional forms, and departures of the empirical F from these benchmarks are taken as indications of 'clustered' or 'ordered' patterns. See Serra (1982, chap. XIII), Stoyan *et al.* (1987, p. 178).

The estimation of F for a random closed set X observed in a window W is an almost trivial extension of the point process case. The border method estimator

$$\widehat{F}^{\mathrm{b}}(r) = \frac{\left| W_{(-r)} \cap X_{(+r)} \right|_d}{\left| W_{(-r)} \right|_d} \tag{2.58}$$

is pointwise unbiased for $F(r)$. Theorem 2.2 extends to the statement that for any stationary random closed set X, the empty space function $F(r)$ is absolutely continuous for $r > 0$, with an atom at $r = 0$ of mass $\mathbb{E}\,|X \cap W|_d / |W|_d$, and has hazard rate

$$\lambda(r) = \frac{\mathbb{E}\left| W \cap \partial\left(X_{(+r)} \right) \right|_{d-1}}{\mathbb{E}\left| W \setminus X_{(+r)} \right|_d}$$

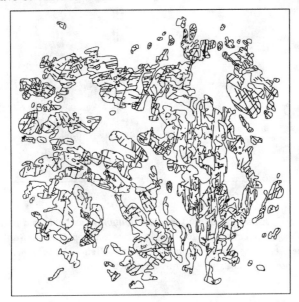

Figure 2.12 *Geological faults (straight lines) observable on exposed granite (region enclosed by curved contours). Square side is 160 metres.*

for almost all r, for any compact window W such that the denominator is positive. The Kaplan–Meier style estimator is

$$\widehat{F}(r) = 1 - \exp\left\{ -\int_0^r \frac{\left|\partial\left(X_{(+s)}\right) \cap W_{(-s)}\right|_{d-1}}{\left|W_{(-s)} \setminus X_{(+s)}\right|_d} \, ds \right\} \qquad (2.59)$$

where $|\cdot|_{d-1}$ denotes $d-1$ dimensional Hausdorff measure. It can also be represented as a window integral

$$\widehat{F}(r) = 1 - \exp\left\{ -\int_W \frac{\mathbf{1}\{t(x) \leq c(x)\} \, \mathbf{1}\{t(x) \leq r\}}{\left|W_{(-t(x))} \setminus X_{(+t(x))}\right|_d} \, dx \right\} \qquad (2.60)$$

where $t(x) = \rho_B(x, X)$ and $c(x) = \rho_B(x, \partial W)$. The hazard rate of $\widehat{F}$ is pointwise ratio-unbiased for the hazard rate of F.

Figure 2.12 represents a pattern of geological fractures in granitic pluton near Lac du Bonnet, Manitoba, Canada, from Stone *et al.* (1984). The faults were mapped only when visible on the surface, the other two-thirds of the granite being covered by soil. In Figure 2.12 the faults are represented as line segments and the boundary of the observable region is indicated as a curved contour.

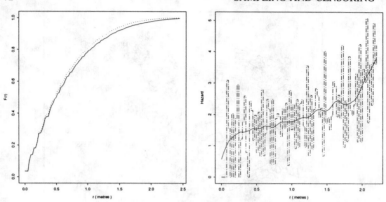

Figure 2.13 *Estimated empty space distribution function F and hazard function* λ *for geological faults*. Left: *Empty space function F.* $--$: *Kaplan–Meier estimate;* $\cdots\cdots$: *reduced sample estimate*. Right: *Hazard function* λ *of F.* $\cdot-\cdot-$: *point estimate;* $--$: *kernel smoothed function*.

Figure 2.13 shows the Kaplan–Meier and reduced sample estimates of the empty space function of the fault pattern, the pointwise Kaplan–Meier estimate of the hazard rate $\lambda(r)$ of F, and a kernel smoothed estimate of $\lambda(r)$.

2.8.2 First contact distribution H_B

More generally, consider the analogue of F

$$F_B(r) = \mathbb{P}\{\rho_B(x, X) \le r\}, \quad r \ge 0$$

where the Euclidean distance $\rho(x, X)$ is replaced by

$$\rho_B(x, X) = \inf\{t \ge 0 \,:\, (tB + x) \cap X \ne \emptyset\}$$

where $B \subset \mathbb{R}^d$ is some fixed 'test set'. Thus $\rho_B(x, X)$ is the earliest time t at which a balloon with shape B will touch X if it begins inflating with zero size at time $t = 0$. The conditional distribution

$$H_B(r) = \mathbb{P}\{\rho_B(x, X) \le r \mid \rho_B(x, X) > 0\} = \frac{F_B(r) - F_B(0)}{1 - F_B(0)}$$

is called the *first contact distribution* with test set B. See Serra (1982), Stoyan *et al.* (1987).

When B is the unit sphere, F_B reduces to the usual empty space function F, and H_B is called the *spherical contact distribution function*.

When B is a line segment, H_B is called the *linear contact distribution function*. When B is convex and contains a neighbourhood of the origin, F_B is a generalised empty space function in which Euclidean distance is replaced by a metric ρ_B on $\mathbb{R}^d$ with unit ball B.

There are various reasons for studying F_B for non-spherical B. Anisotropy (preferential orientation) in a spatial pattern cannot be assessed using the spherical empty space function F, and the usual approach is to use the linear contact distributions for various segments B pointing in different directions Stoyan *et al.* (1987). The linear contact function is not applicable to point patterns; instead one can use F_B for elliptical shapes B at various orientations. When the distance transform algorithm (Borgefors, 1984, 1986) is used to compute approximate distances in a digital image, Euclidean distance has effectively been replaced by a metric with an octagonal unit ball.

Again we have an edge effect, in that $\rho_B(x, X)$ is not determined from information inside W alone. The border method estimator

$$\widehat{F}_B^{\text{b}}(r) = \frac{\left|(W \ominus rB) \cap (X \oplus r\check{B})\right|_d}{|W \ominus rB|_d} \tag{2.61}$$

is a pointwise unbiased estimator of $F_B(r)$.

The Kaplan–Meier estimator of F_B is derived exactly as for point processes. Hansen *et al.* (1996, to appear) treat the cases where B is convex with nonempty interior, and where B is a line segment, respectively.

The 'window integral' representation (2.60) continues to hold for $\widehat{F}_B$, with the modification that $t(x) = \rho_B(x, X)$ and $c(x) = \rho_B(x, \partial W)$. Hence in practice $\widehat{F}_B$ can be obtained using the same technique of discretising the window and computing the discrete Kaplan–Meier estimator.

However, the representation of $\widehat{F}$ as an integral over r now involves a Jacobian:

$$\widehat{F}(r) = 1 - \frac{|W \setminus X|_d}{|X|_d} \tag{2.62}$$

$$\times \exp\left\{-\int_0^r \int_{(W_{(-sB)}) \cap \partial (X_{(+sB)})} \frac{1}{J_1 \rho_B(x, X)} \, d\mathcal{H}^{d-1} x \; ds\right\}$$

where $J_1 \rho_B(x, X)$ is the 1-dimensional approximate Jacobian of the Lipschitz function $\rho_B(\cdot, X)$. If $B = b(0, 1)$ then $J_1 \rho_B(x, X) \equiv 1$ and we recover (2.59).

Let B be compact, convex and contain a neighbourhood of the origin. Then Hansen *et al.* to appear) showed that for any stationary random closed set X, both F_B and $\widehat{F}_B$ are continuous monotone functions, and absolutely continuous for $r > 0$ with an atom at $r = 0$. The hazard rate of F_B equals

$$\lambda_B(r) = \frac{1}{\mathbb{E}\left|W \setminus X_{(+rB)}\right|_d} \mathbb{E} \int_{W \cap \partial(X_{(+rB)})} \frac{1}{J_1 \rho_B(x, X)} \, d\mathcal{H}^{d-1}x$$

for almost all $r > 0$, for any compact window W such that the denominator is positive. Thus the Kaplan–Meier estimator of $\lambda_B(r)$

$$\widehat{\lambda}_B(r) = \frac{1}{\left|W \setminus X_{(+rB)}\right|_d} \int_{W_{(-rB)} \cap \partial(X_{(+rB)})} \frac{1}{J_1 \rho_B(x, X)} \, d\mathcal{H}^{d-1}x$$

is ratio-unbiased for almost all $r > 0$. Examples of applications of the Kaplan–Meier estimator can be seen in Hansen *et al.* (to appear).

A Hanisch-type estimator for F_B for convex B was proposed by Chiu and Stoyan (1995) and Stoyan *et al.* (1995, pp. 138, 215). This is defined by the direct analogue of (2.55), and can be expressed in a form similar to (2.56) with the introduction of the relevant Jacobian. Detailed assessments of performance have not yet been made.

For the linear contact distribution, where B is a line segment, a Kaplan–Meier type estimator was constructed in Hansen *et al.* (1996). This has a form similar to (2.62) but the statements about regularity of F_B no longer hold. Examples of applications can be seen in Hansen *et al.* (1996).

Conclusion

Edge effects for spatial processes exhibit features analogous to sampling bias and other features analogous to random censoring. These analogies suggest estimators for properties of the process. However, there does not seem to be an adequate optimality theory which identifies the best estimator in this general context.

Acknowledgements

I thank many colleagues for their comments, advice and encouragement, especially Drs Katja Schladitz and Aila Särkkä.

References

A.J. Baddeley. Integrals on a moving manifold and geometrical probability. *Advances in Applied Probability*, 9:588–603, 1977.

A.J. Baddeley. Stereology and survey sampling theory. *Bulletin of the International Statistical Institute*, 50, book 2:435–449, 1993.

A.J. Baddeley and R.D. Gill. Kaplan–Meier estimators for interpoint distance distributions of spatial point processes. Research Report BS-R9315, Centrum voor Wiskunde en Informatica, july 1993.

A.J. Baddeley and R.D. Gill. Kaplan–Meier estimators for interpoint distance distributions of spatial point processes. *Annals of Statistics*, 25:263–292, 1997.

A.J. Baddeley and H.J.A.M. Heijmans. Incidence and lattice calculus with applications to stochastic geometry and image analysis. *Applicable Algebra in Engineering, Communication, and Computing*, 6(3):129–146, 1995.

A.J. Baddeley, R.A. Moyeed, C.V. Howard, and A. Boyde. Analysis of a three-dimensional point pattern with replication. *Applied Statistics*, 42(4):641–668, 1993.

L.G. Barendregt and M.J. Rottschäfer. A statistical analysis of spatial point patterns. A case study. *Statistica Neerlandica*, 45:345–363, 1991.

G. Borgefors. Distance transformations in arbitrary dimensions. *Computer Vision, Graphics and Image Processing*, 27:321–345, 1984.

G. Borgefors. Distance transformations in digital images. *Computer Vision, Graphics and Image Processing*, 34:344–371, 1986.

K.R. Brewer and M. Hanif. *Sampling with unequal probabilities*. Number 15 in Lecture Notes in Statistics. Springer Verlag, New York, 1983.

S.N. Chiu and D. Stoyan. Estimation of distance distributions for spatial patterns. Unpublished manuscript, 1995.

W.G. Cochran. *Sampling Techniques*. John Wiley and Sons, 3rd edition, 1977.

D.R. Cox. Appendix to 'The dye sampling method of measuring fibre length distribution' by D.R. Palmer. *Journal of the Textile Institute*, 39:T8–T22, 1949.

D.R. Cox. Some sampling problems in technology. In N.L. Johnson and H. Smith, editors, *New developments in survey sampling*, pages 506–527. John Wiley and Sons, 1969.

N.A.C. Cressie. *Statistics for spatial data*. John Wiley and Sons, New York, 1991.

M.W. Crofton. Sur quelques théorèmes du calcul intégral. *Comptes Rendus de l'Académie des Sciences de Paris*, 68:1469–1470, 1869.

D.J. Daley and D. Vere-Jones. *An introduction to the theory of point processes*. Springer Verlag, New York, 1988.

H.E. Daniels. A new technique for the analysis of fibre length distribution in wool. *Journal of the Textile Institute*, 33:T137–T150, 1942.

R.T. DeHoff. The geometric meaning of the integral mean curvature. In *Microstructural Science*, volume 5, pages 331–348, Amsterdam, 1977. Elsevier.

P.J. Diggle. On parameter estimation and goodness-of-fit testing for spatial point patterns. *Biometrika*, 35:87–101, 1979.

P.J. Diggle. *Statistical analysis of spatial point patterns*. Academic Press, London, 1983.

P.J. Diggle and B. Matérn. On sampling designs for the estimation of point-event nearest neighbour distributions. *Scandinavian Journal of Statistics*, 7:80–84, 1981.

S.I. Doguwa. A comparative study of the edge-corrected kernel-based nearest neighbour density estimators for point processes. *J. Statist. Comp. Simul.*, 33:83–100, 1989.

S.I. Doguwa. On edge-corrected kernel-based pair correlation function estimators for point processes. *Biometrical Journal*, 32:95–106, 1990.

S.I. Doguwa. On the estimation of the point-object nearest neighbour distribution $F(y)$ for point processes. *J. Statist. Comp. Simul.*, 41:95–107, 1992.

S.I. Doguwa and D.N.Choji. On edge-corrected probability density function estimators for point processes. *Biometrical Journal*, 33:623–637, 1991.

S.I. Doguwa and G.J.G. Upton. Edge-corrected estimators for the reduced second moment measure of point processes. *Biometrical Journal*, 31:563–675, 1989.

S.I. Doguwa and G.J.G. Upton. On the estimation of the nearest neighbour distribution, $G(t)$, for point processes. *Biometrical Journal*, 32:863–876, 1990.

Thomas Fiksel. Edge-corrected density estimators for point processes. *Statistics*, 19:67–75, 1988.

E.M. Floresroux and M.L. Stein. A new method of edge correction for estimating the nearest neighbor distribution. *J. Statist. Planning and Inference*, 50:353–371, 1996.

R.D. Gill. Lectures on survival analysis. In P. Bernard, editor, *22e Ecole d'Eté de Probabilités de Saint-Flour 1992*, number 1581 in Lecture Notes in Mathematics. Springer, 1994.

R.D. Gill and S. Johansen. A survey of product-integration with a view toward application in survival analysis. *Annals of Statistics*, 18:1501–1555, 1990.

H.J.G. Gundersen. Notes on the estimation of the numerical density of arbitrary profiles: the edge effect. *Journal of Microscopy*, 111:219–223, 1977.

H.J.G. Gundersen. Estimators of the number of objects per area unbiased by edge effects. *Microscopica Acta*, 81:107–117, 1978.

H.J.G. Gundersen et al. The new stereological tools: disector, fractionator, nucleator and point sampled intercepts and their use in pathological research and diagnosis. *Acta Pathologica Microbiologica et Immunologica Scandinavica*, 96:857–881, 1988a.

H.J.G. Gundersen et al. Some new, simple and efficient stereological methods and their use in pathological research and diagnosis. *Acta Pathologica Microbiologica et Immunologica Scandinavica*, 96:379–394, 1988b.

J. Hájek. Asymptotic normality of simple linear rank statistics under alternatives. *Annals of Mathematical Statistics*, 39:325–346, 1968.

P. Hall. Correcting segment counts for edge effects when estimating intensity. *Biometrika*, 72:459–463, 1985.

Peter Hall. *An introduction to the theory of coverage processes*. John Wiley and Sons, New York, 1988.

K.-H. Hanisch. Some remarks on estimators of the distribution function of nearest neighbour distance in stationary spatial point patterns. *Mathematische Operationsforschung und Statistik, series Statistics*, 15:409–412, 1984.

M.B. Hansen, A.J. Baddeley, and R.D. Gill. First contact distributions for spatial patterns: regularity and estimation. Provisionally accepted for publication.

M.B. Hansen, R.D. Gill, and A.J. Baddeley. Kaplan–Meier type estimators for linear contact distributions. *Scandinavian Journal of Statistics*, 23:129–155, 1996.

L. Heinrich. Asymptotic behaviour of an empirical nearest-neighbour distance function for stationary Poisson cluster processes. *Mathematische Nachrichten*, 136:131–148, 1988a.

L. Heinrich. Asymptotic Gaussianity of some estimators for reduced factorial moment measures and product densities of stationary Poisson cluster processes. *Statistics*, 19:87–106, 1988b.

L. Heinrich. Goodness-of-fit tests for the second moment function of a stationary multidimensional Poisson process. *Statistics*, 22:245–268, 1991.

D.G. Horvitz and D.J. Thompson. A generalization of sampling without replacement from a finite universe. *Journal of the American Statistical Association*, 47:663–685, 1952.

E.B. Jensen and R. Sundberg. Generalized associated point methods for sampling planar objects. *Journal of Microscopy*, 144:55–70, 1986.

E.L. Kaplan and P. Meier. Nonparametric estimation from incomplete observations. *Journal of the American Statistical Association*, 53:457–481, 1958.

P.R. Krishnaiah and C.R. Rao, editors. *Sampling*. Number 6 in Handbook of Statistics. North-Holland, Amsterdam, 1988.

Ch. Lantuéjoul. Computation of the histograms of the number of edges and neighbours of cells in a tessellation. In R.E. Miles and J. Serra, editors, *Geometrical Probability and Biological Structures: Buffon's 200th Anniversary*, Lecture Notes in Biomathematics, No 23, pages 323–329. Springer Verlag, Berlin-Heidelberg-New York, 1978.

Ch. Lantuéjoul. *La squelettisation et son application aux mesures topologiques des mosaiques polycristallines*. Thesis, docteur-ingénieur en Sciences et Techniques Minières, Ecole Nationale Supérieure de Mines de Paris, Fontainebleau, 1978.

G.M. Laslett. Censoring and edge effects in areal and line transect sampling of rock joint traces. *Mathematical Geology*, 14:125–140, 1982a.

G.M. Laslett. The survival curve under monotone density constraints with applications to two-dimensional line segment processes. *Biometrika*, 69:153–

160, 1982b.

R.E. Miles. On the elimination of edge-effects in planar sampling. In E F Harding and D G Kendall, editors, *Stochastic geometry: a tribute to the memory of Rollo Davidson*, pages 228–247. John Wiley and Sons, London-New York-Sydney-Toronto, 1974.

R.E. Miles. The sampling, by quadrats, of planar aggregates. *Journal of Microscopy*, 113:257–267, 1978.

J. Ohser. On estimators for the reduced second moment measure of point processes. *Mathematische Operationsforschung und Statistik, series Statistics*, 14:63–71, 1983.

B.D. Ripley. The second-order analysis of stationary point processes. *jap*, 13:255–266, 1976.

B.D. Ripley. Modelling spatial patterns (with discussion). *Journal of the Royal Statistical Society, Series B*, 39:172–212, 1977.

B.D. Ripley. On tests of randomness for spatial point patterns. *Journal of the Royal Statistical Society, series B*, 41:368–374, 1979.

B.D. Ripley. *Spatial statistics*. John Wiley and Sons, New York, 1981.

B.D. Ripley. *Statistical inference for spatial processes*. Cambridge University Press, 1988.

A. Rosenfeld and J. L. Pfalz. Sequential operations in digital picture processing. *Journal of the Association for Computing Machinery*, 13:471, 1966.

A. Rosenfeld and J. L. Pfalz. Distance functions on digital pictures. *Pattern Recognition*, 1:33–61, 1968.

J. Serra. *Image analysis and mathematical morphology*. Academic Press, London, 1982.

M.L. Stein. A new class of estimators for the reduced second moment measure of point processes. *Biometrika*, 78:281–286, 1991.

M.L. Stein. Asymptotically optimal estimation for the reduced second moment measure of point processes. *Biometrika*, 80:443–449, 1993.

M.L. Stein. An approach to asymptotic inference for spatial point processes. *Statistica Sinica*, 5:221–234, 1995.

D. Stone, D.C. Kamineni, and A.Brown. Geology and fracture characteristics of the Underground Research Laboratory lease near Lac du Bonnet, Manitoba. Technical Report 243, Atomic Energy of Canada Ltd. Research Co., 1984.

D. Stoyan, W.S. Kendall, and J. Mecke. *Stochastic Geometry and its Applications*. John Wiley and Sons, Chichester, 1987.

D. Stoyan, W.S. Kendall, and J. Mecke. *Stochastic Geometry and its Applications*. John Wiley and Sons, Chichester, second edition, 1995.

W. Weil and J.A. Wieacker. A representation theorem for random sets. *Probability and Mathematical Statistics*, 9:147–151, 1987.

B.J. Wijers. *Nonparametric estimation for a windowed line segment process*. PhD thesis, University of Leiden, Leiden, The Netherlands, January 1995.

D. Zimmerman. Censored distance-based intensity estimation of spatial point processes. *Biometrika*, 78:287–294, 1991.

CHAPTER 3

Likelihood inference for spatial point processes

C. Geyer

Department Statistics, Univ Minnesota, Vincent Hall, Minneapolis, MN
55455, USA

3.1 Introduction

This chapter deals with likelihood inference for spatial point processes
using the methods of Moyeed and Baddeley (1991), Geyer and Thompson (1992), Gelfand and Carlin (1993), Geyer (1994), and Geyer and
Møller (1994) using Markov chain Monte Carlo (MCMC). The basic
message is

> If you can write down a model, I can do likelihood inference for it, not
> only maximum likelihood estimation, but also likelihood ratio tests, likeli-
> hood based confidence regions, profile likelihoods, whatever. That includes
> conditional likelihood inference and inference with missing data.

This is overstated, of course. There is no question that one can write
down a model so complicated that Monte Carlo will take so much time
that no one would wait for an answer. But analyses that can be done
are far beyond what is generally recognized.

This does not mean that MCMC likelihood analyses are easy. They
usually require writing of computer code or at least modification of
someone else's code, and they require hours of computer time. Each
analysis is more a mini research project than a routine calculation. But
if an analysis is worth doing, it generally can be done.

The code used to do the analyses in this chapter is available by
anonymous ftp from the server at the School of Statistics, University
of Minnesota

    ftp://stat.umn.edu/pub/points/points.tar.gz

Most of the code is written in the S language except for the samplers,
which are written in C to be dynamically loaded into S or S-plus. One

bit of code uses the S-plus function `acf`. The reader who has S-plus can reproduce the results in this section using this code. The reader who has S can reproduce all but the optimal spacing calculation in Section 3.13.2.

Samplers are provided only for three models. To do a new model, it would be necessary to write new code or modify my code. If there is a way to make samplers both efficient and easy to modify, I do not know it.

Disclaimer *This code is not guaranteed to be bug free. MCMC is not guaranteed to converge. The code may not correctly implement the algorithms described in this chapter. This chapter may contain typographical or conceptual errors.*

Computer code traditionally comes with such a disclaimer, scientific papers traditionally do not. MCMC is a complex mixture of computer programming, statistical theory, and practical experience. When it works, it does things that cannot be done any other way, but it is good to remember that it is not foolproof.

3.2 Families of unnormalized densities

Returning to the overstated claim with which we began, what did the phrase 'write down a model' mean? The models our methods can handle are those specified by families of unnormalized densities.

The notion of an unnormalized density is elementary, occurring in the problem often assigned in introductory probability courses of finding the constant that normalizes a given function to be a probability density. Thus this section contains nothing deep. It merely gives new language for old concepts. But this is important. Language constrains thought, and new language leads to new ways of looking at old problems.

Let $(\Omega, \mathcal{A}, \mu)$ be a measure space and h a nonnegative measurable real function on Ω. If

$$c = \int h(x)\mu(dx) \tag{3.1}$$

is finite and nonzero, then

$$f(x) = \frac{1}{c}h(x), \qquad x \in \Omega$$

defines a probability density f with respect to μ of a measure P. We say that

- h is an *unnormalized density* of P with respect to μ.

- f is the *normalized density* of P with respect to μ.

- c is the *normalizing constant* for h.

Use of the term 'unnormalized density' implies that the normalizing constant (3.1) is finite and nonzero.

This game can be played for families of densities specifying a statistical model. Let $\mathcal{H} = \{\, h_\theta : \theta \in \Theta \,\}$ be a family of unnormalized densities with respect to μ. This terminology implies that

$$c(\theta) = \int h_\theta(x)\mu(dx)$$

is finite and nonzero for all $\theta \in \Theta$. Also

$$f_\theta(x) = \frac{1}{c(\theta)} h_\theta(x), \qquad x \in \Omega \tag{3.2}$$

defines a probability density f_θ with respect to μ of a measure P_θ for all $\theta \in \Theta$. Then $\mathcal{P} = \{\, P_\theta : \theta \in \Theta \,\}$ is a statistical model specified by the family $\mathcal{F} = \{\, f_\theta : \theta \in \Theta \,\}$ of probability densities with respect to μ. We say that

- $\mathcal{H}$ is a *family of unnormalized densities* for the model $\mathcal{P}$.

- $\mathcal{F}$ is the *family of normalized densities* for $\mathcal{P}$.

- $c : \Theta \to (0, \infty)$ is the *normalizing function* for the family $\mathcal{H}$.

This terminology was introduced in Geyer (1994). It arises naturally from consideration of the most general case in which the methods of Monte Carlo maximum likelihood work.

3.3 Examples of families of unnormalized densities

3.3.1 Exponential families

The oldest general class of models specified by unnormalized densities are exponential families. Let $t : \Omega \to \mathbb{R}^d$ be a measurable map, and define

$$c(\theta) = \int e^{\langle t(x), \theta \rangle} \mu(dx), \qquad \theta \in \mathbb{R}^d.$$

Suppose μ is not the zero measure so that $c(\theta) > 0$ for all θ. Define

$$\Theta = \{\, \theta \in \mathbb{R}^d : c(\theta) < \infty \,\}.$$

For each $\theta \in \Theta$, define

$$h_\theta(x) = e^{\langle t(x), \theta \rangle}, \qquad x \in \Omega, \tag{3.3}$$

and define f_θ by (3.2). Then $\mathcal{F} = \{ f_\theta : \theta \in \Theta \}$ is the *full exponential family* of probability densities with respect to μ with *canonical statistic* $t(X)$, *canonical parameter* θ, and *canonical parameter space* Θ (also called *natural* statistic, parameter, and parameter space, respectively).

It is clear that $\mathcal{H} = \{ h_\theta : \theta \in \Theta \}$ is a family of unnormalized densities with respect to μ specifying the exponential family $\mathcal{F}$ and that c is the normalizing function of the family. The term 'normalizing function' is not used in the exponential family literature; c is sometimes called the *Laplace transform* of the measure $\mu \circ t^{-1}$, and $\log c$ is sometimes called the *cumulant function* of the family. These names only apply when the unnormalized densities have the special structure (3.3), so we shall not use them, being interested in general families of unnormalized densities.

Of particular interest in spatial statistics and stochastic geometry are exponential families associated with finite-volume Gibbs distributions, Markov random fields, and Markov point processes.

3.3.2 Conditional families

Suppose we are given a family of normalized probability densities $f_\theta(x, y)$ but are interested in the *conditional family* $f_\theta(x|y)$.

Slogan *A joint density is an unnormalized conditional density. The marginal is its normalizing constant.*

What this means is the following. Say f_θ is a density with respect to $\mu \times \nu$. For fixed y consider the function

$$h_\theta(x) = f_\theta(x, y). \tag{3.4}$$

This is an unnormalized density with respect to μ with normalizing constant

$$c(\theta|y) = \int h_\theta(x)\mu(dx) = \int f_\theta(x, y)\mu(dx) = p_\theta(y)$$

where p_θ is the marginal density with respect to ν of the random variable Y. The normalized density is then

$$\frac{1}{c(\theta|y)} h_\theta(x) = \frac{f_\theta(x, y)}{p_\theta(y)} = f_\theta(x|y).$$

Conditional families arise in conditional likelihood inference and in Bayesian inference. In conditional likelihood inference, although the complete data (x, y) are observed, rather than maximizing the joint likelihood $f_\theta(x, y)$ to estimate θ, one maximizes the conditional likelihood $f_\theta(x|y)$. This is usually done because y is exactly or approximately ancillary. In Bayesian inference the likelihood times the prior is a known normalized joint density for parameter and data, but the object of inference is the conditional distribution of the parameter given the data.

3.3.3 Missing data

Missing data involve the same considerations as conditional families. If x is missing and y is observed and the joint density is $f_\theta(x, y)$, then the likelihood is the marginal density $p_\theta(y)$ considered a function of θ for y fixed at the observed value.

As we observed in the preceding section $p_\theta(y)$ is the normalizing constant for the joint density considered as an unnormalized conditional density (3.4). Thus the family of unnormalized densities is involved in both conditional likelihood inference and likelihood inference with missing data. Latent variables, random effects, and ordinary (non-Bayes) empirical Bayes models all involve missing data of some form, and all give rise to the same considerations.

3.3.4 Unknown normalizing functions and missing data

The situation does not get much worse when the distribution of the complete data (x, y) is specified by a family of unnormalized densities $h_\theta(x, y)$. Let

$$c(\theta) = \int h_\theta(x, y)\mu(dx)\nu(dy) \qquad (3.5)$$

be the normalizing function for the joint density. Following our slogan, h_θ is not only an unnormalized joint density for x and y with normalizing function (3.5) but also an unnormalized conditional density of x given y with normalizing function

$$c(\theta|y) = \int h_\theta(x, y)\mu(dx).$$

Only the normalizing functions are different. Now the likelihood when x is missing and y observed is

$$p_\theta(y) = \int f_\theta(x, y)\mu(dx) = \int \frac{1}{c(\theta)} h_\theta(x, y)\mu(dx) = \frac{c(\theta|y)}{c(\theta)}$$

a ratio of two normalizing functions. As we shall see, even if both normalizing functions are unknown, we can still do Monte Carlo likelihood inference.

3.3.5 Unknown normalizing functions and Bayesian inference

Models specified by unnormalized densities present a problem for Bayesian inference. If the model is specified by a family of unnormalized densities $h(x|\theta)$ having a normalizing function

$$c(\theta) = \int h(x|\theta)\mu(dx)$$

that is analytically intractable, then the likelihood

$$f(x|\theta) = \frac{1}{c(\theta)}h(x|\theta)$$

is also analytically intractable, as is the joint density of the data and the parameter

$$f(x|\theta)\pi(\theta) = \frac{1}{c(\theta)}h(x|\theta)\pi(\theta).$$

This joint density is an unnormalized posterior, when considered as a function of θ for fixed x. The trouble is that since the normalizing function $c(\theta)$ is unknown, this is an unknown function of θ. Dealing with such families satisfactorily seems to be an open research question.

This issue has been obscured by the remarkable recent progress in Bayesian inference using MCMC. No one points out that likelihood inference is still easier than Bayesian inference in most situations. Many models for which exact likelihood inference is easy, generalized linear models, for example, require MCMC to carry out Bayesian inference. Many models for which MCMC likelihood inference is routine, have another level of difficulty when Bayesian inference is tried and are still open research problems. All of the likelihood analysis in this chapter is an example of this phenomenon.

3.3.6 General models specified by unnormalized densities

Going beyond the special cases so far described, the greatest potential for modelling with unnormalized families of densities is achieved by freeing oneself from all limitations. *Any* family of nonnegative functions on the state space can be used to specify a model so long as all of the functions can be proved to be integrable and not almost everywhere zero so that the normalizing function is finite and nonzero. This permits tremendous and largely unexplored flexibility in specifying models with complex dependence.

3.4 Markov chain Monte Carlo

Markov chain Monte Carlo (MCMC), including the Gibbs sampler and the Metropolis, the Metropolis–Hastings, and the Metropolis–Hastings–Green algorithms, permits the simulation of any stochastic process specified by an unnormalized density.

3.4.1 Updates not 'algorithms'

The traditional language of MCMC which talks about 'algorithms' and 'samplers' constrains inventiveness when taken seriously. It is much better to talk about 'updates' or 'update mechanisms'. The point is that updates can be combined in many ways to make an MCMC algorithm. Talking about the 'Gibbs sampler' focuses on an MCMC algorithm composed only of Gibbs updates. Nowadays, the only reason for that is sheer ignorance. Gibbs updates often are 'part of the problem rather than part of the solution'. When asked for advice on an MCMC algorithm, I often find myself saying 'Why are you using that Gibbs update? That's what makes the sampler so slow.' As soon as one thinks about using a non-Gibbs update, the problem becomes easy. A quip in the same vein is Peter Clifford's characterization of the Gibbs sampler as a *cul-de-sac* that many statisticians are now trying to reverse out of (Clifford, 1993).

Even when the terminology of algorithms rather than updates does not lead to construction of inferior Markov chain samplers, it still makes description of algorithms more complicated than need be and makes it harder to talk about samplers. For these reasons we always describe the 'basic update mechanisms' that are combined to make the sampler.

3.4.2 Combination of updates

There are two basic methods of combining update mechanisms, *composition* and *mixing*, corresponding to multiplication and linear combination of Markov transition kernels.

Stationarity

A probability measure π is *stationary* or *invariant* for the kernel P if $\pi P = \pi$ where right multiplication by a measure is defined by

$$(vP)(A) = \int v(dx)P(x, A).$$

This is the basic property required of an MCMC sampler, that it have a specified stationary distribution π.

Composition

If an update step described by a transition kernel P_1 is followed by another step with transition kernel P_2, then the combined step is described by the transition kernel $P_1 P_2$ defined by

$$(P_1 P_2)(x, A) = \int P_1(x, dy)P_2(y, A).$$

When the kernels are thought of as operators on a function space

$$(P_i f)(x) = \int P_i(x, dy)f(y)$$

then this multiplication of kernels corresponds to composition of operators $(P_1 P_2)f = P_1(P_2 f)$.

The notion of functional composition also describes the computer code. The composition of updates is implemented by following the code for the first update by the code for the second. If the code bits are thought of as implementing mathematical operations, then this also is a functional composition.

It is obvious from the definition that if π is stationary for P_i, $i = 1$, $\ldots, d$. Then π is also stationary for the composition $P_1 \ldots P_d$.

Mixing

Another way of combining updates is by choosing an update mechanism at random and then implementing that update, thus producing a statistical mixture of the updates. If update steps described by transition kernels P_i are chosen with probabilities p_i then the transition kernel of the mixed update step is $P = \sum_i p_i P_i$. The sum can be finite or infinite.

In order for the p_i to specify a probability distribution, the p_i must be nonnegative and sum to one. Thus mixing updates corresponds to the operation of forming a convex combination of kernels. Again, it is obvious that if π is stationary for each of the P_i, then π is also stationary for $\sum_i p_i P_i$.

Reversibility

A general, not necessarily stochastic, kernel $P(x, A)$ is said to be *reversible* with respect to a measure π if for every $u, v \in L^2(\pi)$ the integral

$$\iint u(x)v(y)\pi(dx)P(x, dy)$$

is symmetric under the interchange of u and v. Equivalently, the operator T on $L^2(\pi)$ defined by

$$(Tf)(x) = \int P(x, dy)f(y), \qquad f \in L^2(\pi)$$

is self-adjoint. This is the definition of 'reversible' preferred by those who like to think of probabilities as positive linear operators on function spaces rather than as set functions on σ-fields. Those who like probabilities as set functions use the same definition with u and v replaced by indicators of sets, that

$$\iint 1_A(x)1_B(y)\pi(dx)P(x, dy) = \int_A \pi(dx)P(x, B)$$

is symmetric under the interchange of A and B. The two views are, of course, equivalent by extension by linearity and monotone convergence. When P is stochastic, the special case $B = \Omega$ gives

$$\pi(A) = \int \pi(dx)P(x, A)$$

so $\pi = \pi P$ and π is stationary for P. Thus verification of reversibility of a stochastic kernel P with respect to π is a tool for establishing stationarity of π for P.

State-dependent mixing

Green (1995) proposed an algorithm that involves *state-dependent mixing* having mixing probabilities that depend on the current state. There are a finite or infinite set of transition kernels $P_i(x, A), i \in I$, and state-dependent mixing probabilities $p_i(x)$. The overall transition kernel is

$$P(x, A) = \sum_{i \in I} p_i(x)P_i(x, A)$$

To make a move when at x, we choose kernel P_i with probability $p_i(x)$ and then simulate the next state with probability $P_i(x, \cdot)$.

No nice properties are transferred from the P_i to P, so we introduce the substochastic kernels $K_i(x, A) = p_i(x)P_i(x, A)$. Then $P = \sum_{i \in I} K_i$, so if all of the K_i are reversible with respect to π, then so is P, and if P is stochastic, then π is a stationary distribution of P. Note that each K_i determines p_i and P_i through

$$p_i(x) = K_i(x, \Omega) \tag{3.6}$$

and

$$P_i(x, A) = \frac{K_i(x, A)}{K_i(x, \Omega)} \tag{3.7}$$

so we may consider that we have been given the K_i to specify the algorithm.

A simple trick allows us to use the same argument when we are given a set of substochastic kernels $K_i(x, A)$, $i \in I$, that sum to a substochastic kernel, that is

$$\sum_{i \in I} K_i(x, \Omega) \le 1, \qquad \forall x \in \Omega. \tag{3.8}$$

Define the defect

$$d(x) = 1 - \sum_{i \in I} K_i(x, \Omega), \qquad x \in \Omega$$

and a new kernel

$$\widetilde{K}(x, A) = d(x)I(x, A)$$

where I is the identity kernel, defined by

$$I(x, A) = \begin{cases} 1, & x \in A \\ 0, & x \notin A. \end{cases}$$

Then $\widetilde{K}$ is reversible with respect to any distribution π since

$$\iint u(x)v(y)\pi(dx)\widetilde{K}(x, dy) = \int u(x)v(x)d(x)\pi(dx)$$

is trivially symmetric under the interchange of u and v. If we add $\widetilde{K}$ to our set of kernels, then the sum is stochastic.

Thus we have the following formulation of state-dependent mixing. Suppose we are given a set of substochastic kernels satisfying (3.8). Then the following combined update

1. Choose i with probability $p_i(x)$ defined by (3.6). With probability $1 - \sum_{i \in I} p_i(x)$, skip step 2 and stay at the current position.

2. Simulate a new value of x from the probability distribution $P_i(x, \cdot)$ defined by (3.7).

has the stochastic transition kernel $\widetilde{K} + \sum_i K_i$ and is reversible with respect to π if each of the K_i is reversible with respect to π.

3.4.3 The Metropolis–Hastings update

The Metropolis–Hastings update (Metropolis, et al., 1953; Hastings, 1970) simulates a distribution specified by an unnormalized density h with respect to a measure μ. It uses an auxiliary 'proposal' density $q(x, \cdot)$ with respect to μ having the following properties

- $q(x, \cdot)$ must be normalized, $\int q(x, y)\mu(dy) = 1$.
- $q(x, y)$ can be evaluated for all x and y.
- for each x, it is possible to simulate realizations from the distribution having density $q(x, \cdot)$ with respect to μ.

Other than this there is no particular relation between q and h. There is always an infinite variety of proposal densities that will do the job.

The Metropolis–Hastings update changes the state x as follows.

1. Simulate $y \sim q(x, \cdot)$.

2. Evaluate the 'Hastings ratio'

$$R = \frac{h(y)q(y, x)}{h(x)q(x, y)}.$$

3. ('Metropolis rejection') Move to y with probability $\min(1, R)$. Otherwise stay at x.

Note that 'Metropolis rejection' has no relation to the 'rejection sampling' of ordinary independent-sample Monte Carlo. If step 3 'rejects' the 'proposal' y and stays at the current state x, then the Markov chain stays at x for two consecutive iterations (at least it does so if this update is not composed with another update). This is not the case in ordinary rejection sampling, which keeps making proposals until one is accepted and the next sample is the first accepted proposal.

Call a state x *feasible* if $h(x) > 0$. If the sampler is started in a feasible state, the proposal y must be possible under the proposal density, so $q(x, y) > 0$ too. Thus the denominator of the Hastings ratio is not zero. If the proposal y is not feasible, then the Hastings ratio is zero and the proposal is not accepted. Hence the entire sample path of the chain remains in feasible states, and the denominator of the Hastings ratio can never be zero.

In general, the Metropolis–Hastings update makes no sense when x is not feasible. Since infeasible simulations are pointless, it is best to simply require that the sampler be started in a feasible state.

3.4.4 The Metropolis update

If $q(x, y) = q(y, x)$ for all x and y, then the Hastings ratio R becomes the 'odds ratio' $h(y)/h(x)$. This special case is called a 'Metropolis update'. It avoids the requirement of being able to evaluate $q(x, y)$, but has no other virtues. It should be used when convenient and effective and not otherwise.

3.4.5 Variable-at-a-time Metropolis–Hastings updates

When the state x is a vector $x = (x_1, \ldots, x_d)$, a widely used form of Metropolis–Hastings updating involves only one variable at a time. Suppose μ is a product measure $\mu_1 \times \cdots \times \mu_d$. To update the ith variable, we need a proposal density $q_i(x, \cdot)$ with respect to μ_i. The update proceeds as follows.

1. Simulate $y \sim q_i(x, \cdot)$. Note that y has the dimension of x_i not x. Let x_y denote the state with x_i replaced by y

$$x_y = (x_1, \ldots, x_{i-1}, y, x_{i+1} \ldots x_d)$$

2. Evaluate the Hastings ratio

$$R = \frac{h(x_y)q(x_y, x_i)}{h(x)q(x, y)}.$$

3. (Metropolis rejection) Move to x_y with probability $\min(1, R)$. Otherwise stay at x.

In order to make an effective sampler we must, of course, combine these basic update steps, using either composition or mixing, so that the combined update changes all the variables. This update also stays in feasible states if started in a feasible state.

3.4.6 The Gibbs update

The Gibbs update, the elementary update step of the 'Gibbs sampler' (Geman and Geman, 1984; Gelfand and Smith, 1990) is the special case of variable-at-a-time Metropolis–Hastings obtained by making $q(x, \cdot)$ the conditional distribution of x_i given the rest of the variables induced by h, that is, h is an unnormalized version of the proposal density. If c is the normalizing constant, the Hastings ratio becomes

$$R = \frac{h(x_y)q(x_y, x_i)}{h(x)q(x, y)} = \frac{h(x_y)h(x)/c}{h(x)h(x_y)/c} = 1$$

so the Metropolis rejection step always accepts the proposal.

Like the Metropolis update, the Gibbs sampler has no particular virtues. It too should be used when convenient and effective and not otherwise.

3.4.7 The Metropolis–Hastings–Green update

The schemes described up to this point used to have separate theories. Now they are special cases of the Metropolis–Hastings–Green (MHG) algorithm (Green, 1995), which is essentially Metropolis–Hastings with measures rather than densities. Green was not the first to present an algorithm of this type. Multigrid methods in statistical physics are special cases, as are the methods for point processes presented by Geyer and Møller (1994), but Green gave the first general formulation.

The MHG update is best understood by comparison with the Metropolis–Hastings update.

- The unnormalized density h is replaced by an unnormalized measure π on the state space Ω.
- The proposal density $q(x, y)$ is now replaced by a proposal kernel $Q(x, A)$.
- We also need a symmetric measure ξ on $\Omega \times \Omega$ to play the role played by $\mu \times \mu$ in the ordinary Metropolis–Hastings algorithm.

The measure ξ must dominate $\pi(dx)Q(x, dy)$ so that there is a Radon-Nikodym derivative

$$f(x, y) = \frac{\pi(dx)Q(x, dy)}{\xi(dx, dy)}$$

which replaces $h(x)q(x, y)$ in the ordinary Metropolis–Hastings update. Then the Hastings ratio becomes 'Green's ratio'

$$R = \frac{f(y, x)}{f(x, y)}. \tag{3.9}$$

The update is

1. Simulate $y \sim Q(x, \cdot)$.

2. Evaluate Green's ratio (3.9)

3. Accept y with probability $\min(1, R)$.

State-dependent mixing

Green (1995) proposed a more general form of the algorithm using state-dependent mixing. There are a finite or infinite set of proposal kernels $Q_i(x, A)$, $i \in I$, which are permitted to be *substochastic*. The requirements on the proposal kernels are

- $Q_i(x, \Omega)$ is known for all i.
- $\sum_{i \in I} Q_i(x, \Omega) \leq 1, \forall x \in \Omega$
- For all i

$$f_i(x, y) = \frac{\pi(dx) Q_i(x, dy)}{\xi_i(dx, dy)} \tag{3.10}$$

is known, at least up to a constant of proportionality, and it is possible to evaluate $f_i(x, y)$ for all x and y. A different symmetric measure ξ_i may be used for each i.

- for each x and i, it is possible to simulate realizations from the distribution having the normalized proposal distribution

$$P_i(x, \cdot) = \frac{Q_i(x, \cdot)}{Q_i(x, \Omega)}. \tag{3.11}$$

1. Choose a proposal kernel Q_i with probability $p_i(x) = Q_i(x, \Omega)$. With probability $1 - \sum_{i \in I} p_i(x)$, skip the remaining steps and stay at x

2. Simulate $y \sim P_i(x, \cdot)$ defined by (3.11).

3. Evaluate Green's ratio

$$R = \frac{f_i(y, x)}{f_i(x, y)}, \tag{3.12}$$

where f_i is defined by (3.10).

4. Accept y with probability $\min(1, R)$.

All of this is just the MHG update described in preceding section combined with the idea of state-dependent mixing.

3.4.8 Why it works

It is enough to verify that an MHG update with a stochastic kernel Q_i is reversible. Applying state-dependent mixing gives the general case.

Define

$$a_i(x, y) = \min\left(1, \frac{f_i(y, x)}{f_i(x, y)}\right)$$

the probability of accepting a proposed step from x to y and

$$r_i(x) = \int Q_i(x, dy)[1 - a_i(x, y)]$$

the probability of staying at x when the proposal Q_i is used. Then

$$K_i(x, A) = r_i(x)I(x, A) + \int_A Q_i(x, dy)a_i(x, y)$$

is the transition kernel of the Metropolis–Hastings–Green update, and

$$\iint u(x)v(y)\pi(dx)K_i(x, dy) = \int u(x)v(x)r_i(x)\pi(dx)$$

$$+ \iint u(x)v(y)\pi(dx)Q_i(x, dy)a_i(x, y).$$

The first term on the right hand side is obviously symmetric under interchange of u and v. The second term

$$\iint u(x)v(y)\pi(dx)Q_i(x, dy)a_i(x, y)$$

$$= \iint u(x)v(y)f_i(x, y)a_i(x, y)\xi_i(dx, dy),$$

is symmetric under interchange of u and v because of the symmetry of ξ_i and because the MHG update has been defined to make

$$f_i(x, y)a_i(x, y) = f_i(y, x)a_i(y, x) \tag{3.13}$$

hold for all i, x, and y. To see this suppose $R \leq 1$ in (3.12). Then

$$a_i(x, y) = \frac{f_i(y, x)}{f_i(x, y)} \quad \text{and} \quad a_i(y, x) = 1$$

and (3.13) holds. The argument for $R \geq 1$ is analogous.

3.5 Finite point processes

3.5.1 Setup and notation

Let $(S, \mathcal{A}, \lambda)$ be a measure space, with λ an atomless finite measure. We are interested in point processes having densities with respect to the Poisson process with intensity measure λ. The requirement $\lambda(S) < \infty$ gives us finite point processes, and the requirement that λ is atomless gives us simple point processes. For a concrete example, take S to be a bounded region in $\mathbb{R}^d$ and λ proportional to Lebesgue measure.

Let Ω be the disjoint union of all finite Cartesian products of S with itself, $S^k = S \times \cdots \times S$ (k factors), including the empty product $S^0 = \{\varnothing\}$. Let λ^k denote the k-fold product of λ on S^k for $k \geq 1$ and let λ^0 denote the measure on S^0 that gives mass one to the only element of S^0, which is $\varnothing$. Let $\mathcal{B}$ denote the σ-field in Ω inherited from $\mathcal{A}$, that is the family of sets B such that $B \cap S^k$ is an element of $\mathcal{A}^k$. Then μ defined by

$$\mu(B) = \sum_{k=0}^{\infty} \frac{\lambda^k(B \cap S^k)}{k!}, \qquad B \in \mathcal{B}.$$

is a measure on $(\Omega, \mathcal{B})$, which is proportional to the Poisson process with intensity measure λ, the proportionality constant being $\mu(\Omega) = \exp[\lambda(S)]$. An element $x \in S^k$ is visualized as a pattern of k points in S. The empty set, which is the only element of S^0, is interpreted as the pattern with no points. Let the number of points in x be denoted $n(x)$, so n is a map from Ω to $\mathbb{N}$, defined by $n(x) = k$ when $x \in S^k$.

Let h be a nonnegative function on Ω. Then the measure ν having density h with respect to μ is defined by

$$\nu(B) = \int_B h(x)\mu(dx)$$

$$= \sum_{k=0}^{\infty} \frac{1}{k!} \int_{S^k \cap B} h(x)\lambda^k(dx), \qquad B \in \mathcal{B}.$$

If $0 < \nu(\Omega) < \infty$, then ν can be normalized to make a probability measure P defined by $P(B) = \nu(B)/\nu(\Omega)$, $B \in \mathcal{B}$. So h is an unnormalized density of P with respect to μ and $\nu(\Omega)$ its normalizing constant.

Despite the definition of elements of Ω as ordered k-tuples we are only interested in the interpretation of elements of Ω as point patterns, that is, as unordered sets of points. Thus we are only interested in unnormalized densities h that are symmetric under permutation of points in a pattern: $h(x) = h(y)$ if x and y would be identical if considered unordered sets. For a detailed discussion of this issue, see Section 5.3 of Daley and Vere-Jones (1988).

3.5.2 Statistical models

The opportunity to do statistics arises when we consider a statistical model specified by a family $\mathcal{H} = \{h_\theta : \theta \in \Theta\}$ of unnormalized densities with respect to μ. The normalizing function of the family is

$$c(\theta) = \int_\Omega h_\theta(x)\mu(dx)$$
$$= \sum_{k=0}^\infty \frac{1}{k!} \int_{S^k} h_\theta(x)\lambda^k(dx), \qquad \theta \in \Theta. \tag{3.14}$$

Recall that the requirement that $0 < c(\theta) < \infty$ for all $\theta \in \Theta$ is implied by using the terminology 'unnormalized density' for each of the h_θ.

Normalized densities of the family are defined by

$$f_\theta(x) = \frac{1}{c(\theta)} h_\theta(x), \qquad x \in \Omega \tag{3.15}$$

and the corresponding probability measures by

$$P_\theta(B) = \int_B f_\theta(x)\mu(dx)$$
$$= \frac{1}{c(\theta)} \sum_{k=0}^\infty \frac{1}{k!} \int_{S^k \cap B} h_\theta(x)\lambda^k(dx), \qquad B \in \mathcal{B} \tag{3.16}$$

Models of this type include Markov point processes (Ripley and Kelly, 1977) and nearest-neighbour Markov point processes (Baddeley and Møller, 1989). Another broad family is exponential family models in which h_θ has the form $h_\theta(x) = e^{\langle t(x), \theta \rangle}$ where t is a map from Ω to $\mathbb{R}^d$ and the parameter space Θ is a subset of $\mathbb{R}^d$

3.5.3 Strauss processes

A simple example that is both a Markov point process and an exponential family is the Strauss process (Strauss, 1975; Kelly and Ripley, 1976). Let r be a fixed real number, let $d(\cdot, \cdot)$ be a distance function on S, and let $s(x)$ be the number of pairs of points in x that are separated by distance no more than r.

$$s(x) = \frac{1}{2} \sum_{\xi \in x} \sum_{\substack{\eta \in x \\ \eta \neq \xi}} 1_{[0,r]}[d(\xi, \eta)], \tag{3.17}$$

where the notation $\xi \in x$ means ξ ranges over the points in x (this is a slight abuse of notation since x is a k-tuple not a set, but the ordering of the points does not affect the value of the sum, so the meaning is unambiguous). We say that points ξ and η are *neighbours* if $d(\xi, \eta) \leq r$, so $s(x)$ counts the number of pairs of neighbours in the pattern x.

Define $t(x)$ to be the bivariate vector $(n(x), s(x))$. The exponential family having canonical statistic $t(x)$ and unnormalized densities (3.3) is called the unconditional Strauss process. It was pointed out by Kelly and Ripley (1976) that $c(\theta)$ defined by (3.14) is finite if and only if $\theta_2 \le 0$. Thus the natural parameter space of the family is

$$\Theta = \{\theta \in \mathbb{R}^2 : \theta_2 \le 0\}.$$

Although it is of no importance in the sequel, it is interesting to note that since Θ is not an open subset of $\mathbb{R}^2$ this is a *nonregular* exponential family. The Strauss process thus provides an interesting example of a nonregular family not dreamed up just to provide an example of the phenomenon. The nonregularity affects likelihood inference for the model (Geyer and Møller, 1994). The Strauss process is also actually *nonsteep* since $E_\theta s(x)$ is finite for all $\theta \in \Theta$.

Conditioning on the event $\{x : n(x) = k\}$ gives the conditional Strauss process with k points. As in any exponential family, conditioning on a component of the natural statistic (here $n(x)$) eliminates the corresponding parameter (θ_1). Thus this produces a one-parameter exponential family with natural statistic $s(x)$. If we do not consider this a process derived from the unconditional process, but as a simple multivariate exponential family defined on S^k with normalizing function

$$c(\theta) = \int_{S^k} e^{s(x)\theta} \lambda^k(dx), \qquad \theta \in \mathbb{R},$$

then $c(\theta)$ is now finite for all $\theta \in \mathbb{R}$. Thus this gives another interesting property not seen in typical exponential families, that conditioning can increase the natural parameter space.

The title of Strauss (1975) calls this process a 'model for clustering,' but the unconditional Strauss process is not at all a model for clustering. By the usual formulas for exponential families

$$\nabla E_\theta t(X) = \text{Var}_\theta\, t(X)$$

so

$$E_\theta s(X) \le E_{\theta'} s(X), \qquad \text{if } \theta_2 < \theta_2' \text{ and } \theta_1 = \theta_1'$$

and, in particular, every unconditional Strauss process with $\theta_2 < 0$ is less clustered than the Poisson process with $\theta_2 = 0$ and the same value of θ_1. The conditional Strauss process can be a model for clustering, since $\theta_2 > 0$ is allowed, but it is a very poor model. Before explaining this we look at another property of exponential families.

3.5.4 Completion of exponential family models

Another standard property of exponential families gives new processes by taking limits. Let φ be a direction in the parameter space such that $\theta + s\varphi \in \Theta$ for all $\theta \in \Theta$ and for all $s > 0$. Then φ is called a *direction of recession* of Θ (Rockafellar, 1970, p. 61). If Θ is closed, then φ is a direction of recession if there exists even one $\theta \in \Theta$ such that $\theta + s\varphi \in \Theta$ for all $s > 0$ (Rockafellar, 1970, Theorem 8.3). For any direction φ, let m_φ be the essential supremum of the natural statistic $t(X)$, that is the infimum of all real numbers r such that $P_\theta(t(X) > r) = 0$ (since the distributions in the family are absolutely continuous with respect to each other, any $\theta \in \Theta$ can be used here). If m_φ is finite, let

$$H_\varphi = \{x \in \Omega : \langle t(x), \varphi \rangle = m_\varphi\}.$$

Define normalized densities and measures by (3.15) and (3.16) with $h_\theta(x) = e^{\langle t(x), \theta \rangle}$, and if $P_\theta(H_\varphi) > 0$ define

$$f_\theta(x|H_\varphi) = \begin{cases} 0, & \langle t(x), \theta \rangle < m_\varphi \\ \frac{1}{P_\theta(H_\varphi)} f_\theta(x), & \langle t(x), \theta \rangle = m_\varphi \\ +\infty, & \langle t(x), \theta \rangle > m_\varphi. \end{cases}$$

Then $f_\theta(\cdot|H_\varphi)$ is a density with respect to μ of the conditional probability measure $P_\theta(\cdot|H_\varphi)$. (By definition of essential supremum and continuity of measure, the set where $f_\theta(\cdot|H_\varphi)$ is $+\infty$ has μ-measure zero.)

Proposition 3.1 *If φ is a direction of recession, $m_\varphi < \infty$, and $P_\theta(H_\varphi) > 0$, then $f_{\theta+s\varphi}(\cdot)$ converges pointwise to $f_\theta(\cdot|H_\varphi)$ as $s \to \infty$.*

This proposition follows from Barndorff-Nielsen (1978), p. 105.

3.5.5 Limit models from Strauss processes

Conditional Strauss processes

For the conditional Strauss process with k points, every direction is a direction of recession (there just being two directions in the one-dimensional parameter space). Taking the limit as $\theta \to -\infty$ gives the *hard core* process with k points. This is a *binomial process* (k points uniformly distributed in the region) conditioned on each pair of points having separation greater than r. Taking the limit as $\theta \to +\infty$ gives the process conditioned $s(x)$ having its maximum value, which occurs when every pair of points are neighbours so $s(x) = k(k-1)/2$. This rather uninteresting process has not been studied and has no name. Dub it the *one-clump* process. In every realization of the process all k points are in a ball of diameter r.

Even when the limit process is uninteresting in itself, it tells us something about the original process. The reason why the conditional Strauss process is an uninteresting model for clustering is that when $\theta > 0$ all realizations of the process look much like one of the limits $\theta = 0$ or $\theta = +\infty$. The first is the binomial process and the second is the one-clump process. At intermediate values of θ, the conditional Strauss process is bimodal. Realizations in one mode look much like realizations from the binomial process, slightly more clustered, but hardly noticeable to the eye. Realizations in the other mode look like realizations from the one-clump process, slightly less clustered, most but not all of the points in a ball of diameter r. The parameter θ only governs how much probability goes to each of the modes. This behaviour of the Strauss process seems to have been first noticed by Julian Besag (David Strauss, personal communication). A clear demonstration of this bimodality can be seen in Figure 1 of Geyer and Thompson (1995).

It is interesting that the behaviour of the conditional Strauss process is so well described as a mixture of the processes obtained by taking limits in directions of recession. There is, of course, no reason why this phenomenon must always hold. There could be interesting novel behaviour at intermediate parameter values. Nevertheless, it does give a reason to think about all of the limiting processes. If none of the limiting processes are interesting, then the entire model may be uninteresting, as is the case for the attractive conditional Strauss process with $\theta > 0$.

Unconditional Strauss processes

Applying Proposition 3.1 to the unconditional Strauss process, any φ with $\varphi_2 \leq 0$ is a direction of recession. But if $\varphi_1 < 0$ then $H_\varphi = S^0$ and the limit obtained using the proposition is the uninteresting process that has no points with probability one.

If $\varphi_1 = 0$ and $\varphi_2 < 0$, then

$$H_\varphi = \{ x \in \Omega : s(x) = 0 \}.$$

The process obtained by conditioning a Strauss process on H_φ is the general *hard core* process. Note that one obtains the same hard core process by conditioning any Strauss process $P_{\theta + \lambda \varphi}$ for any $\lambda \geq 0$. Thus we obtain the complete family of hard core processes by conditioning Strauss processes P_θ with $\theta_2 = 0$, that is, Poisson processes. This connexion between the Strauss process and hard core process is, of course, well-known, though not usually derived using general properties of exponential families. The reason for drawing the connexion here, is that we will also be interested in other point process models for which there is no existing literature. The point of the proposition is that whenever we consider a point process model that is an exponential family a natural question arises as to what models arise from taking limits in directions of recession using Proposition 3.1.

There is still one kind of direction of recession left to consider. If $\varphi_1 > 0$ and $\varphi_2 < 0$, then the characterization of H_φ is not obvious. When the region S in which the process is defined is bounded, every such vector is a direction of recession. To see this we divide S into m disjoint subregions of radius less than r so that any two points in a subregion are neighbours. If there are k_i points in the ith subregion, then $n(x) = \sum_i k_i$ and $s(x) \geq \sum_i \binom{k_i}{2}$. It is easy to see that the latter is minimized over a real ntuples $(k_1, \ldots, k_m)$ when $k_i = n/m$. Hence

$$s(x) \geq \frac{n(x)[n(x) - m]}{2m^2}, \qquad n(x) \geq m.$$

Since this is quadratic in $n(x)$, it follows that $\varphi_1 n(x) + \varphi_2 s(x)$ is bounded above on Ω. Thus H_φ, the set of x where the least upper bound is achieved, is nonempty, and a limit process in the direction φ can be defined. What it not clear, is what the process or the set H_φ looks like. Perhaps this is related to the strange plots of unconditional Strauss processes exhibited by Professor Møller at the meeting.

3.6 Simulating finite point processes

Finite point processes are easily simulated by a special case of the MHG algorithm described by Geyer and Møller (1994) prior to Green's work. Let h be the unnormalized density of the process with respect to the measure μ of the Poisson process on S with intensity measure λ (as in Section 3.5.1). Consider the points in the pattern to be ordered, as they must be to be stored in a computer: if $n(x) = m$, write $x = (x_1, \ldots, x_m)$. A proposal kernel $Q_m(x, A)$ is defined as follows: when $n(x) = m$, it proposes to add a point x_{m+1} distributed on S with distribution proportional to λ, and when $n(x) = m + 1$ it proposes to delete x_{m+1}. For $x \notin S^m \cup S^{m+1}$ this kernel does nothing; $Q_m(x, A) = 0$ for all A. The algorithm is composed of these kernels Q_m for $m \geq 0$.

In order to fit this algorithm into the MHG framework we need to find a symmetric measure ξ_m on $\Omega \times \Omega$ such that $\pi(dx)Q_m(x, dy)$ has a density $f_m(x, y)$ with respect to $\xi_m(dx, dy)$, where π is the measure having unnormalized density h with respect to μ. The measure $\pi(dx)Q_m(x, dy)$ is concentrated on the set of possible values (x, y) such that x is a realization of the process and y is a proposal made at x. So ξ_m can be concentrated on the same set.

First consider the part of the update that attempts to add a point. Then x is in S^m, and the proposal y is in S^{m+1} The new point y_{m+1} is distributed on S with distribution proportional to λ and the rest of the points are not moved, $x_i = y_i$ for $i \leq m$. Thus the joint distribution of the pair (x, y) is concentrated on the set

$$D_m^+ = \{ (x, y) \in S^m \times S^{m+1} : x_i = y_i, \ i \leq m \}.$$

This set is not symmetric. Interchange of x and y gives

$$D_m^- = \{ (x, y) \in S^{m+1} \times S^m : x_i = y_i, \ i \leq m \}.$$

D_m^+ is isomorphic to S^{m+1} via the map $(x, y) \mapsto y$, and D_m^- is isomorphic to S^{m+1} via the map $(x, y) \mapsto x$.

Now consider the part of the update that attempts to delete a point. Then x is in S^{m+1}, and the proposal y is in S^m, and the joint distribution is concentrated on the set D_m^-. Thus $\pi(dx)Q_m(x, dy)$ and ξ_m are concentrated on the set $D_m^+ \cup D_m^-$ and this set is symmetric. We may take ξ_m to be the measure 'equal' to λ^{m+1} on D_m^+ and D_m^-, where the inverted commas call attention to the isomorphisms just mentioned.

Returning to the consideration of adding a point using the kernel Q_m so that $(x, y) \in D_m^+$, where x is the current state and y the proposal, the distribution of (x, y) is

$$\pi(dx)Q_m(x, dy) = h(x)\mu(dx)I(x, S^m)\frac{\lambda(dy_{m+1})}{\lambda(S)}$$

because x is concentrated on S^m and we must divide by $\lambda(S)$ to make λ a probability measure for the proposal. On S^m the measure μ is equal to $\lambda^m/m!$ (3.16). Hence

$$\pi(dx)Q_m(x, dy) = \frac{h(x)\lambda^m(dx)\lambda(dy_{m+1})}{m!\lambda(S)}$$

$$= \frac{h(x)\lambda^{m+1}(dy)}{m!\lambda(S)}$$

because $x_i = y_i, i \leq m$. Since $\xi_m(dx, dy) = \lambda^{m+1}(dy)$,

$$f_m(x, y) = \frac{\pi(dx)Q_m(x, dy)}{\xi_m(dx, dy)} = \frac{h(x)}{m!\lambda(S)}. \tag{3.18}$$

Now consider deleting a point using the kernel Q_m in which case

$(x, y) \in D_m^-$, where x is the current state and y the proposal, and the distribution of (x, y) is

$$\pi(dx)Q_m(x, dy) = h(x)\mu(dx)I(x, S^{m+1})$$
$$= \frac{h(x)\lambda^{m+1}(dx)}{(m+1)!}$$

because x is concentrated on S^{m+1} and the proposal is deterministic, always deleting x_{m+1}. Since now $\xi_m(dx, dy) = \lambda^{m+1}(dx)$

$$f_m(x, y) = \frac{h(x)}{(m+1)!}. \tag{3.19}$$

This now allows us to calculate Green's ratio. For a proposal going up Green's ratio is (3.19) with x and y interchanged divided by (3.18)

$$R = \frac{\lambda(S)}{m+1} \frac{h(y)}{h(x)}.$$

For a proposal going down, Green's ratio is the reciprocal.

Now consider composing this update mechanism with an update that merely permutes the order of the points. Then at each update step x_{m+1} is a random one of the $m + 1$ points. Hence we use the same Green's ratio whether we always delete the last point or whether we delete a random point. Let $P_m(x, A)$ be the kernel that performs the update just described, proposing a random point to delete.

Finally consider the mixture

$$P(x, A) = \frac{1}{2}I(x, S^0) + \sum_{m=0}^{\infty} \frac{1}{2}P_m(x, A).$$

This proposes half the time to add a point and half the time to delete a point, except when $x = \varnothing$ in which case it is impossible to go down. When $x = \varnothing$, this proposes half the time to add a point and half the time to do nothing.

In summary the algorithm is

Algorithm A.

- With probability $\frac{1}{2}$ propose an upstep and with probability $\frac{1}{2}$ propose a downstep unless $x = \varnothing$, in which case skip the downstep.

- **upstep**

  - Simulate ξ distributed proportional to λ.
  - Calculate
    $$R = \frac{\lambda(S)}{n(x)+1} \frac{h(x \cup \xi)}{h(x)}.$$

    – Set $x := x \cup \xi$ with probability $\min(1, R)$.

- **downstep**

    – Choose ξ uniformly from the points of x.
    – Calculate

$$R = \frac{n(x)}{\lambda(S)} \frac{h(x \setminus \xi)}{h(x)}.$$

    – Set $x := x \setminus \xi$ with probability $\min(1, R)$.

Note that Algorithm A, like the Metropolis–Hastings algorithm, can never move to an infeasible point, an x with $h(x) = 0$. So if started at a feasible point, the entire sample path is feasible.

3.7 Stability conditions

Ruelle (1969, Chapter 3) gives two conditions for an infinite-volume point process to have thermodynamic behaviour. The first condition *temperedness* is a restriction on the range of interactions between points. This will not concern us here, because we are interested only in finite point processes, for which the condition is irrelevant. Ruelle's second condition *stability* does apply to finite point processes.

Condition 3.1 *A process with unnormalized density h with respect to μ is stable in the sense of Ruelle if there exists a real number $M \geq 1$ such that*

$$h(x) \leq M^{n(x)}, \qquad \forall x \in \Omega. \tag{3.20}$$

This condition is clearly sufficient for h to be normalizable, since the normalizing constant is

$$\int h(x)\mu(dx) \leq \int M^{n(x)}\mu(dx) = \sum_{k=0}^{\infty} \frac{M^k}{k!}\lambda(S)^k = e^{M\lambda(S)}.$$

Proposition 3.2.2 of Ruelle shows that this condition is necessary for normalizability of densities determined by lower semicontinuous pair potentials, that is

$$h(x) = \exp\left(\frac{1}{2} \sum_{\xi \in x} \sum_{\substack{\eta \in x \\ \eta \neq \xi}} g(\xi, \eta) \right)$$

where g is a lower semicontinuous function on $S \times S$. That is, Ruelle's condition is necessary and sufficient for thermodynamic behaviour in a certain class of processes.

We now turn to an even stronger stability condition that is only sufficient, not necessary, for a different kind of stability, geometric ergodicity of Markov chain Monte Carlo simulation schemes.

Condition 3.2 *A process with unnormalized density h with respect to μ is stable if there exists a real number M such that*

$$h(x \cup \xi) \leq Mh(x), \qquad \text{for all } x \in \Omega \text{ and } \xi \in S. \qquad (3.21)$$

This condition implies Condition 3.1 since it implies

$$h(x) \leq h(\emptyset)M^{n(x)}, \qquad \text{for all } x \in \Omega.$$

It also implies a *hereditary* condition

$$h(x) = 0 \quad \text{implies} \quad h(y) = 0, \qquad \text{whenever } x \subset y.$$

which makes it possible to define the conditional intensity (see the contribution of Baddeley to this volume)

$$\lambda(\xi|x) = \frac{h(x \cup \xi)}{h(x)},$$

taking $0/0 = 0$. Thus Condition 3.2 can be restated saying that the process has a bounded conditional intensity.

The point of Condition 3.2 is that it permits easy proofs of geometric ergodicity of simple Markov chain Monte Carlo schemes for simulating the process. It is not clear that there are any interesting processes that satisfy Condition 3.1 and fail to satisfy Condition 3.2, but if there are they may require more delicate Markov chain Monte Carlo schemes or they may lead to a generalization of Condition 3.2 that still yields geometric ergodicity. Here we only investigate processes satisfying Condition 3.2.

3.8 Markov Chain Convergence

We consider four aspects of what Meyn and Tweedie (1993) call 'stability' of Markov chains: φ-irreducibility, small sets, Harris recurrence, and geometric ergodicity. They are generally proved in that order, since the simpler notions must be established before the more complex can be considered. For the point process samplers in Algorithm A we really need only two proofs. The same argument shows both φ-irreducibility and that every bounded set is small, and a second argument shows both Harris recurrence and geometric ergodicity.

3.8.1 φ-Irreducibility

Let $L(x, A)$ denote the probability that a Markov chain started at x ever hits the set A. A rather complicated formula for $L(x, A)$ is given by Meyn and Tweedie (1993, p. 72) but is not of interest here.

A Markov chain on a measurable space $(\Omega, \mathcal{B})$ is φ-*irreducible* if φ is a nonzero measure on $(\Omega, \mathcal{B})$ such that $L(x, B) > 0$ for all $x \in \Omega$ and all $B \in \mathcal{B}$ such that $\varphi(B) > 0$. If a chain is φ-irreducible for any φ then it is also π-irreducible if π is a stationary distribution for the chain (Meyn and Tweedie, 1993, Proposition 4.2.2 and Theorem 10.4.9), moreover π must then be the unique stationary distribution. The point of the weaker notion of φ-irreducibility is that it allows one to check fewer sets.

Let $P^n(x, B)$ denote the probability $\Pr\{X_{n+1} \in B | X_1 = x\}$ that the chain started at x is in B after n steps. Then an equivalent definition of φ-irreducibility is that for every $x \in \Omega$ and every $B \in \mathcal{B}$ such that $\varphi(B) > 0$ there exists an n such that $P^n(x, B) > 0$.

3.8.2 Small Sets

A set C is *small* if there exists a nonzero measure ν and an integer n such that

$$P^n(x, B) \geq \nu(B), \qquad \text{for all } x \in C \text{ and all } B \in \mathcal{B}. \tag{3.22}$$

What one typically finds for well-behaved Markov chains on locally compact topological spaces is that every bounded measurable set is small. Chapter 6 of Meyn and Tweedie (1993) gives a number of tools for establishing this property.

Here we characterize φ-irreducibility and small sets directly for our point process samplers. There is a technical detail involved in φ-irreducibility. Since the sampler is confined to the set of feasible states $\Omega^+ = \{x \in \Omega : h(x) > 0\}$, and no dynamics are defined off this set, we must take Ω^+ to be the state space.

Proposition 3.2 *If the unnormalized density satisfies Condition 3.2, then Algorithm A started at any $x \in \Omega^+$ simulates a φ-irreducible Markov chain on Ω^+ and every measurable set on which $n(x)$ is bounded is small.*

Proof. We pick for φ in the proof of φ-irreducibility and for ν in the definition of small set the measure $c\lambda^0$, where c is a constant to be named later. The probability of accepting a downstep is greater than

$$\min\left(1, \frac{n(x)}{\lambda(S)}\frac{h(x \setminus \xi)}{h(x)}\right) \geq \frac{1}{M\lambda(S)} \tag{3.23}$$

where M is the constant in Condition 3.2 assuming that M is chosen large enough so that the right hand side in (3.23) is less than one, which entails no loss of generality. The probability of proposing a downstep is $\frac{1}{2}$. Also $P(\varnothing, S^0) \geq \frac{1}{2}$, hence if $m \geq n(x) = k$

$$P^m(x, S^0) \geq P^k(x, S^0)P^{m-k}(\varnothing, S^0) \geq \left(\frac{1}{2M\lambda(S)}\right)^m \tag{3.24}$$

This calculation shows that the chain is λ^0-irreducible, since the only λ^0-positive set is $S^0 = \{\varnothing\}$. It also shows that $P^m(x, S^0) > 0$ whenever $m \geq n(x)$.

The same calculation shows that every set on which $n(x)$ is bounded is small. Let $C = \{x \in \Omega : n(x) \leq m\}$ be such a set, and let c be the right hand side of (3.24) then $P^m(x, S^0) \geq c\lambda^0(S^0)$ for any $x \in C$, since $\lambda^0(S^0) = 1$, and this establishes (3.22) with $v = c\lambda^0$. $\square$

Meyn and Tweedie also introduce a concept they call a *petite set*. For aperiodic Markov chains there is no difference between small and petite sets. Any MHG sampler that has nonzero rejection probabilities, is aperiodic, including Algorithm A. Hence we shall use only the notion of small sets.

3.8.3 Harris Recurrence

Harris recurrence is the property that $L(x, A) = 1$ for all $x \in \Omega$ and all π-positive A, where π is the stationary distribution. Harris recurrence is a stronger property than π-irreducibility, which only requires that $L(x, A) > 0$ for all x and all π-positive A, although it can be proved that when a chain is π-irreducible, there is a π-null set N such that $L(x, A) = 1$ does hold for all $x \notin N$ and all π-positive A. (Meyn and Tweedie, 1993, Proposition 9.0.1).

So the point of Harris recurrence is to banish the pathological null set. It would be very strange if a Markov chain that is an idealization of a computer simulation would be π-irreducible but not Harris recurrent. If null sets matter when the computer's real numbers are replaced by those of real analysis, then the simulation cannot be well described by the theory.

Fortunately, an irreducible Gibbs or Metropolis sampler is always

Harris recurrent under very weak conditions (Tierney, 1994, Corollaries 1 and 2) and the same is true of variable-at-a-time Metropolis–Hastings (Chan and Geyer, 1994, Theorem 1). Because of the general measures employed in the MHG algorithm there can be no such simple theorem implying Harris recurrence for general MHG samplers, but the same principles can sometimes be used to check Harris recurrence for them too.

A different way to prove Harris recurrence uses a so-called drift condition. Recall that for any function V

$$PV(x) = E\{V(X_{t+1})|X_t = x\} = \int P(x, dy)V(y).$$

The function V is said to be *unbounded* off small sets if for every $\alpha > 0$ the level set $\{x \in \Omega : V(x) \leq \alpha\}$ is small. We say a Markov chain satisfies the *drift condition for recurrence* if there exists a function $V : \Omega \to (0, \infty)$ which is unbounded off small sets and a small set $C \in B$ such that

$$PV(x) \leq V(x), \qquad x \notin C.$$

If a chain satisfies the drift condition for recurrence, then it is Harris recurrent (Meyn and Tweedie, 1993, Theorem 9.1.8).

3.8.4 Geometric Ergodicity

A stronger property is geometric ergodicity. For any signed measure ν on Ω, let $\|\nu\|$ denote the total variation of ν

$$\|\nu\| = \sup_{f:|f|\leq 1} \int f \, d\nu.$$

A Markov chain is *geometrically ergodic* if there exists a constant $r > 1$ such that

$$\sum_{n=1}^{\infty} r^n \|P^n(x, \cdot) - \pi(\cdot)\| < \infty, \qquad \forall x \in \Omega.$$

This is implied (Meyn and Tweedie, 1993, Theorem 15.0.1) by the *geometric drift condition*: there exists a function $V : \Omega \to [1, \infty)$, constants $b < \infty$ and $\lambda < 1$, and a small set $C \in B$ such that

$$PV(x) \leq \lambda V(x) + b1_C(x), \qquad \forall x \in \Omega. \tag{3.25}$$

Any V that satisfies the geometric drift condition is unbounded off small sets (when the chain is aperiodic). Hence the geometric drift condition implies the drift condition for recurrence.

Proposition 3.3 *If the unnormalized density satisfies Condition 3.2, then Algorithm A started at $x \in \Omega^+$ simulates a Markov chain on Ω^+ that is Harris recurrent and geometrically ergodic.*

Proof. Take $V(x) = A^{n(x)}$, where $A > M\lambda(S)$ and $A > 1$. The probability of accepting an upstep is

$$\min\left(1, \frac{h(x \cup \xi)}{h(x)} \frac{\lambda(S)}{n(x)+1}\right) \le \frac{M\lambda(S)}{n(x)+1}$$

For any $\epsilon \in (0, 1)$, this probability is less than ϵ when $n(x) \ge K_\epsilon$ where $K_\epsilon = M\lambda(S)/\epsilon$. The probability of accepting a downstep is

$$\min\left(1, \frac{n(x)}{\lambda(S)} \frac{h(x \setminus \xi)}{h(x)}\right) \ge \min\left(1, \frac{n(x)}{M\lambda(S)}\right) = 1$$

when $n(x) \ge K_\epsilon$. Then

$$PV(x) \le \left\{\frac{1}{2}\frac{1}{A} + \frac{1}{2}(1 - \epsilon) + \frac{1}{2}\epsilon A\right\} V(x), \qquad n(x) \ge K_\epsilon.$$

Since the choice of ϵ is still free and the term in brackets on the right hand side converges to $\frac{1}{2}(\frac{1}{A}+1) < 1$ as $\epsilon \to 0$, we can choose ϵ small enough so that

$$PV(x) \le \lambda V(x), \qquad n(x) \ge K_\epsilon$$

for some $\lambda < 1$. Take $C = \{x \in \Omega : n(x) < K_\epsilon\}$ and the geometric drift condition (3.25) is satisfied for $x \notin C$. For $x \in C$, one step of the chain can increase $n(x)$ by no more than 1, hence $PV(x) \le A^{K_\epsilon+1}$, $x \in C$ and the geometric drift condition holds for $x \in C$ if we take $b = A^{K_\epsilon+1}$. $\square$

Neither of these proofs has any novelty. They follow exactly the logic of the proofs for the Strauss process sampler given in Geyer and Møller (1994). The only point of repeating them here is to show that the same argument works for any point process with unnormalized density having bounded conditional intensity.

3.8.5 Central Limit Theorem

Geometric ergodicity often implies a central limit theorem (CLT). Let g be any π-integrable function,

$$\mu = E_\pi g(X) = \int g(x)\pi(dx)$$

where π is the stationary distribution of a Harris recurrent chain, and let

$$\hat{\mu}_n = \frac{1}{n} \sum_{i=1}^{n} g(X_i)$$

be the sample mean of a run of length n starting at $X_0 = x$, which may be chosen arbitrarily. Then the strong law of large numbers (Meyn and Tweedie, 1993, Theorem 17.1.7) implies $\hat{\mu}_n \to \mu$ with probability one.

If the chain is geometrically ergodic and the function g satisfies a Lyapunov condition

$$\int |g(x)|^{2+\epsilon} \pi(dx) < \infty \tag{3.26}$$

for some $\epsilon > 0$, then the CLT also holds

$$\sqrt{n} \left(\hat{\mu}_n - \mu \right) \xrightarrow{D} N(0, \sigma^2) \tag{3.27}$$

where

$$
\begin{aligned}
\sigma^2 &= \text{Var}_\pi \left(g(X) \right) + 2 \sum_{k=1}^{\infty} \text{Cov}_\pi \left(g(X_t), g(X_{t+k}) \right) \\
&= \int \bar{g}(x)^2 \pi(dx) + 2 \sum_{k=1}^{\infty} \iint \bar{g}(x) \bar{g}(y) \pi(dx) P^k(x, dy)
\end{aligned}
\tag{3.28}
$$

where $\bar{g}(x) = g(x) - \mu$ (Chan and Geyer, 1994, Theorem 2). A different CLT is given by Meyn and Tweedie (1993, Theorem 17.5.4). They replace the Lyapunov condition (3.26) by a condition that relates g to the function V in the geometric drift condition. If $g^2 \le V$, then the CLT (3.27) holds with σ^2 given by (3.28).

When the chain is Harris recurrent, the CLT and the SLLN hold for every starting position, indeed every starting distribution, if they hold for one starting position (Meyn and Tweedie, 1993, Theorem 17.1.6). The idea that one must run the chain until it 'reaches equilibrium' before starting to sample, something that is, strictly speaking, impossible, is not required by the asymptotics.

3.8.6 Comment on Asymptotics

A careful examination of our proofs of irreducibility and geometric ergodicity for our point process samplers shows that they tell us very little about the behaviour of the samplers in actual use.

The irreducibility proof shows that the chain will eventually reach the empty pattern with no points, but in practice we will never see this event of exceedingly small probability. Thus the proof tells us nothing about the behaviour of the sampler during any run we are likely to have patience to endure.

The proof of geometric ergodicity does tell us that if we start the chain at a pattern having far more points than the average number under the stationary distribution, the sampler would move towards the mode of the stationary distribution geometrically fast, losing points until it reaches some bounded set C on which we do not bother to follow its behaviour. The set C could be very large, so large that it contains with high probability the entire sample path of any run we have patience to endure. Thus, while geometric ergodicity tells us something, it does not tell us much.

The main benefit of proving geometric ergodicity seems to be the implication that the CLT holds. Of course one cannot calculate the variance (3.28), but one can estimate it using time series methods or regeneration. Examples will appear later in this chapter. This is not much different from most applications in asymptotics in statistics. The theorems have mainly heuristic value, giving one a calculation that will be approximately valid if n is large enough, but no one can tell how large is large enough.

In early statistical papers on MCMC, such as Gelfand and Smith (1990), the CLT was avoided because simple conditions that would imply it were not known. The connexion between geometric drift and geometric ergodicity was given by Nummelin and Tuominen (1982) and Nummelin (1984). The CLT given by Chan and Geyer (1994) follows from the fact that Harris recurrence implies what is called β-mixing in the stationary stochastic process literature (Bradley, 1986) and that geometric ergodicity implies exponentially fast β-mixing. Although the proofs are almost identical, the latter seems not to have been noticed before Chan and Geyer (1994). That exponentially fast β-mixing implies a CLT is in Ibragimov and Linnik (1971, Theorem 18.5.3). The other version of the CLT involving geometric drift is also recent (Meyn and Tweedie, 1992, 1993). Lacking useful conditions implying a CLT, statisticians avoided it in MCMC, which is exceedingly peculiar given the prominence of the CLT in other areas of statistics. This shows the power of theory to control practice even when the actual relevance of the theory is questionable.

While the theory presented here may be of little comfort to the practical-minded statistician, there is a point to having proofs of what can be proved. Lacking proofs may set people to wondering and worrying unnecessarily. For Markov chain samplers that converge much faster than any known samplers for general point processes there is a theory due to Rosenthal (1995) that gives bounds on total variation distance, but this seems not to be useful for chains that take many tens of iterations to mix well, see the appendix of the contribution of Møller to this volume.

3.9 Two New Point Processes

In this section two new point process are described and shown to satisfy Condition 3.2. No particular virtues are claimed for these processes. The main point of proposing them is to show how easy it is to invent new processes and do statistical inference for them. In an application motivated by a real scientific question, it is likely that no process in the existing literature would be of scientific interest and a model specific to the application would be invented. The two processes studied here are 'models for clustering' that avoid, in different ways, the pitfall that ensnared the Strauss process. They will be fitted to a made-up data set simulated from one of the processes in question.

3.9.1 The Triplets Process

The idea of Markov point processes (Ripley and Kelly, 1977) suggests adding the clique of next higher order to get a process that permits positive attraction of pairs of points. Define

$$w(x) = \frac{1}{6} \sum_{\xi \in x} \sum_{\substack{\eta \in x \\ \eta \neq \xi}} \sum_{\substack{\zeta \in x \\ \zeta \neq \xi \\ \zeta \neq \eta}} 1_{[0,r]}[d(\xi, \eta)] 1_{[0,r]}[d(\eta, \zeta)] 1_{[0,r]}[d(\zeta, \xi)]$$

$$(3.29)$$

the number of triples of points that are mutual neighbours, where 'neighbour' has the same definition as for the Strauss process (separation by distance no greater than r). Define $t(x)$ to be the 3-dimensional vector $(n(x), s(x), w(x))$, where $s(x)$ given by (3.17) is the same as in the Strauss process. The exponential family having canonical statistic $t(x)$ and unnormalized densities (3.3) is called here the *triplets process*.

Proposition 3.4 *The triplets process satisfies Condition 3.2, if $\theta_2 \leq 0$ and $\theta_3 \leq 0$ or if $\theta_3 < 0$.*

Proof. We need to obtain an upper bound that does not depend on x

for

$$\log \frac{h(x \cup \xi)}{h(x)} = \theta_1 + [s(x \cup \xi) - s(x)]\theta_2 + [w(x \cup \xi) - w(x)]\theta_3.$$

Since the factors in brackets are both nonnegative, this is trivial if θ_2 and θ_3 are both nonpositive. Thus the only difficult case is $\theta_2 \geq 0$ and $\theta_3 < 0$, which we now assume. For this case we need an upper bound on $s(x \cup \xi) - s(x)$ and a lower bound on $w(x \cup \xi) - w(x)$. When adding a point ξ, the only points of x that matter in calculating the odds ratio $h(x \cup \xi)/h(x)$ are those in a circle of radius r centred at ξ. Divide the circle into six equal sectors Let k_i, $i = 1, \ldots, 6$ be the number of

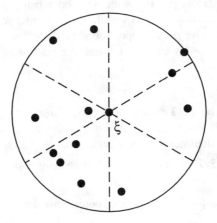

points (other than ξ) in the six sectors. Then

$$s(x \cup \xi) - s(x) = \sum_{i=1}^{6} k_i$$

and

$$w(x \cup \xi) - w(x) \geq \sum_{i=1}^{6} \binom{k_i}{2}$$

because each pair of points in a sector are neighbours and also neighbours of ξ. Hence, since $\theta_3 < 0$,

$$\log \frac{h(x \cup \xi)}{h(x)} \leq \theta_1 + \sum_{i=1}^{6} \left\{ \theta_2 k_i + \theta_3 \frac{k_i(k_i - 1)}{2} \right\}$$

$$\leq \theta_1 + 6 \sup_{0 < z < \infty} \left\{ \theta_2 z + \theta_3 \frac{z(z - 1)}{2} \right\}$$

Since the factor in braces is a negative-definite quadratic form, it has a maximum, and this gives the required upper bound.   □

3.9.2 The Saturation Process

A simpler way of obtaining a 'model for clustering' is the following. For each point $\xi \in x$ define

$$m_\xi(x) = \sum_{\substack{\eta \in x \\ \eta \neq \xi}} 1_{[0,r]}[d(\xi, \eta)].$$

Then $\sum_{\xi \in x} m_\xi(x)$ is just twice the neighbour pair statistic $s(x)$ used in defining the Strauss process and the triplets process. But instead of adding the $m_\xi(x)$ we put an upper bound on the influence of any single point. Let $c > 0$ be an arbitrary constant, and define

$$u(x) = \sum_{\xi \in x} \min\left[c, m_\xi(x)\right].$$

Define $t(x)$ to be the 2-dimensional vector $(n(x), u(x))$. Then the exponential family having canonical statistic $t(x)$ and unnormalized densities (3.3) is called here the *saturation process*.

Proposition 3.5 *The saturation process satisfies Condition 3.2 for all values of θ.*

Proof. We need to obtain an upper bound that does not depend on x for

$$\log \frac{h(x \cup \xi)}{h(x)} = \theta_1 + [u(x \cup \xi) - u(x)]\theta_2.$$

Since the factor in brackets is nonnegative, this is trivial if $\theta_2 \leq 0$. Hence assume $\theta_2 > 0$.

Again consider, as in the preceding proof, the circle of radius r centred at ξ divided into six sectors with $k_i, i = 1, \ldots, 6$ points (other than ξ) in each sector. The contribution to $u(x \cup \xi) - u(x)$ involving $m_\xi(x \cup \xi)$ is at most c. For $\eta \neq \xi$, the contribution

$$\min\left[c, m_\eta(x \cup \xi)\right] - \min\left[c, m_\eta(x)\right]$$

is at most 1 and is nonzero only if $m_\eta(x) < c$, which implies that $k_i - 1 < c$ if η is in the ith sector. Hence

$$\log \frac{h(x \cup \xi)}{h(x)} \leq \theta_1 + [c + 6(c + 1)]\theta_2,$$

and this gives the required upper bound.   □

3.9.3 Limit Models from the Saturation and Triplets Processes

Consider first the saturation process. As with the unconditional Strauss process, the direction of recession $\varphi = (0, -1)$ gives the hard core process as a limiting distribution. The upper bound $u(x) \leq cn(x)$ gives a new limiting process. Consider the direction $\varphi = (-c, 1)$. Then $\varphi_1 n(x) + \varphi_2 u(x) \leq 0$ and is maximized on the set

$$H_\varphi = \{ x \in \Omega : u(x) = cn(x) \}$$

the set of points such that each point has at least c neighbours. The limiting process is the family of Poisson processes conditioned to lie in H_φ.

Directions of recession with $\varphi_1 < 0$ and $\varphi_1 > -c\varphi_2$ give rise to the uninteresting limit distribution conditioned on the empty realization. Directions of recession with $\varphi_1 < c\varphi_2$ give rise to no limit distributions, since H_φ is empty.

As with the unconditional Strauss process, directions of recession with $\varphi_1 > 0$ and $\varphi_1 > -c\varphi_2$ seem difficult to describe. They are, at least models, for repulsion rather than clustering. So we can say that the saturation process has very simple limiting behaviour in the clustering region of the parameter space.

No such simple limits arise with the triplets process. Not only is the parameter space three-dimensional so that we must consider directions of recession in three dimensions, there are no simple linear inequalities involving the canonical statistics $n(x)$, $s(x)$ and $w(x)$. All of the limiting behaviour of the triplets process seems complicated. This is one way to see that the triplets process, despite its simple motivation from the point of view of cliques in Markov point processes, is actually a much more complicated statistical model than the saturation process.

3.10 Monte Carlo Likelihood Inference

There are a number of different approaches to likelihood inference for models specified by families of unnormalized densities. The main approach used here will be the Monte Carlo likelihood (MCL) of Geyer and Thompson (1992) and Geyer (1994). This will be compared with the Monte Carlo Newton–Raphson (MCNR) approach of Penttinen (1984) and the stochastic approximation approach of Younes (1988) and Moyeed and Baddeley (1991). For other approaches, see the introduction and references of Geyer and Thompson (1992).

3.10.1 Monte Carlo Likelihood

With the normalized densities of a model given by (3.2), the log likelihood for an observation x is

$$l(\theta) = \log h_\theta(x) - \log c(\theta).$$

It turns out to be more convenient to use the log likelihood ratio against
a fixed point $\psi \in \Theta$

$$l(\theta) = \log \frac{h_\theta(x)}{h_\psi(x)} - \log \frac{c(\theta)}{c(\psi)}. \tag{3.30}$$

The first term, involving the unnormalized densities, is known in closed
form, the second, involving the normalizing function, is not. But, if
$h_\theta(x) = 0$ whenever $h_\psi(x) = 0$, then

$$
\begin{aligned}
\frac{c(\theta)}{c(\psi)} &= \frac{1}{c(\psi)} \int h_\theta(x)\mu(dx) \\
&= \int \frac{h_\theta(x)}{h_\psi(x)} \frac{h_\psi(x)}{c(\psi)} \mu(dx) \\
&= \int \frac{h_\theta(x)}{h_\psi(x)} f_\psi(x)\mu(dx) \\
&= E_\psi \frac{h_\theta(X)}{h_\psi(X)}.
\end{aligned}
\tag{3.31}
$$

This permits calculation of the log likelihood by MCMC. Let X_1, X_2,
... be simulations from P_ψ. Then

$$l(\theta) = \log \frac{h_\theta(x)}{h_\psi(x)} - \log \left(E_\psi \frac{h_\theta(X)}{h_\psi(X)} \right) \tag{3.32}$$

is approximated by

$$l_n(\theta) = \log \frac{h_\theta(x)}{h_\psi(x)} - \log \left(\frac{1}{n} \sum_{i=1}^{n} \frac{h_\theta(X_i)}{h_\psi(X_i)} \right) \tag{3.33}$$

Maximizing (3.33) gives a Monte Carlo approximation $\hat{\theta}_n$ to the MLE
$\hat{\theta}$, which maximizes (3.32).
   The gradient of (3.33) is

$$\nabla l_n(\theta) = \nabla \log h_\theta(x) - \frac{\sum_{i=1}^{n} \frac{\nabla h_\theta(X_i)}{h_\theta(X_i)} \frac{h_\theta(X_i)}{h_\psi(X_i)}}{\sum_{i=1}^{n} \frac{h_\theta(X_i)}{h_\psi(X_i)}} \tag{3.34}$$

which can be recognized as a case of importance sampling. Define

$$w_{n,\theta,\psi}(x) = \frac{h_\theta(x)/h_\psi(x)}{\sum_{i=1}^n h_\theta(X_i)/h_\psi(X_i)} \tag{3.35}$$

and for any function g

$$\mathbb{E}_{n,\theta,\psi} g(X) = \sum_{i=1}^n g(X_i) w_{n,\theta,\psi}(X_i). \tag{3.36}$$

Then (3.36) is the Monte Carlo approximation of $E_\theta g(X)$ given by the importance sampling formula using normalized importance weights (3.35). Using this notation, we get

$$\nabla l_n(\theta) = \nabla \log h_\theta(x) - \mathbb{E}_{n,\theta,\psi} \nabla \log h_\theta(X). \tag{3.37}$$

Similarly,

$$\nabla^2 l_n(\theta) = \nabla^2 \log h_\theta(x) - \mathbb{E}_{n,\theta,\psi} \nabla^2 \log h_\theta(X) \\ - \mathbb{V}\mathrm{ar}_{n,\theta,\psi} \nabla \log h_\theta(X) \tag{3.38}$$

where for a vector-valued function g

$$\mathbb{V}\mathrm{ar}_{n,\theta,\psi} g(X) = \mathbb{E}_{n,\theta,\psi} g(X) g(X)^T - \left[\mathbb{E}_{n,\theta,\psi} g(X)\right]\left[\mathbb{E}_{n,\theta,\psi} g(X)\right]^T$$

This all simplifies considerably in the exponential family case where we have

$$\log h_\theta(x) = \langle t(x), \theta \rangle$$
$$\nabla \log h_\theta(x) = t(x)$$
$$\nabla^2 \log h_\theta(x) = 0.$$

and normalized importance weights are

$$w_{n,\theta,\psi}(x) = \frac{e^{\langle t(x), \theta - \psi \rangle}}{\sum_{i=1}^n e^{\langle t(X_i), \theta - \psi \rangle}}.$$

Giving estimates of the score

$$\nabla l_n(\theta) = t(x) - \mathbb{E}_{n,\theta,\psi} t(X) \tag{3.39}$$

and Fisher information

$$-\nabla^2 l_n(\theta) = \mathbb{V}\mathrm{ar}_{n,\theta,\psi} t(X) \tag{3.40}$$

just what one expects, since the exact score and Fisher information are obtained by replacing Monte Carlo expectations by exact expectations.

3.10.2 Monte Carlo Newton–Raphson

If one uses the formulas (3.37) and (3.38) or (3.39) and (3.40) in the special case $\theta = \psi$, there is considerable simplification. In this case, all the importance weights are $\frac{1}{n}$. Hence, for example, (3.39) and (3.40) become

$$\nabla l_n(\psi) = t(x) - \frac{1}{n}\sum_{i=1}^{n} t(X_i)$$

and

$$-\nabla^2 l_n(\psi) = \sum_{i=1}^{n} t(X_i)t(X)^T - \left[\frac{1}{n}\sum_{i=1}^{n} t(X_i)\right]\left[\frac{1}{n}\sum_{i=1}^{n} t(X_i)\right]^T$$

and one can use these approximations to do a Newton–Raphson update of the Monte Carlo approximation of the MLE.

Denote the current MCNR iterate by ψ_k, then the Newton–Raphson update

$$\psi_{k+1} = \psi_k - \left[\nabla^2 l_n(\psi_k)\right]^{-1}\nabla l_n(\psi_k)$$

gives the next iterate. There are, however, three problems with this approach, compared to the MCL approach.

- MCNR must iterate indefinitely, 'until convergence' which is ill-defined. MCL requires only one sample if ψ is close enough to $\hat{\theta}$ so that the importance weights behave.

- Geyer (1994) gives an asymptotic approximation for the Monte Carlo error $\sqrt{n}(\hat{\theta}_n - \hat{\theta})$ using MCL. No analogous formulas exist for MCNR.

- When ψ is not close to $\hat{\theta}$ neither approach works, but MCL can be fixed up by using a trust region in the maximization of the l_n (Geyer and Thompson, 1992). No analogous fix-up exists for MCNR.

Thus MCNR has no benefits over MCL except that the calculations of the Monte Carlo approximation are slightly simpler. MCNR only makes sense as an approximation to MCL, which is the right thing.

3.11 Stochastic Approximation

The method of stochastic approximation, used to obtain Monte Carlo maximum likelihood estimates by Younes (1988) and Moyeed and Baddeley (1991) is not useful for obtaining estimates of even moderate precision, nor is there a good method of estimating the accuracy of its estimates. It may be used to get a starting point for MCL methods.

Here we use a very simplified version of stochastic approximation. The idea is to run a Markov chain with nonstationary transition probabilities, adjusting the parameter θ in each iteration moving it towards the MLE. We thus obtain a sequence (x_n, θ_n) of point patterns and parameter values. If the current position is x_n and x is the observed point pattern in an exponential family model, then a very crude estimate of the score is $t(x) - t(x_n)$. Moving θ in that direction will, on average, move it towards the MLE. Hence for some $\epsilon > 0$, we update θ using

$$\theta_{n+1} = \theta_n + \epsilon[t(x) - t(x_n)].$$

In classical stochastic approximation, ϵ is also a function of n, decreasing with time so that θ_n converges to the MLE. The usual practice is to use

$$\epsilon_k = \frac{\epsilon_0}{1 + \rho k}$$

in iteration k, where $\rho > 0$ and $\epsilon_0 > 0$ are constants. This form is implemented in the computer code described in Section 3.13, but since there are no guidelines for choosing ρ or ϵ_0, we prefer a more 'hands-on' approach in which $\rho = 0$ and $\epsilon_0 = \epsilon$ is chosen by looking at plots.

There is no known way to choose ϵ except by experiment. If ϵ is too large, the sampler may react too rapidly to the initial position, running away from the MLE. If ϵ is too small, the sampler makes no appreciable progress. The sampler should remain in approximate equilibrium, so that x_n is an approximate realization from θ_n, and ϵ should be set small enough so that the sampler stays in approximate equilibrium. One must not change θ so fast that the sampler can't change x rapidly enough to stay in approximate equilibrium.

3.12 Reverse Logistic Regression

While (3.33) and similar formulas work well in calculating log likelihood ratios for points θ and ψ that are not widely separated, they fail miserably when θ and ψ are far apart, as may occur when they are the estimates in the null and alternative of a hypothesis test. Some other method is needed.

Geyer (1991) proposed the method of 'reverse logistic regression' in a tech report that has never appeared in print. (The report was revised in 1994, but Theorem 4 added in the revision was wrong.) As applied to the problem of estimating normalizing constants and log likelihood ratios, the method is sound. It has been used to calculate likelihood ratio test statistics by Thompson, Lin, Olshen and Wijsman (1993).

Suppose we have samples from independent runs for m models with unnormalized densities h_j, $j = 1, \ldots, m$ and unknown normalizing constants $c_j = \int h_j \, d\mu$. We want to estimate these normalizing constants, up to an overall constant of proportionality, or equivalently we want to estimate c_j/c_1 for all j. How this is used in calculating likelihood ratio tests and how one chooses the densities h_j and the number m is explained in Section 3.14.3.

Suppose X_{ij}, $i = 1, \ldots, n_j$ are samples from an MCMC sampler for the distribution with unnormalized density h_j. Write

$$\psi_j = -\log c_j$$

so the normalized densities have the form $h_j(x)e^{\psi_j}$. Reverse logistic involves the following curious scheme for estimating the vector ψ. If we are given a point x that is one of the X_{ij} but we are not told which one, the probability that it belongs to the jth sample is

$$\frac{n_j h_j(x)e^{\psi_j}}{\sum_k n_k h_k(x)e^{\psi_k}}. \tag{3.41}$$

We simplify this by absorbing the factor n_j into the definition of the 'parameter' ψ, writing

$$\eta_j = \psi_j + \log n_j$$

so that (3.41) becomes

$$p_j(x, \eta) = \frac{h_j(x)e^{\eta_j}}{\sum_k h_k(x)e^{\eta_k}}. \tag{3.42}$$

We propose to estimate η by maximizing the quasi-likelihood

$$l(\eta) = \sum_{j=1}^{m} \sum_{i=1}^{n_j} p_j(X_{ij}, \eta). \tag{3.43}$$

It is not really a likelihood because in defining (3.42) we pretend we don't know which sample x belongs to, but in (3.43) we do use the knowledge of which sample X_{ij} belongs to. Or perhaps we don't in some sense, because the quasi-score is

$$\frac{\partial l(\eta)}{\partial \eta_r} = n_r - \sum_{j=1}^{m} \sum_{i=1}^{n_j} p_r(X_{ij}, \eta). \tag{3.44}$$

For each r the sum on i and j runs over all the sample points, not just over the points in the r-th sample. Thus (3.44) does not use information about which sample a data point belongs to.

We shall not try to further explain the philosophy of this method further here, but merely note that (3.43) is arithmetically equivalent to the log likelihood for a multivariate logistic regression (also called a multinomial response model). Statistically, it is not a logistic regression, because the regression is reversed. This is clear from (3.44) which has the usual 'observed minus expected' form of an exponential family. This makes n_r the 'response' and X_{ij} the 'predictor' when we think of this as a logistic regression. That makes the 'response' a fixed known quantity and the 'predictor' the random quantity. So let us just say that (3.44) defines a system of estimating equations used to estimate the vector ψ. It is clear from (3.42) that the value of the quasi-likelihood or quasi-score is unchanged by adding a constant to all of the η_j. Hence η is estimated only up to an additive constant, as is ψ, and the c_j are estimated only up to a multiplicative constant, but that is all we want.

We know from the properties of exponential families that the estimator of ψ will be unique up to an additive constant if ψ is identifiable up to an additive constant. Geyer (1991) shows this happens if the densities h_j cannot be divided into two nonempty subsets such that for each data point $h_j(x) = 0$ for all j in one of the subsets. The quasi-score is the average of a bounded function $p_r(X, \eta)$. Hence it satisfies a Lyapunov condition, hence a sufficient condition for a CLT (Theorem 2 in Geyer, 1991) is that the Markov chain samplers be geometrically ergodic. From the CLT we can obtain a Monte Carlo standard error for $\hat{\psi}_n - \psi$. Geyer (1991) gives the details, which we shall pass over here. The computer code described in Section 3.13 does this calculation.

A possible alternative to the method of reverse logistic regression for calculating log likelihood ratios for widely separated points is the method of umbrella sampling (Torrie and Valleau, 1977; Geyer and Thompson, 1995), but we shall not use it in this chapter.

3.13 Fitting the Saturation Model

The point pattern in Figure 3.1 is made-up data. It is a simulation of the saturation model with $r = .05$, $c = 4.5$, $\theta_1 = 4.0$, and $\theta_2 = 0.4$. The simulation was actually done in a toroidal region with side 1.2, and the points outside the unit square were thrown away. We pretend that Figure 3.1 is a data set of unknown origin to which we will fit several models.

The first model we will fit is the saturation model with $r = .05$ and

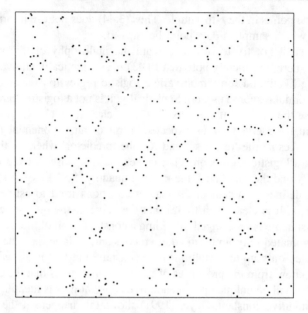

Figure 3.1 *Simulated point pattern*.

$c = 4.5$ (the same as the simulation) with θ an unknown parameter. In this section we consider the process to be defined in the unit square with free boundary conditions, which is not the same as the simulation. The observed value of the canonical statistic $t(x)$ for the pattern in Figure 3.1 is (372.0, 1321.5).

We start with a Monte Carlo sample of size 10^4 with spacing between samples 100. At the MLE, the observed and expected values of the canonical statistic are equal, so the mean number of points is 372. A spacing of 100 only attempts to update about one quarter of the points between sampled point patterns. This spacing is not optimal. We shall see how to choose better spacing presently.

Figure 3.2 is a scatter plot of the values of the canonical statistic $t(x) = (n(x), u(x))$ for this simulation. The large dot is the observed value of the canonical statistic. If the simulation parameter $\psi = (4.0, 0.4)$ were the MLE, the scatter plot would be centered at the observed value. If the observed value were outside the convex hull of the simulated values, the Monte Carlo likelihood l_n would have no maximum, and if the observed value were inside the convex hull but very close to the boundary, the MCL approximation would be very bad.

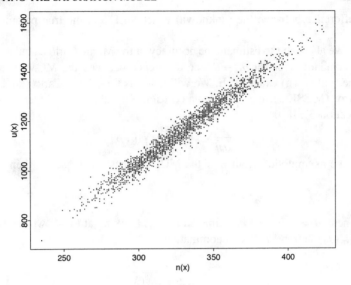

Figure 3.2 *Simulated values of the canonical statistic.*

In either case, we would have to use stochastic approximation (Section 3.11) or a trust region (Geyer and Thompson, 1992) to get closer to the MLE. For the simulation in Figure 3.2 ψ is close enough to the true MLE $\hat{\theta}$ so that the MCL approximation gives a good estimate $\hat{\theta}_n$ of the MLE.

Using the code described earlier in this section, we obtain $\hat{\theta}_n = (4.23, 0.359)$ as the MCMLE, the maximizer of the Monte Carlo likelihood l_n. For comparison the MCNR update gives $(4.21, 0.364)$.

3.13.1 Standard Errors

The MCL approximation of the observed Fisher information is $B_n = -\nabla^2 l_n(\hat{\theta}_n)$ given by (3.38) or (3.40). For these data we get

$$B_n = \begin{pmatrix} 885 & 4060 \\ 4060 & 19300 \end{pmatrix} \quad \text{and} \quad B_n^{-1} = \begin{pmatrix} 0.0339 & -0.00715 \\ -0.00715 & 0.00156 \end{pmatrix}$$

If the usual asymptotics of maximum likelihood hold (this appears to be an open research question) B_n^{-1} would be the asymptotic variance of the MLE. The square root of the diagonal elements $(0.18, 0.039)$ would be the estimated sampling error of the MLE. This estimates $\hat{\theta} - \theta_0$, the difference between the (unknown) exact MLE and the true parameter value.

We also want to estimate the accuracy of the Monte Carlo calculation, a standard error for $\hat{\theta}_n - \hat{\theta}$, the difference between the MCMLE and the (unknown) exact MLE. We will call this the Monte Carlo standard error (MCSE). It can be calculated using formulas from Geyer (1994, Section 3). If

$$\frac{1}{\sqrt{n}} \nabla l_n(\hat{\theta}) \overset{D}{\to} N(0, A)$$

is the asymptotic variance of the MCL estimate of the score, then

$$\sqrt{n}\left(\hat{\theta}_n - \hat{\theta}\right) \overset{D}{\to} N\left(0, B^{-1}AB^{-1}\right)$$

where $B = -\nabla^2 l(\hat{\theta})$ is estimated by B_n. Looking at (3.34) we see that $\nabla l_n(\hat{\theta}_n)$ is a ratio. The denominator

$$d_n = \frac{1}{n} \sum_{i=1}^{n} \frac{h_{\hat{\theta}_n}(X_i)}{h_\psi(X_i)}$$

estimates $c(\hat{\theta})/c(\psi)$, and the numerator is the sample mean of the vector time series

$$Z_{n,i} = \left(\nabla \log h_\theta(x) - \nabla \log h_{\hat{\theta}_n}(X_i)\right) \frac{h_{\hat{\theta}_n}(X_i)}{h_\psi(X_i)} \tag{3.45}$$

$i = 1, \ldots, n$. This time series has theoretical mean zero (see Geyer, 1994) and also has a sample mean exactly zero when $\hat{\theta}_n$ is the MCMLE. If C_n is the estimated asymptotic variance of the sample mean of this series, then A is estimated by

$$A_n = \frac{C_n}{d_n^2}$$

(Geyer, 1994, equation 20). We estimate C_n using the method of overlapping batch means (Meketon and Schmeiser, 1984) with a batch length of 200.

Using all of this we obtain an estimate of the MC error variance

$$\frac{1}{\sqrt{n}} B_n^{-1} A_n B_n^{-1} = \begin{pmatrix} 3.77 & -0.703 \\ -0.703 & 0.140 \end{pmatrix} \times 10^{-4}$$

and the square root of the diagonal elements gives $(0.019, 0.0037)$ for the MCSE.

One gets an approximate MCSE for MCNR by using the same formulas as for MCL except setting all the importance weights to $1/n$.

In our example, this gives $(0.0062, 0.0012)$ for the MCSE. Although, this would give the correct answer for a very small Newton step, it generally underestimates the true error, since it estimates the difference between the true Newton step and the Monte Carlo Newton step, and the true Newton step may still leave us far from the true MLE. The MCSE using the importance weights described above should be used instead.

3.13.2 Choosing the Spacing

We now turn to the question of choosing the optimal spacing as described by Geyer (1992, Section 3.6). The optimal spacing depends on the use made of the samples, in particular of the ratio of the cost of *generating* a sample to the cost of *using* a sample in subsequent calculations. The computer on which all of the computations for this chapter were done (an HP 715/100) took 2.4 seconds in calculating the MCMLE and 0.7 seconds in calculating the MCSE for a total of 3.1 seconds using the samples. It took 255.7 seconds to run the MCMC sampler for 10^6 iterations (10^4 samples spaced 100 iterations apart). Thus the cost ratio is $R = 3.1/255.7 = .012$.

The cost ratio is very small despite the code for the sampler being written in C and fairly efficient and the code for the maximization of the likelihood being written in S and very slow. Cost ratios this low or lower are typical of most applications.

Let $A_{n,m}$ denote an estimate of the variance of the mean of the time series (3.45) subsampled at spacing m, estimated by summing autocovariances at lags that are multiples of m. Then $\Sigma_m = B_n^{-1} A_{n,m} B_n^{-1} / \sqrt{n}$ is the estimate of the MC error variance, and the relative efficiency of spacing m is proportional to $(m + R)\Sigma_m$. This goes up linearly in m for large m, so large spacing is definitely bad (Geyer, 1992). Applied to our example, this method says that spacings 1 or 2 times the spacing of 100 used for samples at hand are about equally efficient, and any larger spacing wastes time.

In order to have any idea what is a useful spacing, one must do a calculation like this. Heuristic arguments don't help. Note that a spacing of 200 does not allow every point to be changed between samples, because the points are selected for attempted deletion only half the time, and it takes roughly $n \log n$ steps to visit each of n points at least once when a random point is visited each step. The naive notion that the spacing should be large enough so that most points are changed between samples is wrong.

3.13.3 A Final Estimate

As a check that these results are sensible, a sample of size 10^5 was collected using spacing 200. This sample gave an estimate $\hat{\theta} = (4.2129, 0.36222)$ with MCSE $(0.0015, 0.00029)$. Recall that our preliminary estimate $(4.23, 0.359)$ using a much smaller sample, had MCSE $(0.019, 0.0037)$, more than ten times as large as the MCSE for our final estimate. The difference between the two estimates, $(0.017, -0.0032)$, is just about the size of the MCSE of the cruder estimate. Thus the MCSE calculation seems to have provided a reasonable estimate of the Monte Carlo error.

The final estimate of the inverse Fisher information is

$$B_n^{-1} = \begin{pmatrix} 0.0274 & -0.00571 \\ -0.00571 & 0.00124 \end{pmatrix}. \tag{3.46}$$

Taking square roots of the diagonal elements gives $(0.170, 0.035)$ for the standard errors of the components of the MLE. Thus the Monte Carlo error in the final estimate is negligible, only one-tenth the error due to stochastic variability under the assumed model.

3.13.4 Stochastic Approximation

We can easily simulate a realization of the Poisson process conditioned on $n = 372$, i. e. a binomial process, so we start stochastic approximation in equilibrium with $\theta = (5.92, 0)$, the MLE for the Poisson model, and a simulation with exactly 372 points uniformly distributed in the unit square. The starting ϵ was 10^{-6}, a wild guess that happened to work. Two runs of 10^5 and 10^6 iterations, taking a total of five and a half minutes, got close enough to the MLE to switch to MCL methods. The results of the two runs were.

ϵ	iterations	θ (at end)		time (sec)
10^{-6}	10^5	4.296	0.4271	31.6
10^{-7}	10^6	4.218	0.3818	316.2

Whether stochastic approximation does this well in general is an open question. For a bit more difficult example, see Section 3.16

3.14 Hypothesis Tests

Consider a test of the null hypothesis that the data in Figure 3.1 is Poisson versus the alternative hypothesis that it comes from a saturation model. Of course, tests of Poissonness are a traditional topic in point processes. The conventional methodology is a nonparametric test using Ripley's K function (see Ripley, 1988, Chapter 3). Here we focus on parametric inference. We calculate the test statistics for the likelihood ratio test and the associated Wald and Rao tests (Rao, 1973, Section 6e3).

3.14.1 Wald Test

A Wald test can be performed with the samples already done to determine the MLE. If the constraints imposed in the null hypothesis are of the form $R(\theta) = 0$ and $T(\theta) = \nabla R(\theta)$, and $I(\theta)$ is the (observed or expected) Fisher information, then the Wald test statistic is

$$Q_W = R(\hat{\theta})^T \left(T(\hat{\theta})^T I(\hat{\theta})^{-1} T(\hat{\theta}) \right)^{-1} R(\hat{\theta})$$

where $\hat{\theta}$ is the MLE in the alternative hypothesis.

In our example, the constraint is simply $\theta_2 = 0$, so $T(\theta)^T = (0, 1)$. So the formula simplifies to

$$Q_W = \frac{\hat{\theta}_2^2}{(I(\hat{\theta})^{-1})_{22}}$$

where $\hat{\theta}_2 = 0.36222$ is the second component of the MLE and $(I(\hat{\theta})^{-1})_{22} = 0.00124$ is the 2, 2 component of the inverse Fisher information (3.46). This gives $Q_W = 105.5$.

3.14.2 Rao Test

A Rao test can be performed using samples from the null model, the Poisson process with rate 372. We could write a programme to produce independent samples for this process, but here we just use the MCMC sampler for the saturation process with parameter $\theta = (5.92, 0)$.

The Rao test statistic is

$$Q_R = \left(\nabla l(\theta^*) \right)^T I(\theta^*)^{-1} \nabla l(\theta^*)$$

where θ^* is the MLE in the null hypothesis. From computations done the same way as for the saturation model, we obtain

$$\nabla l_n(\theta^*) = (-0.3950 \quad 347.6)$$

and

$$I(\theta^*)^{-1} = \begin{pmatrix} 243.0 & -45.87 \\ -45.87 & 9.716 \end{pmatrix} \times 10^{-4}$$

with $I(\theta^*)$ estimated by $\nabla^2 l_n(\theta^*)$ as usual, which gives $Q_R = 118.7$.

3.14.3 Likelihood Ratio Test

The likelihood ratio test is harder to calculate and cannot be done using samples from one model. It can easily be done using the method of reverse logistic regression (Section 3.12) to estimate the normalizing constants.

Having estimated the normalizing constants, the likelihood ratio test statistic is

$$Q_L = 2\left[\langle t(x), \hat{\theta} - \theta^* \rangle - \log \frac{c(\hat{\theta})}{c(\theta^*)} \right]$$

$$= 2\langle t(x), \hat{\theta} - \theta^* \rangle + 2(\psi_m - \psi_1)$$

if we choose the sequence of distributions so that h_1 is the unnormalized density for the parameter value θ^* which is the MLE in the null hypothesis and h_m is the unnormalized density for the parameter value $\hat{\theta}$ which is the MLE in the alternative hypothesis.

Although the method is defined when there are just two samples, one for the null and one for the alternative hypothesis, it does not work unless the samples overlap. There are no guidelines for how much overlap is needed. One must try and see. We start with seven samples for distributions with parameter values evenly spaced along a line between the null and alternative

5.919	0.0000
5.635	0.0604
5.350	0.1207
5.066	0.1811
4.782	0.2415
4.497	0.3019
4.213	0.3622

Then we try four evenly spaced distributions using every other line above (and the same samples for those parameters). The results are

distributions	Q_L	MCSE
7	121.9	0.76
4	121.2	1.2

The samples all have size 1000 and spacing 200. With larger samples, we would get more accuracy. It is clear that the more overlap the more accuracy. It was not possible to do reverse logistic regression with only three samples. The optimization code crashed when the Hessian became singular to the precision of computer arithmetic. If we had obtained estimates, the MCSE would have been huge. The MCSE here is understated, since it only estimates the error of the log likelihood ratio assuming the two parameter values being compared are known. If we took proper account of the effect of errors in the parameter estimates, the error would be larger. MCSE formulas for the Wald and Rao test statistics have not been worked out.

All three test statistics are to be compared to the chi-squared distribution on one degree of freedom and give P-values essentially zero. The test statistics agree as well as one might expect given their size. As mentioned in connexion with confidence intervals for the MLE, it is not clear that the usual asymptotics are valid, whether there exist infinite-volume limiting distributions for the saturation model and whether the usual asymptotics obtain in the limit. One would expect from the elliptical shape of scatter plots of the canonical statistic that these asymptotics give reasonable answers. If one were very worried about the validity of asymptotics, one could do a parametric bootstrap, like Geyer and Møller (1994). One might find, like they did, that the bootstrap gave the same answer as the asymptotics and hence was unnecessary except for whatever reassurance it provided.

3.15 Missing Data and Edge Effects

To a likelihood person the problem of edge effects is a missing data problem. The process is observed in a window, but exists in a larger region. The process outside the window is missing data. The way to handle this is to use standard procedures for likelihood inference with missing data. One might jump to the conclusion that this means using the EM algorithm, but that would be wrong. The EM algorithm is so slow that when combined with Monte Carlo, making MCEM, it is terribly inefficient. Much better methods exist.

3.15.1 Monte Carlo Likelihood with Missing Data

The general method for MCL analysis with missing data for statistical models specified by families of unnormalized densities was laid out by Gelfand and Carlin (1993); see also the introduction of Geyer (1994).

Suppose now that x is missing data and y is observed data for a model specified by unnormalized joint densities $h_\theta(x, y)$ for the complete data (x, y). Then the normalizing function for the complete data model is

$$c(\theta) = \int h_\theta(x, y)\mu(dx)\nu(dy).$$

Also recall from Sections 3.3.2 and 3.3.4 that $h_\theta(x, y)$ is also an unnormalized conditional density of x given y and that the normalizing function for the conditional family is

$$c(\theta|y) = \int h_\theta(x, y)\mu(dx).$$

The likelihood is the ratio of these normalizing functions

$$L(\theta) = p_\theta(y) = \int f_\theta(x, y)\mu(dx) = \frac{1}{c(\theta)} \int h_\theta(x, y)\mu(dx) = \frac{c(\theta|y)}{c(\theta)}.$$

As usual, we write the log likelihood as a ratio against a fixed parameter point ψ

$$l(\theta) = \log \frac{p_\theta(y)}{p_\psi(y)}$$
$$= \log \frac{c(\theta|y)}{c(\psi|y)} - \log \frac{c(\theta)}{c(\psi)}.$$

The same argument (3.31) that gave us ratios of normalizing constants in the non-missing-data case now says

$$\frac{c(\theta)}{c(\psi)} = E_\psi \frac{h_\theta(X, Y)}{h_\psi(X, Y)}$$

when applied to the complete data model and

$$\frac{c(\theta|y)}{c(\psi|y)} = E_\psi \left\{ \frac{h_\theta(X, Y)}{h_\psi(X, Y)} \middle| Y = y \right\}$$

when applied to the conditional family. Hence

$$l(\theta) = \log E_\psi \left\{ \frac{h_\theta(X, Y)}{h_\psi(X, Y)} \middle| Y = y \right\} - \log E_\psi \frac{h_\theta(X, Y)}{h_\psi(X, Y)}. \qquad (3.47)$$

In order to approximate this by Monte Carlo we need samples from both the joint distribution of X and Y and the conditional distribution of X given Y, both for the parameter value ψ. We can use MCMC for both. Both have the same unnormalized density h_θ. The only difference is that we leave y fixed at its observed value in one of the samplers. Hence suppose that (X_i, Y_i), $i = 1, 2, \ldots$ are samples from the joint distribution and X_i^*, $i = 1, 2, \ldots$ are samples from the conditional distribution and y is the observed value of Y. Then

$$l_n(\theta) = \log\left(\frac{1}{n}\sum_{i=1}^{n}\frac{h_\theta(X_i^*, y)}{h_\psi(X_i^*, y)}\right) - \log\left(\frac{1}{n}\sum_{i=1}^{n}\frac{h_\theta(X_i, Y_i)}{h_\psi(X_i, Y_i)}\right) \quad (3.48)$$

is a Monte Carlo approximation to the exact log likelihood (3.47).

3.15.2 Comparison of MCL and MCEM

The gradient of (3.48) is

$$\nabla l_n(\theta) = \mathbb{E}_{n,\theta,\psi}\{\log\nabla h_\theta(X, Y)|Y = y\}$$
$$- \mathbb{E}_{n,\theta,\psi}\{\log\nabla h_\theta(X, Y)\} \quad (3.49)$$

where the empirical expectation operator $\mathbb{E}_{n,\theta,\psi}$ for the distribution with unnormalized density h_θ using samples having unnormalized density h_ψ is defined by (3.36). MCL solves $\nabla l_n(\theta) = 0$ with ∇l_n given by (3.49) to determine the MCMLE $\hat{\theta}_n$.

MCEM uses the same formula, but much less efficiently. It only uses the special case where in $\mathbb{E}_{n,\theta,\psi}$ we set $\theta = \psi = \theta_k$ the current iterate of the MCEM algorithm. This simplifies the formulas by eliminating the importance weights. But the cost of this simplification is that (3.49) becomes

$$\mathbb{E}_{n,\theta_k,\theta_k}\{\log\nabla h_\theta(X, Y)|Y = y\} - \mathbb{E}_{n,\theta_k,\theta_k}\{\log\nabla h_\theta(X, Y)\}$$

and only the θ's inside the integrands are considered variables. The θ's subscripting the expectation operators should also be variable, but EM leaves them fixed. The result is that MCEM takes hundreds of iterations, each involving the collection of two Monte Carlo samples to do what MCL does with no iteration as long as ψ is close enough to the exact MLE $\hat{\theta}$ so that the importance weights behave. The advantages of MCL methods over MCEM are

- MCL methods are faster, requiring many fewer iterations.

- MCL methods permit calculation of everything. MCEM only calculates MLEs, not likelihood ratios, not Fisher information.

- MCL methods permit calculation of Monte Carlo standard errors, MCEM doesn't.

The advantages of MCEM are weak

- MCEM allows one to avoid the importance sampling theory presented in the preceding section.
- Most statisticians have heard of EM, so MCEM sounds reasonable to people unfamiliar with the area.
- MCEM is better than MCNR because it is more stable.

But there are no real advantages of MCEM over MCL. The best that can be said for it is that it might be reasonable to do a few MCEM iterations to get close to MLE before switching to MCL. However, it must compete with stochastic approximation in that role, and it is not clear that MCEM has any advantages over stochastic approximation in providing crude estimates.

3.15.3 Missing Data in Point Processes

Suppose that as in Section 3.5 we are interested in a point process in a region S having unnormalized density h_θ with respect to a Poisson process with intensity measure λ, and $\lambda(S)$ is finite so we have a finite point process. Suppose that $S = W \cup V$ with $W \cap V = \emptyset$, and we see the process only inside W. We call W the observation *window* and V the *guard region*. We want to do likelihood inference for the parameter θ having observed the process in W.

This is a missing data problem. Let x denote the missing data, the process in V, and y the observed data, the process in W. Let $z = x \cup y$ denote the complete data, the process in S. In order to carry out MCL with missing data we need to understand the conditional distribution of x given y. By the independence properties of the Poisson process and our slogan from Section 3.3.2, this conditional distribution has the unnormalized density $x \mapsto h_\theta(x, y)$ with respect to the Poisson process on V with intensity measure λ restricted to V.

Let us again fit the saturation model to the point pattern in Figure 3.1, but this time we consider the observed pattern to be a window surrounded by a guard region. Here we take the complete region $S = W \cup V$ to be a torus that is a square of side 1.4 with edges pasted together. The reason for using a torus is to eliminate the boundary in the hope of getting something resembling an infinite-volume process. We do not know whether or not the infinite-volume process exists, but even if it does we cannot simulate it. Why 1.4? This almost doubles the area, which in some respects quadruples the work, since we need to do calculations involving pairs of points. We could try a larger guard region if we had the patience to do so.

We again start with simulations from the point $\psi = (4.0, 0.4)$ which was the \true" simulation parameter value for the data in Figure 3.1. We take two samples, one from the unconditional distribution on S and one for the conditional distribution of the process on the guard region V given data in the window W, both samples of size 10^4 with spacing 200. The Monte Carlo MLE was $\hat{\theta}_n = (3.948, 0.4148)$ with MCSE $(0.014, 0.0026)$. If we compare this with the MCMLE $\hat{\theta}_n = (4.2129, 0.36222)$ with MCSE $(0.0015, 0.00029)$ that we obtained when we assumed no guard region and free boundary conditions, we see that the estimates are clearly different.

There is no point in doing a statistical hypothesis test to determine which model, with or without guard region, is correct. In a real application only one model would make scientific sense. Either the process exists outside the observation window or it does not. If the data are positions of trees in a city park surrounded by asphalt, we use a model without a guard region. If the data are positions of trees in a quadrat in a forest, we must use a model with a guard region.

The point of the calculation here is that it makes a big difference which model we use. It is of course well known that it is necessary to deal with boundaries. That is the point of the edge-correction literature. The point of the example here is to show that the same phenomenon arises in likelihood inference. Though this notion of guard regions is very important, being perhaps appropriate for the majority of real applications, we leave it and return to analyses without guard regions.

3.16 Fitting the Triplets Process

Let us now consider fitting the triplets process to the same data (Figure 3.1) to which we fit the saturation process. We start with stochastic approximation doing the same as we did with the saturation process. Start at the Poisson model with parameter $(5.92, 0, 0)$, making a short run with $\epsilon = 10^{-6}$. Here we are a bit more careful, plotting the θ values over the course of the run. Values of θ_3 are shown in Figure 3.3 (following page). This run was followed by three more short runs with ϵ values of 10^{-7}, 10^{-8}, and 10^{-9}. All four runs are shown in the

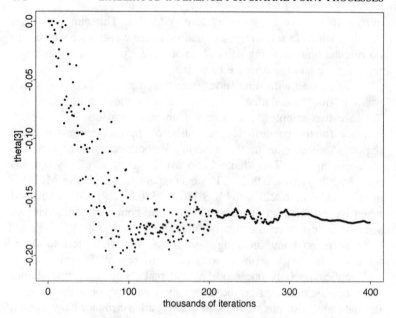

Figure 3.3 *Time Course of Stochastic Approximation. The stochastic approxi-mation parameter ϵ was started at 10^{-6}. It was lowered to 10^{-7} at iteration 100 thousand, to 10^{-8} at 200 thousand, and to 10^{-9} at 300 thousand.*

figure. The lowering of ϵ over the course of the runs seems useful. At $\epsilon = 10^{-6}$ the θ_3 values are all over the lot, at $\epsilon = 10^{-9}$ the θ_3 convergence is very slow.

The parameter values after each of the four short runs shown in Figure 3.3 and a longer run were

ϵ	iterations		θ (at end)		time (sec)
10^{-6}	10^5	4.506	0.6479	−0.1506	35.4
10^{-7}	10^5	4.500	0.5957	−0.1786	33.5
10^{-8}	10^5	4.509	0.6197	−0.1646	32.4
10^{-9}	10^5	4.508	0.6146	−0.1714	35.3
10^{-9}	10^6	4.509	0.6142	−0.1721	345.1

We follow this with two MCL steps to obtain an accurate estimate.

samples	spacing		θ (MCSE below)		time (sec)
10^3	200	4.494	0.6182	−0.1688	62.5
		0.014	0.0058	0.0026	
10^5	200	4.527	0.6072	−0.1673	6946.8
		0.0016	0.00072	0.00028	

In each pair of rows the upper line gives the estimate and the lower gives the Monte Carlo standard error. Our final estimate (4.527, 0.6072, −0.1673) has three to four significant figures of accuracy in the Monte Carlo. The inverse Fisher information is

$$B_n^{-1} = \begin{pmatrix} 0.0216 & -0.0103 & 0.00396 \\ -0.0103 & 0.00573 & -0.00252 \\ 0.00396 & -0.00252 & 0.00126 \end{pmatrix}$$

and the square root of the diagonal elements is (0.15, 0.076, 0.035), which is about 100 times the Monte Carlo error. So the Monte Carlo error is negligible.

3.17 Comparing the Saturated and Triplets Models

Which model, saturation or triplets, fits the data better? Since this is made-up data, simulated from a saturation model, we know that the saturation process should fit better. But is the difference between the two models enough so that we can tell using statistics? And what statistical procedures do we use?

A method of testing nonnested hypotheses was proposed by Cox (1961, 1962). Kent (1986) compares Cox's test with other procedures. Here we follow a different notion also suggested by Cox (1962) and followed up by Atkinson (1970) of embedding the two models in a supermodel that contains both. This is particularly easy when both models are exponential families. The supermodel is just the family having the vector canonical statistic that is the union of the statistics for the two models. The supermodel for the saturation and triplets models is the exponential family with the four-dimensional canonical statistic $t(x) = (n(x), s(x), w(x), u(x))$. The saturation model is the submodel obtained by setting $\theta_2 = \theta_3 = 0$, and the triplets model is the submodel obtained by setting $\theta_4 = 0$.

The first order of business is to find the MLE in the combined model. We start at the MLE in the saturation model, which is (4.212, 0, 0, 0.3626), and do two MCL iterations

samples	spacing	θ (MCSE below)			
10^3	200	4.179	0.1113	−0.03072	0.3177
		0.015	0.013	0.0041	0.0067
10^5	200	4.207	0.1069	−0.02894	0.3128
		0.0016	0.0013	0.00039	0.00061

giving three to four significant figures. For comparison, the parameter values for the three models are

model	θ			
saturation	4.213	0	0	0.3622
combined	4.207	0.1069	−0.02894	0.3128
triplets	4.527	0.6072	−0.1673	0

As expected, the fit in the combined model is much closer to the fit in the saturated model than the fit in the triplets model, when we make the comparison in the canonical parameter space.

We will test saturated versus combined and triplets versus combined. If one null hypothesis can be rejected and the other not, then we declare the null hypothesis that cannot be rejected to be the one that fits. If both null hypotheses can be rejected, then we declare that neither model fits. If neither null hypothesis can be rejected, then we declare that both models fit well, and there is no statistically significant difference between them. This is common garden variety statistics in action, but lest the reader think this inference is easy, Figures 3.4 and 3.5 (following pages) invite an attempt at doing the same inference by eye. It's not an easy task without statistics.

We now must collect three new samples using the sampler for the combined process for each of these parameter values in order to use reverse logistic regression. We need new samples because we need the four-dimensional canonical statistic output by the sampler for the combined process. Collecting samples of size 10000 at spacing 200 and running reverse logistic regression gives $\psi = (2.060, 5.599, -7.659)$ for the log inverse normalizing constants of the three distributions with estimated Monte Carlo error variance

$$V = \begin{pmatrix} 15.1 & 12.8 & -27.9 \\ 12.8 & 12.4 & -25.2 \\ -27.9 & -25.2 & 53.1 \end{pmatrix} \times 10^{-4} \qquad (3.50)$$

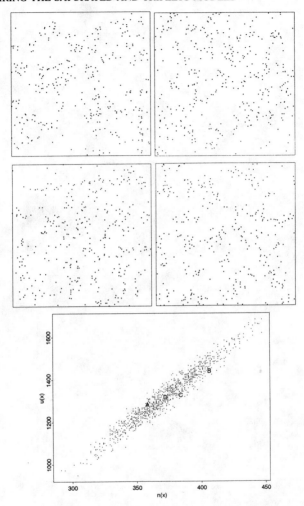

Figure 3.4 *Simulated point patterns from the saturation process and scatter plot of the distribution of the canonical statistics $n(x)$ and $u(x)$. Three of the point patterns are simulations from the maximum likelihood model; the lower right pattern is the observed data (Figure 3.1). Letters in the scatter plot mark the four patterns, D is the observed data.*

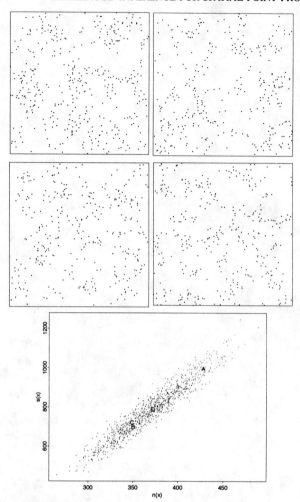

Figure 3.5 *Simulated point patterns from the triplets process and scatter plot of the distribution of the canonical statistics $n(x)$ and $s(x)$. Three of the point patterns are simulations from the maximum likelihood model; the lower right pattern is the observed data (Figure 3.1). Letters in the scatter plot mark the four patterns, D is the observed data.*

now the log likelihood ratio for two models with parameters θ_i and θ_j and log normalizing constants ψ_i and ψ_j estimated by reverse logistic regression is $\langle t(x), \theta_i - \theta_j \rangle + (\psi_i - \psi_j)$, and the MC error variance is estimated by the delta method using (3.50) and

$$V_{ii} + V_{jj} - 2V_{ij}.$$

The deviance, twice the log likelihood ratio, has an asymptotic chi-square distribution if the usual asymptotics hold. Assuming they do hold, the results are

test	deviance	MCSE	d.f.	P-value
saturated vs. combined	0.607	0.028	2	0.74
triplets vs. combined	17.4	0.22	1	0.000030

Again we find the result we expected. The saturation model fits the data well. Adding the other two canonical statistics to the model improves the fit no more what one expects whenever two parameters are added to a model. The triplets model does not fit.

3.18 Conclusion

Exotic areas of statistics, such as spatial statistics in general and spatial point processes in particular often recapitulate the history of statistics. The first formal inference is nonparametric, using method of moments estimators. At this stage there is no modelling. Ordinary statistics was in this phase a century ago with Pearson families of curves fit by method of moments. Time series was in this phase in the 1950s with spectral analysis. Spatial statistics was in this phase in the 1970s. Ripley's K-functions are an example. So is spatial autocorrelation analysis of lattice processes.

The next phase involves the introduction of parametric statistical models, efficient estimation methods, and hypothesis tests, which happened in ordinary statistics in the 1920s and 1930s, in time series in the 1960s, and is only now happening in spatial statistics. As this phase is entered, there are many special methods developed that are later replaced by the methods commonplace in ordinary statistics. An example of this would be the replacement of pseudolikelihood by ordinary likelihood, which is not yet complete.

Some readers, no doubt, will object to this story. Spatial statistics, they will say, is special and needs its own special methods. This is partly true. Aspects of spatial statistics that have no analogue in ordinary statistics must use special methods, but aspects that are universal, stochastic modelling and estimation of parameters are not without their analogues in ordinary statistics. It stands to reason that what works well in general should also work well in spatial statistics. If my work in spatial statistics has any clear goal, it is to make spatial statistics as much like ordinary statistics as possible. I hope readers will agree that this chapter shows we are advancing in that direction.

References

Atkinson, A.C. (1970). A method for discriminating between models (with discussion). *Journal of the Royal Statistical Society, Series B*, **32**, 323–353.

Baddeley, A. and Møller, J. (1989). Nearest-neighbour Markov point processes and random sets. *International Statistical Review*, **57**, 89–121.

Barndorff-Nielsen, O. (1978). *Information and Exponential Families in Statistical Theory*. Wiley, New York.

Bradley, R.C. (1986). Basic properties of strong mixing conditions. In *Dependence in Probability and Statistics: A Survey of Recent Results (Oberwolfach, 1985)*, (eds. E. Eberlein and M. S. Taqqu), 165–192. Birkhäuser, Boston.

Chan, K.S. and Geyer, C.J. (1994). Discussion of the paper by Tierney. *Annals of Statistics*, **22**, 1747–1758.

Clifford, P. (1993). Discussion on the meeting on the Gibbs sampler and other Markov chain Monte Carlo methods. *Journal of the Royal Statistical Society, Series B*, **55**, 53–54.

Cox, D.R. (1961). Tests of separate families of hypotheses. *Proceedings of the Fourth Berkeley Symposium on Mathematical Statistics and Probability*, **1**, 105–123. University of California Press, Berkeley, CA.

Cox, D.R. (1962). Further results on tests of separate families of hypotheses. *Journal of the Royal Statistical Society, Series B*, **24**, 406–424.

Daley, D.J. and Vere-Jones, D. (1988). *An Introduction to the Theory of Point Processes*. Springer-Verlag, New York.

Gelfand, A.E. and Carlin, B.P. (1993). Maximum-likelihood estimation for constrained- or missing-data models. *Canadian Journal of Statistics*, **21**, 303–311.

Gelfand, A.E. and Smith A.F.M. (1990). Sampling-based approaches to calculating marginal densities. *Journal of the American Statistical Association*, **85**, 398–409.

Geman, S. and Geman, D. (1984). Stochastic relaxation, Gibbs distributions, and the Bayesian restoration of images. *IEEE transactions on pattern analysis and machine intelligence*, **6**, 721–741.

Geyer, C.J. (1991). Reweighting Monte Carlo mixtures. Technical Report No. 568, School of Statistics, University of Minnesota, `http://stat.umn.edu/PAPERS/tr568.html`. Revised version (1994) `http://stat.umn.edu/PAPERS/tr568r.html`.

Geyer, C.J. (1992). Practical Markov chain Monte Carlo (with discussion). *Statistical Science*, **7**, 473–511.

Geyer, C.J. (1994). On the convergence of Monte Carlo maximum likelihood calculations. *Journal of the Royal Statistical Society, Series B*, **56**, 261–274.

Geyer, C.J. and Møller, J. (1994). Simulation and likelihood inference for spatial point processes. *Scandinavian Journal of Statistics*, **21**, 359–373.

Geyer, C.J. and Thompson, E.A. (1992). Constrained Monte Carlo maximum likelihood for dependent data, (with discussion). *Journal of the Royal Statistical Society, Series B*, **54**, 657–699.

Geyer, C.J. and Thompson, E.A. (1995). Annealing Markov chain Monte Carlo with applications to ancestral inference. *Journal of the American Statistical Association*, **90**, 909–920.

Green, P.J. (1995). Reversible jump Markov chain Monte Carlo computation and Bayesian model determination. *Biometrika*, **82**, 711–732.

Hastings, W.K. (1970). Monte Carlo sampling methods using Markov chains and their applications. *Biometrika*, **57**, 97–109.

Ibragimov, I.A. and Linnik, Yu. V. (1971). *Independent and stationary sequences of random variables*. Groningen: Wolters-Noordhoff (English translation).

Kelly, F.P. and Ripley, B.D. (1976). A note on Strauss's model for clustering. *Biometrika*, **63**, 357–360.

Kent, J.T. (1986). The underlying structure of nonnested hypothesis tests. *Biometrika*, **73**, 333–343.

Meketon, M.S. and Schmeiser B.W. (1984). Overlapping batch means: something for nothing? In *Proceedings of the 1984 Winter Simulation Conference*, (eds. S. Sheppard, U. Pooch, and D. Pegden), 227–230.

Metropolis, N., Rosenbluth, A.W., Rosenbluth, M.N., Teller, A.H., and Teller, E. (1953). Equation of state calculations by fast computing machines. *Journal of Chemical Physics*, **21**, 1087–1092.

Meyn, S.P. and Tweedie, R.L. (1992). Stability of Markov processes I: Discrete time chains. *Advances in Applied Probability*, **24**, 542–574.

Meyn, S.P. and Tweedie, R.L. (1993). *Markov Chains and Stochastic Stability*. Springer-Verlag, London.

Moyeed, R.A. and Baddeley, A.J. (1991). Stochastic approximation of the MLE for a spatial point pattern. *Scandinavian Journal of Statistics*, **18**, 39–50.

Nummelin, E. (1984). *General Irreducible Markov Chains and Non-Negative Operators*. Cambridge University Press.

Nummelin, E. and Tuominen, P. (1982). Geometric ergodicity of Harris recurrent Markov chains with applications to renewal theory. *Stochastic Processes and their Applications*, **12**, 187–202.

Penttinen, A. (1984). *Modelling Interaction in Spatial Point Patterns: Parameter Estimation by the Maximum Likelihood Method*. Number 7 in

140   LIKELIHOOD INFERENCE FOR SPATIAL POINT PROCESSES

Jyväskylä Studies in Computer Science, Economics, and Statistics. University of Jyväskylä.

Preston, C.J. (1975). Spatial birth-and-death processes. *Bulletin of the International Statistical Institute*, **46**, vol. 2, 371–391.

Rao, C.R. (1973). *Linear Statistical Inference and Its Applications* (2nd ed.) Wiley, New York.

Ripley, B.D. (1977). Modelling spatial patterns (with discussion). *Journal of the Royal Statistical Society, Series B*, **39**, 172–212.

Ripley, B.D. (1979). Simulating spatial patterns: Dependent samples from a multivariate density (Algorithm AS 137). *Applied Statistics*, **28**, 109–112.

Ripley, B.D. (1988). *Statistical Inference for Spatial Point Processes*. Cambridge University Press.

Ripley, B.D. and Kelly, F.P. (1977). Markov point processes. *Journal of the London Mathematical Society*, **15**, 188–192.

Rockafellar, R.T. (1970). *Convex Analysis*. Princeton University Press.

Rosenthal, J. (1995). Minorization conditions and convergence rates for Markov chain Monte Carlo. *Journal of the American Statistical Association*, **90**, 558–566.

Ruelle, D. (1969). *Statistical Mechanics: Rigorous Results*. W. A. Benjamin, Reading, Massachusetts.

Strauss, D.J. (1975). A model for clustering. *Biometrika*, **62**, 467–75.

Thompson, E.A., Lin, S., Olshen, A.B. and Wijsman, E.M. (1993). Monte Carlo analysis on a large pedigree. In 'Genetic Analysis Workshop 8,' *Genetic Epidemiology*, **10**, 677–682.

Tierney, L. (1994). Markov chains for exploring posterior distributions (with discussion). *Annals of Statistics*, **22**, 1701–1762.

Torrie, G.M. and Valleau, J.P. (1977). Nonphysical sampling distributions in Monte Carlo free-energy estimation: Umbrella sampling. *Journal of Computational Physics*, 23, 187–199.

Younes, L. (1988). Estimation and annealing for Gibbsian fields. *Annales de l'Institut Henri Poincaré. Probabilités et Statistiques*, **24**, 269–294.

CHAPTER 4

Markov chain Monte Carlo and spatial point processes

J. Møller

Department of Mathematics, Aalborg University, Fredrik Bajers Vej 7E,
DK-9220 Aalborg, Denmark

4.1 Introduction

Currently, Markov chain Monte Carlo methods attract much attention among statisticians, cf. e.g. Smith and Roberts (1993), Besag and Green (1993), Besag *et al.* (1995), Tierney (1994) and the accompaning discussions and references. The genesis of most ideas lies in statistical physics following the early work by Metropolis *et al.* (1953). In that paper the first Markov chain Monte Carlo algorithm for simulating a Gibbsian point process with a fixed and finite number of points was developed. Hastings (1970) introduced a general class of Markov chain Monte Carlo algorithms which covers nearly any algorithm considered so far. In statistics some of the earliest and most important applications of Markov chain Monte Carlo appear to be within spatial statistics, see e.g. the discussion in Besag (1974), and especially the Gibbs sampler (Geman and Geman, 1984) have been frequently used. The Gibbs sampler has earlier been introduced in statistical physics (e.g. Creutz, 1979) where it is known as the 'heat bath algorithm'; it appeared implicit in spatial statistics in Suomela's (1976) thesis and in Ripley (1977, 1979) too. Following Preston (1977), Ripley (1977) discussed also other 'birth-death' techniques for simulating finite point processes with a fixed or random number of points.

Indeed Markov chain Monte Carlo methods apply on a broad spectrum of complex problems in statistics, particularly in Bayesian inference, but in this contribution we restrict attention to spatial (marked) point processes as used in spatial statistics and stochastic geometry, see e.g. Diggle (1983), Ripley (1988), Baddeley and Møller (1989), and Stoyan, Kendall and Mecke (1995). The objective is both to review those algorithms which are used for simulating point processes with a finite and either fixed or random number of points, and to discuss the problem of modelling spatial (marked) point processes which exhibit regular or clustered patterns.

This contribution is a revised version of Møller (1994b). Section 4.2 provides a general setup for finite processes which covers ordinary spatial point processes and marked point processes. In Section 4.3 we study general Metropolis–Hastings single point updating type algorithms following Hastings (1970) and relate these algorithms to particular examples like the Metropolis algorithm, the Gibbs sampler, and various spatial birth-and-death process techniques. Section 4.4 is concerned with some practical aspects of using such algorithms in cases where different types of Gibbsian or Markovian point processes are used as models for either regular or clustered patterns. In particular, in the case of clustering, it is argued that the classical Gibbsian or Ripley and Kelly (1977) Markov point processes or at least the frequently used pairwise interaction point processes do not provide very flexible and satisfactory models and the single point updating algorithms for simulating such models become inefficient due to an extremely strong attraction — see also Gates and Westcott (1986). In such situations multiple point updating algorithms (Besag and Green, 1993; Møller, 1992, 1993; Hurn and Jennison, 1993; Häggström, Lieshout and Møller, 1996), Metropolis-coupled Markov chains (Geyer, 1991), and simulated tempering schemes (Marinari and Parisi, 1992; Geyer and Thompson, 1995) may prove to be more efficient. However, it is demonstrated that the class of nearest-neighbour Markov point processes introduced in Baddeley and Møller (1989), which includes the Ripley–Kelly Markov point processes as special cases, may provide better models for clustering and in turn they seem more feasible for simulation based on the single point updating algorithms. This is illustrated by considering certain models of disc processes which can produce clustered as well as regular patterns. Incidentally, we study also some Markov properties of nearest-neighbour Markov point processes, in particular the spatial Markov property (Kendall, 1990) which becomes important for statistical applications where point patterns are only partly observed so that edge effects are of great significance. We shall also very briefly mention some recent research on exact simulation (Propp and Wilson, 1996; Kendall, 1996; Häggström, Lieshout and Møller, 1996).

Some general background material on Markov chain Monte Carlo is given in Appendix A, while Appendix B contains a further discussion on quantitative bounds for the rate of convergence. Finally, in

Appendix C a characterization result for the continuum random cluster model considered in Section 4.4 is established. For background material and references on statistical inference for spatial point processes based on Markov chain Monte Carlo methods the reader is refered to Geyer and Møller (1994). Further background material can be found in Geyer's contribution in this volume.

4.2 Setup

We will use a general setup, namely finite point processes defined on the exponential space over a measure space $(S, \mathcal{B}, \lambda)$ where $0 < \lambda(S) < \infty$ and $\mathcal{B}$ is separable and contains all singletons, cf. Carter and Prenter (1972). This covers both ordinary finite spatial point processes and marked point processes such as multitype point processes, line segment processes, discs processes, and other processes of geometrical objects, see e.g. Baddeley and Møller (1989). A brief description follows below.

For $n = 1, 2, \dots$ let Ω_n be the space of all finite point configurations $x = \{x_1, \dots, x_n\}$ or more precisely all counting measures $\varphi = \sum_{i=1}^n \delta_{x_i}$ of total mass n where δ_{x_i} denotes the Dirac measure with mass 1 concentrated at x_i. For convenience we may identify φ with x and think of x as a finite subset of S, though x may consist of multiple points; in fact the ordinary spatial point processes and marked point processes considered below do not have multiple points almost surely. Thus we write $x \cup \xi = \{x_1, \dots, x_n, \xi\}$ for $\varphi + \delta_\xi$ and $x \setminus x_i = \{x_1, \dots, x_{i-1}, x_{i+1}, \dots, x_n\}$ for $\varphi - \delta_{x_i}$, etc. Furthermore, Ω_n is equipped with the σ-field $\mathcal{F}_n = \omega_n(\mathcal{B}^n)$ where $\mathcal{B}^n$ is the product σ-field for S^n and $\omega_n : S^n \to \Omega_n$ is the projection of 'ordered' tuplets $(x_1, \dots, x_n)$ into point configurations $\{x_1, \dots, x_n\}$. The empty point configuration, i.e. the null measure, is denoted $\emptyset$ and we set $\Omega_0 = \{\emptyset\}$ and equip Ω_0 with the corresponding trivial σ-field $\mathcal{F}_0$. Then a finite point process X is a measurable mapping defined on some probability space and taking values in $(\Omega, \mathcal{F})$ where $\Omega = \cup_{n=1}^\infty \Omega_n$ and $\mathcal{F}$ is the smallest σ-field containing $\mathcal{F}_0, \mathcal{F}_1, \dots$.

Henceforth, the induced distribution of X on $\mathcal{F}$ is assumed to be absolutely continuous with respect to the measure μ defined by

$$\mu(F) = e^{-\lambda(S)}\{1[\emptyset \in F] + \sum_{n=1}^\infty \mu_n(F)/n!\}$$

where

$$\mu_n(F) = \int \cdots \int 1[\{x_1, \ldots, x_n\} \in F]\lambda(dx_1)\ldots\lambda(dx_n)$$

and $1[\cdot]$ denotes indicator function. The density of X with respect to μ is denoted by f. The measure μ is the distribution of a Poisson point process on S with intensity measure λ, i.e. the random number N of points under μ is Poisson distributed with mean $\lambda(S)$, and conditional on $N = n$ with $0 < n < \infty$, $\mu(\cdot \mid N = n) = \mu_n(\cdot)/\lambda(S)^n$ forms a binomial point process of n i.i.d. points in S with density $\lambda(\cdot)/\lambda(S)$.

Now, let A be an open bounded subset of $\mathbb{R}^d$ equipped with the Borel σ-field $\mathcal{A}$ and the Lebesgue measure ν. Taking $(S, \mathcal{B}, \lambda) = (A, \mathcal{A}, \nu)$ gives the setup for an ordinary spatial point process and μ becomes a homogeneous Poisson process on S with intensity 1. As a standard example of a non-trivial process under this setup we shall consider Strauss' (1975) model

$$f(x) \propto \beta^{n(x)}\gamma^{s(x)}. \tag{1}$$

Here $n(x)$ denotes the number of points in x and $s(x)$ is the number of pairs $\{\xi, \eta\} \subseteq x$ with distance $\|\xi - \eta\| < R$ for some parameter $R > 0$. Further, $\beta > 0$ and $\gamma \in [0, 1]$ is an interaction parameter which roughly speaking controls the degree of repulsion between the points. Especially, if $\gamma = 0$ then (1) becomes a hard-core model where no points are allowed to be closer than R units apart, whereas $\gamma = 1$ gives a homogeneous Poisson process on S with intensity β. The case $\gamma > 1$ was proposed by Strauss (1975) as a model of clustering, but (1) is not integrable in that case, cf. Kelly and Ripley (1976). Simulated examples for $0 < \gamma \leq 1$ are shown in Figure 4.1, Section 4.2.

Finally, if $(A, \mathcal{A}, \nu)$ is as before and $(M, \mathcal{M}, Q)$ is some probability space we get a setup for a marked point process with 'points' in A and 'marks' in M by setting $(S, \mathcal{B}, \lambda) = (A \times M, \mathcal{A} \otimes \mathcal{M}, \nu \otimes Q)$. Then under μ the points constitute a homogeneous Poisson process on A, while the marks given the number of points are i.i.d. with distribution Q. For instance, consider a random disc process $X = \{(C_1, R_1), \ldots, (C_n, R_n)\}$ where (C_i, R_i) is identified with the disc with center C_i and radius R_i, so let $M = [0, \infty)$ and let $\mathcal{M}$ be the corresponding Borel σ-field. Then a simple model is obtained from (1) if $s(x)$ is now defined as the number of pairs of overlapping discs, see Baddeley and Møller (1989). In Section 4.4.3 we study other models for such disc processes.

Further examples of ordinary point processes and marked point processes can be found in Baddeley and Møller (1989).

4.3 Metropolis–Hastings algorithms for finite point processes

All known Markov chain Monte Carlo algorithms for simulating finite point processes appear to be of Hastings (1970) type. The most widely used algorithms are described by three types of possible transitions: One point of the current point configuration is displaced or removed or a new point is added. In this section we describe such single point updating algorithms and study some of their convergence properties. For simplicity and specificity only random updating schemes are considered; systematic versions can be found in e.g. Ripley (1979) and Allen and Tildesley (1987). Some background material on Markov chain Monte Carlo is given in Appendix A and B.

4.3.1 The fixed number case

Consider a finite point process X with density

$$f(x) \propto \alpha(n(x))g(x) \tag{2}$$

where $n(x)$ is the number of (possible multiple) points in x, $\alpha(\cdot) \geq 0$ is some measurable function, and $g(\cdot) \geq 0$ is another measurable function describing the 'interaction structure' of the process in some way; for example, in the Strauss process (1), we can take $\alpha(n) = \beta^n$ and $g(x) = \gamma^{s(x)}$. It is often claimed that $n(x)$ provides little information about the interaction structure which is usually the feature of most interest, cf. Gates and Westcott (1986) and Ripley (1988). In order to get rid of the 'nuisance' term in (2) we may condition on $n(X) = n$ if n is the observed number of points. Then the conditional density with respect to μ_n becomes

$$f_n(x) \propto g(x). \tag{3}$$

It is in general not clear whether $n(X)$ is (approximately) ancillary, but Geyer and Møller (1994) demonstrate that the unconditional and conditional maximum likelihood estimators of γ in the Strauss process appear to be approximately identical.

Now, suppose we want to construct a Markov chain $\{X_t\}$ with state space $E = \{x \in \Omega_n \mid g(x) > 0\}$ and equilibrium distribution (3). In general this may simply be constructed using the projection $\omega_n :$ $S^n \to \Omega_n$ on any Markov chain Monte Carlo algorithm for simulating a vector process with state space $(S^n, \mathcal{B}^n)$ and density $f^n(x_1, \ldots, x_n) \propto g(\{x_1, \ldots, x_n\})$ with respect to the product measure λ^n. In the case of a single point updating Metropolis–Hastings algorithm an update consists of the (possible) displacement of one point: Suppose $X_t = x$ is the current state with $x \in E$. Then according to some rule it is proposed to make a transition to $X_{t+1} = (x \cup \xi) \setminus \eta$ where $\eta \in x \cup \xi$ and $\xi \in S$. The proposal is accepted with some probability $a(x, \eta, \xi)$ which is zero if $(x \cup \xi) \setminus \eta \notin E$; otherwise we retain $X_{t+1} = x$. Here the proposal may be generated in one of two ways: (i) Either first a point $\eta \in x$ is picked at random with some probability possibly depending on x, and then a new point ξ is generated from some density with respect to λ which may depend on both x and η. (ii) Or first an extra point ξ is generated from a density with respect to λ which possibly depends on x, and secondly a point $\eta \in x \cup \xi$ is picked at random with some probability which may depend on both x and ξ. In both cases we have a joint density $q(x, \cdot, \cdot)$ for (η, ξ) so that

$$\int \sum_{\eta \in x \cup \xi} q(x, \eta, \xi) \lambda(d\xi) = 1$$

where in the case (ii) $q(x, \xi, \xi) = 0$. Thus the transition kernel of the chain is given by

$$P(x, F) = R(x, F) + 1[x \in F]r(x), \quad x \in E, \ F \in \mathcal{E}, \qquad (4)$$

where $\mathcal{E} = \mathcal{F}_n \cap E$ is the induced σ-field on E and

$$R(x, F) = \int \sum_{\eta \in x \cup \xi} q(x, \eta, \xi) a(x, \eta, \xi) 1[(x \cup \xi) \setminus \eta \in F] \lambda(d\xi)$$

is the sub-stochastic kernel describing accepted proposals which are different from the current state x, while

$$r(x) = 1 - R(x, E)$$

is the probability that the chain remains in x in the next step.

The above construction leaves us with a considerable degree of freedom on how to choose the proposal density and the acceptance probability. Following Hastings (1970) let us suppose the following detailed balance condition is imposed:

$$\begin{aligned} &q(x, \eta, \xi) a(x, \eta, \xi) g(x) \\ &= q((x \cup \xi) \setminus \eta, \xi, \eta) a((x \cup \xi) \setminus \eta, \xi, \eta) g((x \cup \xi) \setminus \eta) \end{aligned} \qquad (5)$$

for all $\xi, \eta \in S$ and $x \in E$ with $(x \cup \xi) \setminus \eta \in E$. Combining this with (4) it is easily seen that $\{X_t\}$ is reversible with invariant distribution (3). Moreover, (5) is equivalent to set

$$a(x, \eta, \xi) = q((x \cup \xi) \setminus \eta, \xi, \eta) g((x \cup \xi) \setminus \eta) s(x \setminus \{\eta, \xi\}, \eta, \xi)$$

for some non-negative function $s(x\setminus\{\eta, \xi\}, \eta, \xi)$ which is symmetric in η and ξ and satisfies that $0 \leq a(x, \eta, \xi) \leq 1$. It may be natural to specify this function so that the acceptance probability becomes maximal, cf. Peskun (1973). Hence we arrive with

$$a(x, \eta, \xi) = \min\{1, h(x, \eta, \xi)\}$$
$$\text{for } q(x, \eta, \xi) > 0, \ x \in E, \ (x \cup \xi)\setminus\eta \in E \tag{6}$$

where

$$h(x, \eta, \xi) = \frac{q((x \cup \xi)\setminus\eta, \xi, \eta)g((x \cup \xi)\setminus\eta)}{q(x, \eta, \xi)g(x)} \tag{7}$$

is the so-called Metropolis–Hastings ratio. Henceforth, we let the acceptance probability be specified by (6). According to (5)–(7) we do not need to specify $a(x, \eta, \xi)$ when $q(x, \eta, \xi) = 0$.

The Metropolis *et al.* (1953) algorithm is the first example of this procedure: Firstly, a point $\eta \in x$ is randomly selected with probability $1/n$ and secondly it is proposed to replace η by ξ which is generated from a density which is proportional to $1[\xi \in \partial\eta]$ where $\partial\eta$ denotes some 'neighbourhood' to η (typically a square in the case of ordinary planar point processes). Allen and Tildesley (1987, pp. 121) discuss how to specify such neighbourhoods, but with the fast computers which are commonly used today this does not appear to be so important. Thus we may take $\partial\eta = S$ whereby the Metropolis–Hastings ratio simply becomes $h(x, \eta, \xi) = g((x \cup \xi)\setminus\eta)/g(x)$.

Other examples include situations where we always get acceptance, i.e. when

$$q(x, \eta, \xi)g(x) = q((x \cup \xi)\setminus\eta, \xi, \eta)g((x \cup \xi)\setminus\eta).$$

The Markov chain can then be regarded as the jumps of a spatial birth-and-death process in which births and deaths alternate. Ripley (1977) considered two examples of this: One is obtained by first deleting a point $\eta \in x$ selected with probability $1/n$ and then adding a newborn point ξ to $x\setminus\eta$ where the conditional density of ξ given $x\setminus\eta$ is proportional to $g((x\setminus\eta) \cup \xi)$. This is just the Gibbs sampler with random updating. The other algorithm consists of generating first a 'uniform' point ξ, i.e. its density with respect to λ is constant, and then deleting a point $\eta \in x \cup \xi$ selected with a probability which is proportional to $g((x \cup \xi)\setminus\eta)$. Though the latter algorithm does not involve rejection sampling, Ripley (1977) remarks that it may converge very slowly as it may tend to pick $\eta = \xi$ with a high probability.

In general the convergence properties of the Markov chain $\{X_t\}$ specified by (4) and (6) are inherited by the Markov chain $\{Y_t\}$ obtained by using only the proposal procedure, so $\{Y_t\}$ has transition kernel (4) with $a(\cdot, \cdot, \cdot) = 1$. This is shown in the following statements:

(A) If $\{X_t\}$ is irreducible it becomes Harris recurrent too.

(B) Suppose $q(x, \eta, \xi) > 0$ if and only if $q((x \cup \xi)\backslash\eta, \xi, \eta) > 0$ for $\xi, \eta \in S$ and $x \in E$ with $(x \cup \xi)\backslash\eta \in E$. Then irreducibility of $\{Y_t\}$ implies Harris recurrence of $\{X_t\}$.

(C) If $\{Y_t\}$ is aperiodic or $r(\cdot) > 0$, then $\{X_t\}$ is aperiodic.

(D) Suppose the acceptance probability (6) is bounded below by a positive constant. Then, if $F \in \mathcal{E}$ is a small set for $\{Y_t\}$, it is also a small set for $\{X_t\}$. Especially, uniform ergodicity of $\{Y_t\}$ implies uniform ergodicity of $\{X_t\}$.

Modifying the proof of Corollary 2 in Tierney (1994), we get (A) whereby (B) is easily verified. Finally, (C)–(D) are straightforwardly verified.

Thus under mild conditions the Gibbs sampler and the Metropolis algorithm become Harris ergodic, i.e. Harris recurrent and aperiodic whereby they converge towards (3). For example, suppose $g(\cdot) > 0$ on Ω_n. Then the Gibbs sampler is Harris ergodic and it becomes in fact uniformly ergodic when $g(\cdot)$ is bounded below and above by some positive constants; this is e.g. satisfied for the Strauss process (1) with $g(x) = \gamma^{s(x)}$ and $\gamma > 0$. Similar properties hold for the Metropolis algorithm provided the neighbourhoods are sufficiently large to ensure irreducibility; clearly $\partial\eta = S$ for all $\eta \in S$ is a sufficient condition. However, in the case of a hard-core process, any single point updating algorithm becomes reducible if the packing density is sufficiently high, i.e. whenever n is sufficiently large. In this and other situations (faster) convergence may be obtained by using combinations or mixtures of several chains. Especially, the method in Geyer (1991) of Metropolis-coupled chains and the simulated tempering schemes introduced in Marinari and Parisi (1992) and Geyer and Thompson (1995) may be very efficient. Alternatively, algorithms which update larger subconfigurations of points in each transition may be developed, see e.g. Sokal (1989), Besag and Green (1993), Besag et al. (1994), Møller (1992, 1993), and Hurn and Jennison (1993).

4.3.2 The random number case

We now turn to the case of a random number of points, i.e. when unconditional simulation under (2) is wanted. Hence the state space

$E = \{f > 0\}$ is of varying dimension and the Gibbs sampler and other classsical Metropolis–Hastings algorithms do not handle this. The best known algorithms are based on running a spatial birth-and-death process. This is a time-homogeneous Markov process where a transition consists of either the removal of an existing point or the addition of a new-born point, see Kelly and Ripley (1976), Preston (1977), Ripley (1977), Møller (1989), Baddeley and Møller (1989), and Clifford and Nicholls (1994) for details and examples. An alternative class of Metropolis–Hastings type algorithms for generating reversible Markov chains with equilibrium density f is proposed in Geyer and Møller (1994). These algorithms which are briefly described below may be considered as a special case of a more general algorithm due to Green (1995).

We assume that the density f is hereditary, i.e.

$$f(x \cup \xi) > 0 \Rightarrow f(x) > 0 \text{ for } x \in \Omega \text{ and } \xi \in S, \qquad (8)$$

and define the Papangelou conditional intensity

$$\lambda^*(x, \xi) = \begin{cases} f(x \cup \xi)/f(x) & \text{if } f(x) > 0 \\ 0 & \text{else.} \end{cases} \qquad (9)$$

Hence there is a one-to-one correpondence between f and λ^*. The conditional intensity plays the same role as 'local characteristics' do for Markov random fields in the construction of Markovian models and statistical inference, and it becomes computationally simple for Markovian types of models; we return to this in Section 4.4.

Now, if x is the current state of the Markov chain, we first attempt either (a) with some probability $p(x)$ to add a point ξ generated from some density $b(x, \cdot)$ with respect to λ, or (b) with probability $1 - p(x)$ to delete a point $\eta \in x$ which is selected with some probability $d(x \setminus \eta, \eta)$ (if $n(x) = 0$ we do nothing). Here it is assumed that for $x \in \Omega$ and $\xi \in S$,

$$f(x \cup \xi) > 0 \Rightarrow p(x) > 0, \ p(x \cup \xi) < 1, \ d(x, \xi) > 0, \ b(x, \xi) > 0.$$

Secondly, defining the 'Metropolis–Hastings' ratio

$$r(x, \xi) = \lambda^*(x, \xi) \frac{1 - p(x \cup \xi)}{p(x)} \frac{d(x, \xi)}{b(x, \xi)} \quad \text{for } f(x \cup \xi) > 0 \qquad (10)$$

the transition $x \to x \cup \xi$ in (a) is accepted with probability $\min\{1, r(x, \xi)\}$, while the transition $x \to x \setminus \eta$ in (b) is accepted with probability $\min\{1, 1/r(x \setminus \eta, \eta)\}$. As shown in Geyer and Møller (1994) the acceptance probabilities then become maximal and the procedure generates a reversible and Harris recurrent Markov chain with invariant density $f(\cdot)$. If e.g. $p(\emptyset) < 1$ the chain is also aperiodic and hence Harris ergodic. For instance, one may simply take $p(x) = 0.5, b(x, \xi) \propto 1$, and $d(x\backslash \eta, \eta) = 1/n(x)$. Using this for the Strauss process (1), we have in fact geometrical ergodicity for all $\gamma \in [0, 1]$, cf. Geyer and Møller (1994); see also Geyer's contribution in this volume where it is observed that a certain stability condition, viz. that λ^* is bounded by a constant, together with the hereditary condition imply geometrical ergodicity. But it can easily be shown that uniform ergodicity holds only when $\gamma = 0$, i.e. in the case of a hard-core process. This is due to the fact that for any single point updating algorithm if the state space E is small, then for some $m < \infty$, we have that $f(x) = 0$ whenever $n(x) > m$. For a further discussion of the rate of convergence, see Appendix B.

In conclusion this Metropolis–Hastings algorithm is usually easier to programme and simpler to analyse than the usual spatial birth-and-death process techniques, cf. Geyer and Møller (1994). Clifford and Nicholls (1994) compared certain implementations of the two techniques in the case of the Strauss process. They conclude also that the Metropolis–Hastings algorithm is simplest and most efficient except if β is large and γ is small, i.e. when there is a high degree of regularity and the intensity of points is large.

The above Metropolis–Hastings algorithm may be combined with the algorithms used for moving points (Section 4.3.1), but for simulating the conditional Strauss process (see (11) below) Geyer and Møller (1994) found that this was not useful.

4.4 Empirical results and Markovian models for clustered and regular patterns

In this section we report on some empirical results for simulating models for either clustered or regular point patterns. Though a Markov chain algorithm may have good convergence properties in theory, it is pointed out that it may converge extremely slowly in practise in the case of strong interaction. In particular, it is demonstrated that a certain subclass of the nearest-neighbour Markov processes introduced in Baddeley and Møller (1989), as opposed to pairwise interaction point processes and to some extent the Ripley and Kelly (1977) Markov processes, provide a rich class of models for clustering which may be feasible for simulation based on the methods described in Section 4.3. Some further properties of these processes are also examined, in particular the spatial

Markov property discussed in Kendall (1990). Incidentally, it should be mentioned that there exist indeed other models for clustering like Poisson cluster processes and Cox processes which may be simulated exactly and straightforwardly avoiding Markov chain Monte Carlo, see e.g. Diggle (1983), Stoyan *et al.* (1995), and Møller *et al.* (1996); see also Section 4.4.3.

4.4.1 Pairwise interaction point processes

The perhaps most popular class of models is provided by pairwise interaction point processes, i.e. when the interaction term $g(x)$ in (2) and (3) is of the form

$$g(x) = \prod_{\{\xi,\eta\} \subseteq x} \phi(\{\xi, \eta\})$$

for some function $\phi : \Omega_2 \to [0, \infty)$. We assume also that the density f is hereditary, cf. (8). Then the Metropolis–Hastings ratios (7) and (10) depend on f through the conditional intensity (9) only. This becomes considerably simpler for pairwise interaction point processes as

$$\lambda^*(x, \xi) = \frac{\alpha(n + 1)}{\alpha(n)} \prod_{\eta \in x} \phi(\{\xi, \eta\})$$

with $n(x) = n$.

For example, for the Strauss process (1), we have $\phi(\{\xi, \eta\}) = \gamma^{1[\|\xi - \eta\| < R]}$ and so

$$\lambda^*(x, \xi) = \beta \gamma^{s(x;\xi)} \quad \text{with} \quad s(x; \xi) = \sum_{\eta \in x} 1[\|\eta - \xi\| < R].$$

In particular, Gibbs sampling depends only on the conditional intensity, and in the case of the conditional Strauss process, the Gibbs sampler may be based on rejection sampling: If η is deleted from the current state x, then generate a point ξ which is uniformly distributed in S, and with probability $\gamma^{s(x \setminus \eta; \xi)}$ accept $(x \setminus \eta) \cup \xi$ as the next state; otherwise repeat and generate a new ξ until acceptance. Except when n is very large or γ is too small rejection sampling becomes feasible because $\gamma \leq 1$, cf. Lotwick (1982).

Simulation studies based on the Gibbs sampler for the conditional Strauss process with $\gamma \leq 1$ confirm Ripley's rule of thumb (Ripley, 1987, p. 113): For moderate values of γ and R the chain appears to be in equilibrium after $10n$ transitions, and estimated autocorrelations for $s(X_t)$ and $s(X_{t+m})$ are approximately zero for $t \geq 10n$ and $m \geq 4n$.

The Metropolis algorithm applied on the same model converges a bit slower, but it runs somewhat faster on the machine as it does not involve rejection sampling. From a practical point of view it appears only to be a matter of taste which algorithm one should prefer when $\gamma \leq 1$.

As remarked in Section 4.4.2, Strauss (1975) wanted a model for clustering, so he considered the case where $\gamma > 1$. In fact,

$$f_n(x) \propto \gamma^{s(x)} \tag{11}$$

is integrable with respect to the binomial point process of n points for all $\gamma \geq 0$, but as remarked in Kelly and Ripley (1976) the underlying unconditional density has to be modified when $\gamma > 1$: The term $\beta^{n(x)}$ in (1) should then be replaced by some other function $\alpha(n(x))$ in order to get a well-defined density with respect to the Poisson point process.

Simulation under (11) with $\gamma \gg 1$ is not an easy task: In principle Gibbs sampling based on rejection sampling as described in the case where $\gamma \leq 1$ is still possible, but now it becomes very timeconsuming because of low acceptance rates, see Ripley (1979). Further, single point updating algorithms like the Gibbs sampler and the Metropolis algorithm get stuck in a highly clustered pattern which typically consists of one clump of points, although the algorithms are uniformly ergodic in theory. This is due to the fact that all pairs of points may interact and so the attraction between the points becomes extremely strong. This observation is also in accordance with the theoretical considerations in Gates and Westcott (1986). They conclude that the conditional Strauss process with $\gamma > 1$ as well as many other pairwise interaction processes which aim at modelling clustering violate a stability condition of statistical physics implying that tightly clustered patterns may appear with a very high probability. Such models need therefore very considerable caution. In fact, Ripley (1977) and Ogata and Tanemura (1981, 1984) simulated such models using either the Gibbs sampler or the Metropolis algorithm, but as remarked in Gates and Westcott (1986) the shown simulated patterns are very far from being 'typical'. Recent experiments based on the more efficient method based on simulated tempering as described in Geyer and Thompson (1995) show that the conditional Strauss model has an abrupt transition from 'binomial-like' patterns (γ close to 1) to the tightly clustered pattens rather than exhibiting intermediate moderately clustered patterns.

In order to model clustering with a softer attraction between the points it has also been suggested to combine interaction terms which model repulsion between the points at a small scale with other interaction terms which model attraction at a larger scale. For instance, the

Strauss density (1) may be replaced by

$$f(x) \propto \beta^n \gamma^{s(x)} 1[\| \xi - \eta \| > \epsilon \text{ for all } \{\xi, \eta\} \subseteq x] \qquad (12)$$

where $\epsilon > 0$ is a hard-core parameter, whereby (12) becomes integrable for all $\gamma \geq 0$. This model was proposed in Hanisch and Stoyan (1983) (see also Stoyan, Kendall and Mecke, 1995), but the estimated parameters they use produce in fact much more clustered patterns than the spatial point data considered, and single point updating algorithms get stuck in one clump of points as before when $\gamma \gg 1$. Moreover, simulations show an abrupt transition as in the conditional Strauss process, but now from 'hard-core-like' patterns (γ close to 1) to tightly clustered patterns (γ large).

4.4.2 Ripley–Kelly Markov point processes

The pairwise interaction processes considered above are special examples of the Markov point processes introduced in Ripley and Kelly (1977).

A point process X is Markov in the Ripley–Kelly sense if its density f is hereditary and for $f(x) > 0$ the conditional intensity $\lambda^*(x, \xi)$ depends only on x through $x \cap \partial\{\xi\}$ where $\partial\{\xi\}$ denotes the neighbourhood of ξ. More precisely, for subsets $T \in \mathcal{B}$, the neighbourhood of T,

$$\partial T = \{\xi \in S \backslash T : \xi \sim \eta \text{ for some } \eta \in T\}$$

is defined with respect to some symmetric and reflexive relation $\sim$ on S which is typically of bounded range. The Strauss process (1) is Markov with respect to $\sim$ defined by $\xi \sim \eta$ if and only if $\| \xi - \eta \| < R$. Ripley and Kelly (1977) prove a Hammersley–Clifford type theorem: X is Markov if and only if its density is of the form

$$f(x) = \prod_{y \subseteq x} \phi(y) \qquad (13)$$

where $\phi(\cdot) \geq 0$ with $\phi(y) = 1$ if $\xi \not\sim \eta$ for some $\xi, \eta \in y$. In order to have a well-defined density in (13) conditions on ϕ must be imposed: For instance, it may be assumed that either $\phi(y) \leq 1$ whenever $n(y) \geq 2$ or $\phi(y) = 0$ if y violates a certain condition (typically a hard-core condition), e.g. compare with (1) and (12), respectively. Moreover, interaction of finite order is usually assumed, i.e. $\phi(y) = 1$ whenever $n(y) > n_0$ for some fixed $n_0 < \infty$ (in most cases $n_0 = 2$ as in Section 4.4.1)

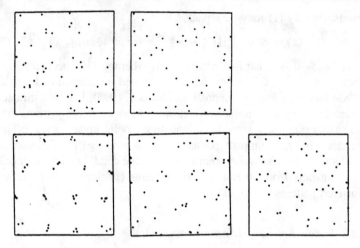

Figure 4.1 *Simulated patterns for a conditional Strauss process with 50 points in the unit square. From left to right:* $\gamma = 1$ *(binomial point process);* $\gamma = 0.02$, $R = 0.05$; $\gamma = 0.02$, $R = 0.25$; $\gamma = 0.05$, $R = 0.25$; $\gamma = 0.5$, $R = 0.05$.

Apart from the computational aspects and the intuitive appeal of these models, (13) implies a spatial Markov property: The conditional distribution of $X_T = X \cap T$ given $X \backslash T$ for subsets $T \in \mathcal{B}$ depends only on $X_{\partial T} = X \cap \partial T$, and the conditional density becomes

$$f(x_T | x_{\partial T}) \propto \prod_{\emptyset \neq y \subseteq x_T} \prod_{z \subseteq x_{\partial T}} \phi(y \cup z). \qquad (14)$$

Here the normalizing constant depends on $x_{\partial T}$ and the conditional density is with respect to the Poisson process μ_T with intensity measure λ_T, the restriction of λ to T. The spatial Markov property becomes important since most practical applications in spatial statistics involve a relatively low number of data points observed through (typically bounded) windows $W \in \mathcal{B}$ — usually $n(X)$ is at most a few hundred — and so the problem of edge-effects is likely to be of great significance, cf. Ripley (1988) and Stoyan, Kendall and Mecke (1995). Therefore, statistical inference as well as simulations may be performed on (14) with $T = W \backslash \partial(S \backslash W)$ and $\partial T = \partial(S \backslash W)$.

It is well-known that the Ripley–Kelly models provide a rich class of models for regular patterns — but probably not so much for modelling clustering. This claim is supported by the discussion in Section 4.4.1. Indeed the conditional Strauss process with $\gamma > 0$ small and R or n

large produces clustered but also highly regular patterns, cf. Figure 4.1. Other special examples of Ripley–Kelly models which exhibit clustering caused by repulsion can be constructed, e.g. if we replace $s(x)$ in (1) by $\sum_{\{\xi,\eta\}\subseteq x} 1(R_1 < \|\xi - \eta\| < R_2)$ for some $R_2 > R_1 \gg 0$ and let $\gamma \geq 0$ be small. Of course, Ripley–Kelly models which are not pairwise interaction point processes may also be considered, e.g. interaction terms for triplets of points may be included too; Geyer shows an example in his contribution, where the density of the Strauss process is modified by including the number of triplets of neighbouring points. Geyer's saturation process is also Markov in Ripley–Kelly's sense, but it exhibits interactions of infinite order as the interaction function $\phi(x)$ is not identical to one whenever $n(x)$ is greater than some constant. Other interesting special models with interactions of infinite order are the the penetrable spheres model (Widom and Rowlinson, 1970) and the area-interaction model (Baddeley and Van Lieshout, 1995); we return to this in Section 4.4.3.

4.4.3 Nearest-neighbour Markov point processes

Alternatively, we may turn attention to the nearest-neighbour Markov point processes introduced in Baddeley and Møller (1989). More precisely, we shall consider a particular class of nearest-neighbour Markov processes which is briefly described below. These include as special cases (a) the Ripley–Kelly Markov point processes, (b) Poisson cluster processes with uniformly bounded clusters, and (c) more general cluster processes where the point process and the clusters are uniformly bounded and nonempty, cf. Baddeley, Lieshout and Møller (1994). This suggests that nearest-neighbour Markov point processes can be used as flexible models for clustered as well as regular point patterns. At the end of this section we return to specific examples which support this conjecture. Moreover, we shall demonstrate that single point updating algorithms may be feasible for simulating certain nearest-neighbour Markov point processes.

We shall consider nearest-neighbour Markov processes defined with respect to the *connected component relation* $\underset{x}{\sim}$. This is a relation between points ξ and η of $x \in \Omega$ so that $\xi \underset{x}{\sim} \eta$ if and only if $\xi = x_1 \sim x_2 \sim \cdots \sim x_m = \eta$ for some $x_1, x_2, \ldots, x_m \in x$ where $\sim$ is some given symmetric and reflexive relation on S. In other words, if $x_{C_1}, \ldots, x_{C_k}$ denote the (maximal) connected components of the finite graph on x whose edges connect every pair of points in x which are related by $\sim$, then $\xi \underset{x}{\sim} \eta$ if and only if $\xi, \eta \in x_{C_j}$ for some j. By Definition 4.9 in Baddeley and Møller (1989) a point process X is then Markov with respect to the connected component relation if its density f is hereditary and for $f(x) > 0$ we have that $\lambda^*(x, \xi)$ depends only on ξ, on $N(\xi|x \cup \xi) = \{\eta \in x : \xi \underset{x \cup \xi}{\sim} \eta\}$, and on the relations $\underset{x}{\sim}$ and $\underset{x \cup \xi}{\sim}$ restricted to $N(\xi|x \cup \xi)$. Equivalently, by Lemma 1 in Baddeley, Lieshout and Møller (1996) the density is of the form

$$f(x) \propto \prod_{j=1}^{k} \Phi(x_{C_j}) \tag{15}$$

where $\Phi(\cdot) \geq 0$ satisfies the following property which is due to the hereditary condition: If $y \in \Omega$ and $z \subseteq y$ do not split into two or more connected components with respect to $\underset{y}{\sim}$ and $\underset{z}{\sim}$, respectively, then $\Phi(y) > 0$ implies $\Phi(z) > 0$. Note that if $\Phi(\cdot)$ is bounded above by a constant, then (15) becomes integrable and the density satisfies the stability condition of statistical physics mentioned in Gates and Westcott (1986). In the special case of the Ripley–Kelly model (13) we can simply take $\Phi(y) = \prod_{\emptyset \neq z \subseteq y} \phi(z)$.

From (15) we can derive a spatial Markov property: Suppose we can only observe the points $y = x \cap W$ of a pattern $x \in \Omega$ where $W \in \mathcal{B}$ is some window. Further, let $y_{C_1}, \ldots, y_{C_l}$ denote those connected components y_C defined by $\underset{y}{\sim}$ which satisfy the following property: There exists some $\eta \in y_C$ with $\partial\{\eta\}\backslash W \neq \emptyset$ where ∂ denotes neighbourhood with respect to $\sim$. Letting

$$B_y(W) = \left[W \cap \bigcup_{j=1}^{l} \bigcup_{\xi \in y_{C_j}} \partial\{\xi\} \right] \cup \partial(S\backslash W)$$

and taking $T = W\backslash B$ for any $B \in \mathcal{B}$ with $B_y(W) \subseteq B \subseteq W$ we have that the conditional density of $X_T = X \cap T$ given $X\backslash T = x_{-T}$ depends only on x_{-T} through $x_{\partial_y T} = B \cap y$. By (15) this conditional density becomes

$$f(x_T|x_{\partial_y T}) \propto \prod_{j=1}^{m} \Phi(z_{D_j}) \tag{16}$$

with respect to μ_T where $z_{D_1}, \ldots, z_{D_m}$ denote the connected components defined by $z = x_T \cup x_{\partial_y T}$ and where the normalizing constant depends on $x_{\partial_y T}$. In the terminology of Kendall (1990), $B_y(W)$ is the splitting boundary of W with respect to $y = x \cap W$. Clearly, $B_y(W) = B_{\partial_y T}(W)$. Note that if $B_y(W)$ is measurable and $B = B_y(W)$, then $x_{\partial_y T} = \bigcup_{j=1}^{l} y_{C_j}$ is minimal in the sense that the conditional density (16) depends on all the points in $x_{\partial_y T}$.

It may be illuminating to consider the specific example of a disc process of closed discs $b(c, r)$ with centers c in an open planar region A and radii $r \leq R$ for a given $R < \infty$, so we can take $S = A \times [0, R]$, cf. Section 4.4.2. Then $\sim$ may be defined by that $(c_1, r_1) \sim (c_2, r_2)$ if and only if $\|c_1 - c_2\| \leq r_1 + r_2$, i.e. when the discs $b(c_1, r_1)$ and $b(c_2, r_2)$ intersect each other. The connected components defined by $\underset{x}{\sim}$ correspond then to the path-connected components in the setunion $\bigcup_{i=1}^{n} b(c_i, r_i)$ of all the discs in $x = \{(c_1, r_1), \ldots, (c_n, r_n)\}$ (Baddeley and Møller, 1989, Examples 4.6 and 5.10 and Appendix A3). Such path-connected components are called clusters. Suppose $K_1, \ldots, K_l$ denote those clusters associated with x which intersect both V and $A \backslash V$ where V is some measurable subset of A; e.g. V may be some planar observation window. Let $W = \{(c, r) \in S : b(c, r) \cap V \neq \emptyset\}$ be the set of discs hitting V and set $y = x \cap W$. Then the splitting boundary $B_y(W)$ consists of all those $(c, r) \in W$ so that the disc $b(c, r)$ intersects either $\bigcup_{j=1}^{l} K_j \cap V$ or $A \backslash V$ or both. Hence $T = W \backslash B_y(W)$ is given by those discs contained in V which do not intersect $\bigcup_{j=1}^{l} K_j$, i.e.

$$T = \left\{ (c, r) \in S : b(c, r) \subseteq V \backslash \bigcup_{j=1}^{l} K_j \right\}$$

Consequently, it follows from (16) that

$$f(x_T | x_{\partial_y T}) \propto f(x_T)$$

where the normalizing constant depends only on T. Therefore, in order to take care of boundary effects we need, roughly speaking, only to condition on the set $\bigcup_{j=1}^{l} K_j \cap V$ which consists of those parts of the clusters which are both 'observed' within the window V and hit the boundary of V.

A simple model for a disc process of the above type is obtained by assuming that the conditional density $\lambda^*(x, \xi)$ depends only on the number $c(x, \xi)$ of clusters associated with x which intersects the disc associated with ξ, i.e.

$$\lambda^*(x, \xi) = g(c(x, \xi)) \tag{17}$$

for some function $g(\cdot) > 0$. As shown in Appendix C (cf. Häggström,

Van Lieshout and Møller, 1996), (17) is equivalent to state that f is of the form

$$f(x) \propto \beta^{n(x)} \gamma^{-c(x)} \qquad (18)$$

for some parameters $\beta > 0$ and $\gamma > 0$ where $c(x)$ is the number of clusters (for a similar characterization result for the Strauss process, see Kelly and Ripley, 1976). In fact,

$$\lambda^*(x, \xi) = (\beta/\gamma)\gamma^{c(x,\xi)}.$$

For $\gamma > 1$ we get models for clustered patterns, while $0 < \gamma < 1$ gives more regular patterns if β is not too large, see Figures 4.2–4.4. For sufficiently large values of β and $\gamma < 1$ fixed a phase transition behaviour may occur so that the high density of discs implies that typical realizations consist of one big cluster and many small clusters; we return to this at the end of this section. The model (18) was introduced independently in Møller (1994a, b) and for the special case $\gamma = 1/2$ and all r_i equal in Chayes et al. (1995). As in Häggström, Lieshout and Møller (1996) we call (18) the *continuum random cluster model*. Notice that if γ is very small (as in Figure 4.3) we have essentially a classical hard-core process as $c(X) = n(X)$ with a high probability, while if γ is very large we get mostly one cluster as $c(X) = 1$ with a high probability. Simulations with a fixed number of points have been performed for varying values of γ showing no abrupt transition as in the case of the conditional Strauss process and other pairwise interaction point processes studied in Section 4.4.1 (however, as remarked at the end of this section, a phase transition behaviour occurs in the unconditional case when $\gamma = \exp(\beta)$ is sufficiently large). This may be explained by a much softer interaction in (18) so that compared to e.g. the Strauss process the density is less peaked. For example, $c_n(\gamma) = \sup_{\Omega_n} f / \inf_{\Omega_n} f$ with $\gamma > 1$ is given by $c_n(\gamma) = \gamma^{n-1}$ under (18) whereas $c_n(\gamma) = \gamma^{n(n-1)/2}$ under the conditional Strauss process (provided all n points or discs may have no neighbours, i.e. when n is not too large).

The Metropolis algorithm was used for the simulations shown in Figures 4.2–4.4. The Markov chains appear to reach equilibrium very fast, though some autocorrelation is evident in the time series shown in Figure 4.4. Whether this remains true for simulating other types of nearest-neighbour Markov processes given by (15) or (16) depend probably much on the interaction function Φ. For instance, for a disc process where Φ depends much on the size and shape of the cluster such algorithms may be expected to be less efficient. Alternatively, one may try to develop algorithms where an update consists of the deletion of (possible large) parts of a cluster or the generation of an entire new cluster.

Baddeley and Van Lieshout (1995) consider another example of a process with density of the form (18), but with $c(x)$ replaced by the area $|U_x|$ of the set union of the clusters, i.e.

$$f(x) \propto \beta^{n(x)} \gamma^{-|U_x|} .$$

They call this the area-interaction process; for $\gamma > 1$ and all r_i equal this is also the penetrable sphere model introduced in Widom and Rowlinson (1970). The process is in fact Markov in the Ripley–Kelly sense, but it exhibits interactions of infinite order, cf. Baddeley and Van Lieshout (1995). For $0 < \gamma < 1$ the area-interaction process produces regular patterns, while $\gamma > 1$ correspond to clustered patterns. Simulated patterns are shown in Baddeley and Van Lieshout (1995), Kendall (1996), and Häggström, Van Lieshout and Møller (1996). In the latter paper a characterisation result similar to (17)–(18) is established for the area-interaction process.

The algorithms studied in Häggström, Van Lieshout and Møller (1996) utilise a correspondance to a mixture model with two types of points $X_1 = \{x_{11}, \ldots, x_{1n_1}\}$ and $X_2 = \{x_{21}, \ldots, x_{2n_2}\}$: The density of (X_1, X_2) with respect to the product measure of the unit Poisson is proportional to $\beta_1^{n_1} \beta_2^{n_2}$ if discs of radii $r > 0$ associated to different types of points do not overlap; otherwise it is 0. Suppose $|\cdot|$ means area within the planar region A or let A be a square identified with a torus. The marginal distribution of X_1 is then a penetrable sphere model with $\beta = \beta_1$ and $\gamma = \exp(\beta_2)$ (Widom and Rowlinson, 1970). Simulation of the mixture model is easy using Gibbs sampling: The conditional distribution of X_1 given $X_2 = x_2$ is a homogeneous Poisson process on $A \setminus U_{x_2}$ with intensity β_1 (this may easily be generated by thinning a homogeneous Poisson process on A with intensity β_1 to U_{x_2}); and similarly for X_2 given $X_1 = x_1$. In the symmetric case $\beta_1 = \beta_2 = \beta$ we can collect samples from both X_1 and X_2, and the superposition $X = X_1 \cup X_2$ becomes the simple component model (18) with the same β but $\gamma = 1/2$ and all radii $r_i = r$.

This is another way to construct models for clustering by the marginal distribution of X_i, $i = 1, \ldots, k$, where $(X_1, \ldots, X_k)$ is a multitype point process with repulsion between points of different types but no (or a softer) interaction between points of the same type. Such multitype point process may be considered as continuum versions of the Ising or Potts model, cf. Georgii and Häggström (1995). For the immediate

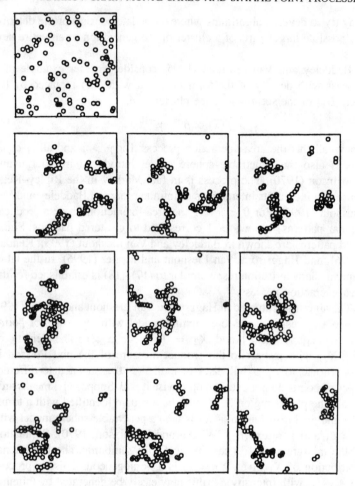

Figure 4.2 *Simulations based on the Metropolis algorithm of the continuum random cluster model (18) with centers in the unit square, $\gamma = \exp(5)$, and fixed $n = 100$ and $r = 0.02$. From left to right: Initial pattern and patterns after every $10\,000$ iterations. The centers of the initial pattern were generated by a binomial process.*

generalization of the Widom and Rowlinson model in the symmetric case $\beta_1 = \cdots = \beta_k = \beta$ we get (18) with $\gamma = 1/k$ for the superposition $X = X_1 \cup \cdots \cup X_k$.

Ruelle (1971) has shown that a phase transition behaviour occurs

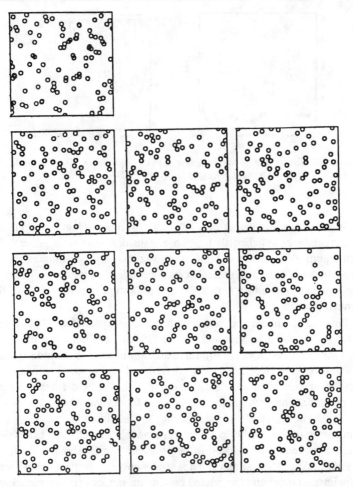

Figure 4.3 *Simulations based on the Metropolis algorithm of the continuum random cluster model (18) with centers in the unit square, $\gamma = \exp(-5)$, and fixed $n = 100$ and $r = 0.02$. From left to right: Initial pattern and patterns after every $10\,000$ iterations. The centers of the initial pattern were generated by a binomial process.*

in the mixture model of Widom and Rowlinson for large values of $\beta_1 = \beta_2 = \beta$, but so far it is not known if there exists a critical point for β (though this is expected). The phase transition is due to the possible existence of an infinite cluster as the domain A entends

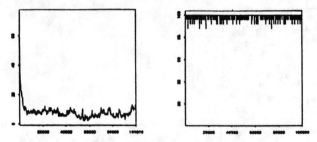

Figure 4.4 *Timeseries of the sufficient statistic* $c(x)$ *after* 100 000 *iterations.
Left: As in Figure 4.2. Right: As in Figure 4.3.*

to infinity, see Chayes *et al.* (1995). Georgii and Häggström (1995)
has recently established that a phase transition occurs in general for
continuum Potts models. Simulation studies may be illuminating in
order to see if there is a critical point (depending on whether β_1 and β_2
are equal or not) and whether there is an abrupt transition like in the
Strauss process.

Finally, it should be noticed that recently Kendall (1996) and
Häggström, Van Lieshout and Møller (1996) have extended the ideas
of Propp and Wilson (1996) for exact simulation to include certain
point processes. Kendall (1996) describes how a coupling construction of spatial birth-and-death processes can be used for simulating
in particular the area-interaction process for all values of $\beta > 0$ and
$\gamma > 0$, but Kendall's algorithm may be modified to include exact
simulation of some other point process models as well. The exact simulation procedure introduced in Häggström, Van Lieshout and Møller
(1996) appears to be simpler, but it seems only to apply on the Widom-
Rowlinson model and the related continuum random cluster model and
area-interaction process. A Swendsen–Wang type algorithm for these
particular models are also introduced in Häggström, Van Lieshout and
Møller (1996).

Acknowledgements

I thank Julian Besag, Charles Geyer, Olle Häggström, Marie-Colette
van Lieshout, Jeffrey Rosenthal, and Rasmus Waagepetersen for helpful comments, and Kenneth Berg and Erik Petersen for programming
assistance.

Appendix A: Some convergence results for stationary Markov chains

Consider a probabilistic model $(E, \mathcal{E}, \pi)$ which is perhaps not tractable for mathematical analysis, but at least we can generate a time homogeneous Markov chain $X_0, X_1, \ldots$ with equilibrium distribution π, i.e.

$$P^t(x, A) \to \pi(A) \text{ as } t \to \infty$$

where $P^t(x, A) = P(X_t \in A | X_0 = x)$ is the t-step transition probability. Constructions of such Markov chains when π is the distribution for a spatial point process are studied in Section 4.3. Below we briefly discuss some general concepts and results for convergence. Most results are based on Orey (1971), Nummelin (1984), and the review in Tierney (1994); a comprehensive collection of results can be found in the excellent monograph by Meyn and Tweedie (1993). As in Nummelin's book the σ-algebra $\mathcal{E}$ is assumed to be separable, i.e. countable generated; all σ-algebras considered in the previous sections are separable too. Finally, some practical aspects are discussed at the end of this appendix.

Henceforth, for ease of presentation, π is assumed to be invariant, i.e.

$$\pi(A) = \int P(x, A)\pi(dx)$$

for all $A \in \mathcal{E}$ where $P(\cdot, \cdot) = P^1(\cdot, \cdot)$ denotes the transition kernel of the chain. Of course, π is invariant if it is the equilibrium distribution. Therefore, in practice any Markov chain Monte Carlo algorithm is constructed so that π becomes invariant. In fact, for most algorithms (including all those considered in this paper) reversibility holds, that is

$$\int_B P(x, A)\pi(dx) = \int_A P(x, B)\pi(dx)$$

for all $A, B \in \mathcal{E}$, whereby π is clearly invariant.

Irreducibility and aperiodicity are essentially equivalent to convergence of the transition probabilities with respect to the total variation norm defined by

$$\|v\| = [\sup\{v(A) : A \in \mathcal{E}\} - \inf\{v(A) : A \in \mathcal{E}\}]/2$$

where v is a bounded signed measure. By definition the chain is irreducible if $P^t(x, A) > 0$ for some $t = t(x, A)$ and all $x \in E$ and all $A \in \mathcal{E}$ with $\pi(A) > 0$. In that case π becomes the unique invariant distribution, cf. Tierney (1994, Theorem 1). Moreover, the chain is said to be aperiodic if there are no disjoint sets $A_0, \ldots, A_{d-1} \in \mathcal{E}$ with $d \geq 2$ so that $P(x, A_{j(i)}) = 1$ for all $x \in A_i$ and $i = 0, \ldots, d - 1$ where $j(i) \equiv i + 1 \pmod{d}$. We have that

$$\| P^t(x, \cdot) - \pi(\cdot) \| \to 0 \text{ for } \pi - a.a. \, x$$

if the chain is irreducible and aperiodic.

In order to get rid of the nullset above Harris recurrence is needed. This means that $P(X_t \in A$ for some $t \mid X_0 = x) = 1$ for all $x \in E$ and $A \in \mathcal{E}$ with $\pi(A) > 0$. Equivalently, as in the definition used in Nummelin (1984) and Tierney (1994), $P(X_t \in A$ for infinitely many $t \mid X_0 = x) = 1$ if $\pi(A) > 0$, cf. Orey (1971, p. 22). Clearly, Harris recurrence implies irreducibility. By Orey's theorem,

$$\| \mu P^t - \pi \| \to 0$$

for any initial distribution μ on $\mathcal{E}$ if and only if the chain is Harris ergodic, i.e. Harris recurrent and aperiodic, cf. Nummelin (1984, Propositions 6.3 and 6.7). Here μP^t is the marginal distribution of X_t when X_0 has distribution μ.

For spatial point processes a stronger form of ergodicity called geometrical ergodicity may often be established, cf. Section 4.3. The chain is geometrically ergodic if

$$\| P^t(x, \cdot) - \pi \| \leq K(x) r^t \tag{A.1}$$

for all t and $x \in E$ where $r < 1$ is some constant and $K : E \to [0, \infty)$ is some measurable function with $\pi K < \infty$. Here

$$\pi K = \int K(x) \pi(dx)$$

denotes the mean of K with respect to π. If the function K in (A.1) is bounded above, then the chain is uniformly ergodic. Usually, this holds only for spatial point processes when the number of points is limited almost surely by a constant like in the hard-core case or if one has conditioned on the number of points as in Section 4.3.1.

In order to verify geometric or uniform ergodicity the concept of a small set is useful. A set $A \in \mathcal{E}$ is small if $\pi(A) > 0$ and the following minorization condition holds: $P^{t_0}(x, \cdot) \geq \epsilon Q(\cdot)$ for all $x \in A$, some t_0, some $\epsilon > 0$, and some probability measure Q on $\mathcal{E}$. Proposition 5.21 in Nummelin (1984) states that geometric ergodicity holds if there exists a measurable function $g : E \to [0, \infty)$, a small set A, and a constant

$c > 1$ so that

$$\sup_{x \notin A} E[cg(X_1) - g(X_0) \mid X_0 = x] < 0$$

and

$$\sup_{x \in A} E[g(X_1)1\{X_1 \notin A\} \mid X_0 = x] < \infty.$$

Especially, uniform ergodicity holds if and only if the state space E is small, cf. Proposition 2 in Tierney (1994). In this case the convergence rate bound r in (A.1) can be chosen so that $r^{t_0} = 1 - \epsilon$ with ϵ and t_0 as in the minorization condition for E, and we can further take $K(\cdot) = 1$ in (A.1); see also Appendix B where other quantitative bounds are discussed.

Now, let us turn to the limit behaviour of the Monte Carlo approximation

$$\bar{g}_t = \frac{1}{t} \sum_{s=1}^{t} g(X_s)$$

of the mean πg where $g : E \to \mathbb{R}$ is a measurable function with $\pi|g| < \infty$. The initial distribution of the chain is arbitrary in the following. By Birkhoff's ergodic theorem $\bar{g}_t \to \pi g$ almost surely when the chain is just Harris recurrent, cf. Tierney (1994, Theorem 3). If the chain is also geometrically ergodic and $\pi(g^2) < \infty$, then

$$\sqrt{n}(\bar{g}_n - \pi g) \xrightarrow{D} N(0, \sigma(g)^2), \tag{A.2}$$

cf. Chan and Geyer (1994, Theorem 1) and Roberts and Rosenthal (1996, Corollary 3). Here

$$\sigma(g)^2 = \gamma_0(g) + 2 \sum_{s=1}^{\infty} \gamma_s(g)$$

with the lag s autocovariance

$$\gamma_s(g) = \mathrm{cov}_\pi(g(x_0), g(x_s))$$

calculated for the stationary chain.

Let us conclude with a few comments on some practical aspects. MCMC algorithms can fail to converge in practise, and exploratory analysis (perhaps in terms of multiple starts) can be used to try and detect badly mixing chains (see e.g. Brooks and Roberts, 1996). In order to consider the chain as being effectively in equilibrium, it is customary to use a 'burn-in' of transitions before collecting samples.

The number needed to be thrown away is usually estimated by investigating estimated autocorrelations and time series plots of different statistics, see e.g. Sokal (1989), Geyer (1991, 1992), Green and Han (1991), Geyer and Møller (1994), Cowles and Carlin (1995) and Brooks and Roberts (1996). Another question is whether a subsample $X_m, X_{m+k}, X_{m+2k}, \ldots$ should be used where m is the burn-in and k is the spacing. Of course this is again a Markov chain satisfying the same conditions as the original chain, but the asymptotic variance in (A.2) now becomes

$$\sigma(g, k)^2 = \gamma_0(g) + 2 \sum_{i=1}^{\infty} \gamma_{ik}(g).$$

We may consider $c(g, k) = k\sigma(g, k)^2$ as a measure of the cost of generating the subsample. As shown in Geyer (1992, Theorem 3.3) $c(g, k)$ is minimal for $k = 1$ if the chain is irreducible and reversible. Geyer (1992) considers also the cost of using samples and remarks that even with a large cost for using samples the optimal spacing may be very low.

Appendix B: Quantitative bounds for convergence rates

Geometrical and uniform ergodicity as discussed in Appendix A provide only qualitative bounds on the rate of convergence. There exist various papers giving quantitative bounds on convergence rates for specific models, see Rosenthal (1993) and the references therein. Below we concentrate on a general result in Rosenthal (1993) and relate this to the Metropolis–Hastings algorithm in Section 4.3.2.

Consider a time homogeneous Markov chain $X_0, X_1, \ldots$ with invariant distribution π and t-step transition probabilities $P^t(\cdot, \cdot)$ defined on an arbitrary measure space $(E, \mathcal{E})$. Suppose the chain satisfies the drift condition

$$E(V(X_{t_0})|X_0 = x) \le aV(x) + b, \ x \in E \tag{B.1}$$

for some measurable nonnegative function V and some constants t_0, $a < 1, b < \infty$. Suppose also that it satisfies the minorization condition

$$P^{t_0}(x, \cdot) \ge \epsilon Q(\cdot) \text{ whenever } V(x) \le d, \tag{B.2}$$

for some $\epsilon > 0$, some probability measure Q on $\mathcal{E}$, and some $d > \frac{2b}{1-a}$. Then according to Theorem 12 in Rosenthal (1993) we have for any

$0 < r < 1, k = 1, 2, \ldots$, and any initial distribution ν,

$$\|\nu P^{kt_0} - \pi\| \leq (1 - \epsilon)^{rk} + \alpha^{(1-r)k} A^{rk} \left(1 + \frac{b}{1-a} + \int V(x)\nu(dx) \right)$$

$$(B.3)$$

where

$$\alpha = \frac{1 + 2b + ad}{1 + d} < 1 , \quad A = 1 + 2(ad + b).$$

Here $\| \cdot \|$ and νP^t are defined as in Appendix A.

We apply now this result on the Metropolis–Hastings algorithm for the Strauss process in the random number case as described at the end of Section 4.3.2, i.e. when $p(x) = 0.5$, $b(x, \xi) \propto 1$, and $d(x \backslash \eta, \eta) = 1/n(x)$. Without loss of generality we assume that S has volume 1. Let $V(x) = n(x)$ and consider any positive integer d. Letting

$$\epsilon = \inf_{n(x)=d} P^d(x, \{\emptyset\}) = 2^{-d} \prod_{j=1}^{d} \min \left\{ 1, \frac{j}{\beta \gamma^{j-1}} \right\}$$

and if Q is the probability measure concentrated at the empty point configuration $\emptyset$, then (B.2) holds with $t_0 = d$. In order to verify the drift condition we use the following: Since $\gamma \leq 1$ we find that

$$E(n(X_1)|X_0 = x) = \frac{n(x) - 1}{2n(x)} \sum_{\eta \in x} \min \left\{ 1, \frac{n(x)}{\beta \gamma^{s(x \backslash \eta; \eta)}} \right\}$$

$$+ \frac{n(x) + 1}{2} \int_S \min \left\{ 1, \frac{\beta \gamma^{s(x; \xi)}}{n(x) + 1} \right\} d\xi$$

$$\leq 1[n(x) \geq \beta]\{[n(x) - 1] + \beta\}/2$$

$$+ 1[n(x) < \beta]\{[n(x) - 1] + [n(x) + 1]\}/2$$

with $s(\cdot; \cdot)$ defined as in Section 4.4.1. Thus

$$E(n(X_t)|n(X_{t-1})) \leq \{n(X_{t-1}) + \delta\}/2 \qquad (B.4)$$

where $\delta = \beta - 1$ if $\beta > 0$ is an integer, and δ denotes the integer part of β otherwise. Conditioning on first X_{d-1}, then X_{d-2}, and so on, we obtain from (B.4) that

$$E(n(X_d)|X_0) = E[E(n(X_d)|X_{d-1})|X_0]$$

$$\leq E(n(X_{d-1}) + \delta|X_0)/2$$

$$\leq \cdots$$

$$\leq a \times n(x) + b$$

with

$$a = 2^{-d} , \ b = \delta(1-a)/2 .$$

Hence taking $d > \delta$, (B.1) and hence (B.3) holds.

In the above situation it is appropriate to start the chain in the empty point configuration, whereby (B.3) gives

$$\| P^{kd}(\emptyset, \cdot) - \pi \| \leq (1-\epsilon)^{rk} + \alpha^{(1-r)k} A^{rk}(1+\delta/2) \qquad (B.5)$$

where π is the distribution of the Strauss process. Here we may take $d = \beta$ if β is integer, in which case

$$\epsilon = 2^{-\beta} \prod_{j=1}^{\beta} \min\left\{1, \frac{j}{\beta\gamma^{j-1}}\right\} , \ \alpha = \frac{\beta + 2^{-\beta}}{\beta + 1} , \ A = \beta + (\beta + 1)2^{-\beta} .$$

For example, if $\gamma = 0.5$, we have $\epsilon \approx 3.7 \times 10^{-36}$, $\alpha \approx 0.99$, and $A \approx 100$ if $d = \beta = 100$, whereas $\epsilon = 0.125$, $\alpha = 0.75$, and $A = 2.75$ if $d = \beta = 2$. Thus we get very weak bounds except for small values of β, but such cases appear not to be very interesting since $En(X) \leq \beta$. Of course better bounds may be obtained by another choice of the function V, e.g. $V(x) = |n(x) - En(X)|$ with the expectation calculated under the Strauss model, but then it becomes much more difficult to verify (B.1) and (B.2) (Since $En(X)$ is unknown we may replace it by an approximate value obtained e.g. by simulations). In any case the bounds which may be obtained will probably be too weak for real applications.

Appendix C: Characterization result for the continuum random cluster model

For the continuum random cluster model studied in Section 4.4.3 we prove below that

$$\lambda^*(x, \xi) = g(c(x \cup \xi) - c(x)) > 0 \text{ for all } x \in \Omega, \xi \in S \qquad (C.1)$$

is equivalent to (18), that is

$$f(x) = \alpha\beta^{n(x)}\gamma^{-c(x)} \text{ for all } x \in \Omega \qquad (C.2)$$

and some parameters $\beta, \gamma > 0$, where $\alpha = f(\emptyset)$ is a normalizing constant. Note that

$$c(x, \xi) = 1 + c(x) - c(x \cup \xi)$$

is the number of clusters generated by x which are intersected by the disc associated with ξ, so (C.1) is obviously equivalent to (17).

Clearly, (C.2) implies (C.1). Assuming (C.1) holds we prove (C.2) by induction on $n(x)$ as follows. Setting

$$\beta = g(0) \quad \text{and} \quad \gamma = g(0)/g(1) \tag{C.3}$$

then (C.2) trivially holds for $n(x) \leq 1$. Letting $(x) \geq 2$, $\xi \in x$, and $k = c(x) - c(x\backslash\xi)$, then $k \leq 1$ and the induction hypothesis and (C.1) imply that

$$f(x) = f(x\backslash\xi)\lambda^*(x\backslash\xi, \xi) = \alpha\beta^{n-1}\gamma^{-c(x\backslash\xi)}g(k).$$

Combining this with (C.3) give (C.2) if $k = 0$ or $k = 1$. Suppose $k < 0$. Then the disc associated with ξ intersects at least two of the clusters generated by $x\backslash\xi$, and it is easily seen that we can delete a disc from one of these clusters without changing the total number of clusters generated by x, i.e.

$$c(x) = c(x\backslash\eta) \quad \text{for some} \quad \eta \in x\backslash\xi.$$

Consequently, by the induction hypothesis and using (C.1) and (C.3),

$$f(x) = f(x\backslash\eta)\lambda^*(x\backslash\eta, \eta) = \alpha\beta^{n-1}\gamma^{-c(x\backslash\eta)}g(0) = \alpha\beta^n\gamma^{-c(x)}$$

whereby (C.2) is verified.

References

Allen, M.P. and Tildesley, D.J. (1987). *Computer Simulation of Liquids*. Oxford: Oxford University Press.

Baddeley, A.J. and Van Lieshout, M.N.M. (1995). Area-interaction point processes. To appear in *Ann. Inst. Statist. Math.* **46**, 601–619.

Baddeley, A.J., Van Lieshout, M.N.M. and Møller, J. (1996). Markov properties of cluster processes. *Adv. Appl. Prob.* **28**, 346–355.

Baddeley, A.J. and Møller, J. (1989). Nearest-neighbour Markov point processes and random sets. *Int. Statist. Rev.* **57**, 90–121.

Besag, J. (1974). Spatial interaction and the statistical analysis of lattice systems (with discussion). *J. R. Statist. Soc.* **B 36**, 192–236.

Besag, J. and Green, P. (1993). Spatial Statistics and Bayesian computation. *J. R. Statist. Soc.* **B 55**, 25–38.

Besag, J., Green, P., Higdon, D. and Mengersen, K. (1995). Bayesian computation and stochastic systems. *Statistical Science* **10**, 3–66.

Brooks, S. and Roberts, G. (1996). Diagnosing convergence of Markov chain Monte Carlo algorithms. Technical Report, Statistical Laboratory, University of Cambridge.

Carter, D.S. and Prenter, P.M. (1972). Exponential spaces and counting processes. *Z. Wahr. verw. Geb.* **21**, 1–19.

Chan, K.S. and Geyer, C.J. (1994). Discussion of the paper 'Markov chains for exploring posterior distributions' by Luke Tierney. *Ann. Statist.* **22**, 1747–1758.

Chayes, J.T., Chayes, L. and Kotecky, R. (1995). The analysis of the Widom-Rowlinson model by stochastic geometric methods. *Commun. Math. Phys.* **172**, 551–569.

Clifford, P. and Nicholls, G. (1994). Comparison of birth-and-death and Metropolis–Hastings Markov chain Monte Carlo for the Strauss process. Department of Statistics, Oxford University. (Manuscript).

Cowles, M.K. and Carlin, B.P. (1995). Markov chain Monte Carlo convergence diagnostics: A comparative review. Div. of Biostatistics, University of Minnesota. (Manuscript).

Creutz, M. (1979). Confinement and the critical dimensionality of space-time. *Phys. Rev. Lett.*, **43**, 553–556.

Diggle, P.J. (1983). *Statistical Analysis of Spatial Point Patterns*. London: Academic Press.

Gates, D.J. and Westscott, M. (1986). Clustering estimates for spatial point distributions with unstable potentials. *Ann. Inst. Statist. Math.* **A 38**, 123–135.

Georgii, H.-O. and Häggström, O. (1995) Phase transition in continuum Potts models. *Comm. Math. Phys.* **181**, 507–528.

Geman, S. and Geman, D. (1984). Stochastic relaxation, Gibbs distributions and the Bayesian restoration of images. *IEEE Transactions on Pattern Analysis and Machine Intelligence* **6**, 721–741.

Geyer, C. (1991). Markov chain Monte Carlo maximum likelihood. In: *Computer Science and Statistics, Proceedings of the 23rd Interface* (Ed. Keramidas). Fairfax Station: Interface Foundation.

Geyer, C. (1992). Practical Markov chain Monte Carlo (with discussion). *Statist. Sci.* **7**, 473–511.

Geyer, C. and Møller, J. (1994). Simulation procedures and likelihood inference for spatial point processes. *Scand. J. Statist.* **21**, 359–373.

Geyer, C. and Thompson, E.A. (1995). Annealing Markov chain Monte Carlo with applications to ancestral inference. *JASA* **90**, 909–920.

Green, P (1995). Reversible jumb MCMC computation and Bayesian model determination. *Biometrika* **82**, 711–732.

Green, P and Han, X. (1991). Metropolis methods, Gaussian proposals and antithetic variables. In: *Stochastic Models, Statistical Methods and Algorithms in Image Analysis*. (Eds. Barone, Frigessi and Piccioni). Berlin: Springer-Verlag.

Häggström, O., Van Lieshout, M.N.M. and Møller, J. (1996). Characterisation results and Markov chain Monte Carlo algorithms including exact simulation for some spatial point processes. Research Report R-96-2040, Department of Mathematics, Aalborg University. To appear in *Bernoulli*

Hanisch, K.-H. and Stoyan, D. (1983). Remarks on statistical inference and prediction for a hard-core clustering model. *Math. Operations. Statist., ser. statist.* **14**, 559–567.

Hastings, W. (1970). Monte Carlo sampling methods using Markov chains and their applications. *Biometrika* **57**, 97–109.

Hurn, M. and Jennison, C. (1993). Multiple-site updates in maximum a posteriori (MAP) and marginal posterior models (MPM) image estimation. In: *Statistics and Images: 1* (Eds. K.V. Mardia and G.K. Kanji). Oxford: Carfax Publishing Company. Pp. 155–186.

Kelly, F.P. and Ripley, B.D. (1976). A note on Strauss' model for clustering. *Biometrika* **63**, 357–360.

Kendall, W.S. (1990). A spatial Markov property for nearest-neighbour Markov point processes. *J. Appl. Prob.* **28**, 767–778.

Kendall, W.S. (1996). Perfect simulation for the area-interaction point process. To appear in *Proceedings of the Symposium on Probability Towards the Year 2000*. (Eds. L. Accardi and C. Heyde). Springer-Verlag.

Lotwick, H.W. (1982). Simulation of some spatial hard core models, and the complete packing problem. *J. Statist. Comput. Simulation* **15**, 295–314.

Marinari, E. and Parisi, G. (1992). Simulated tempering: a new Monte Carlo scheme. *Europhys. Lett.* **19**, 451–458.

Meyn, S.P. and Tweedie, R.L. (1993). *Markov Chains and Stochastic Stability*. London: Springer-Verlag.

Metropolis, N., Rosenbluth, A., Rosenbluth, M., Teller, A. and Teller, E. (1953). Equations of state calculations by fast computing machines. *J. Chem. Phys.* **21**, 1087–1092.

Møller, J. (1989). On the rate of convergence of spatial birth-and-death processes. *Ann. Inst. Statist. Math.* **41**, 565–581.

Møller, J. (1992). Discussion contribution. *J. R. Statist. Soc.* **B 54**, 692–693.

Møller, J. (1993). Discussion contribution. *J. R. Statist. Soc.* **B 55**, 84–85.

Møller, J. (1994a). Discussion contribution. *Scand. J. Statist.* **21**, 346–349.

Møller, J. (1994b). Markov chain Monte Carlo and spatial point processes. Research Report 293, Department of Theoretical Statistics, University of Aarhus.

Møller, J., Syversveen, A.R. and Waagepetersen, R. (1998). Log Gaussian Cox processes. To appear in *Scand. J. Statist.*.

Nummelin, E. (1984). *General Irreducible Markov Chains and Non-negative Operators*. Cambridge: Cambridge University Press.

Ogata, Y. and Tanemura, M. (1981). Estimation of interaction potentials of spatial point pattens through the maximum likelihood procedure. *Ann. Inst. Statist. Math.* **33 B**, 315–338.

Ogata, Y. and Tanemura, M. (1984). Likelihood analysis of spatial point patterns. *J. R. Statist. Soc.* **B 46**, 496–518.

Orey, S. (1971). *Limit Theorems for Markov Chain Transition Probabilities*. London: Van Nostrand.

Peskun, P. (1973). Optimum Monte-Carlo sampling using Markov chains.

Biometrika **60**, 607–612.

Preston, C.J. (1977). Spatial birth-and-death processes. *Bull. Int. Statist. Inst.* **46**, 371–391.

Propp, J. G. (1996). Exact sampling with coupled Markov chains and applications to statistical mechanics. *Random Structures and Algorithms* **9**, 223–252.

Ripley, B.D. (1977). Modelling spatial patterns (with discussion). *J. R. Statist. Soc.* **B 39**, 172–212.

Ripley, B.D. (1979). Simulating spatial patterns: dependent samples from a multivariate density. Algorithm AS 137. *Appl. Statist.* **28**, 109–112.

Ripley, B.D. (1987). *Stochastic Simulation.* New York: Wiley.

Ripley, B.D. (1988). *Statistical Inference for Spatial Processes.* Cambridge: Cambridge University Press.

Ripley, B.D. and Kelly, F.P. (1977). Markov point processes. *J. Lond. Math. Soc.* **15**, 188–192.

Roberts, G.O. and Rosenthal, J.S. (1996). Geometric ergodicity and hybrid Markov chains. Technical Report, Statistical Laboratory, University of Cambridge.

Rosenthal, J. S. (1993). Minorization conditions and convergence rates for Markov chain Monte Carlo. Technical Report 9321, Department of Statistics, University of Toronto.

Ruelle, D. (1971). Existence of a phase transition in a continuous classical system. *Phys. Rev. Lett.* **27**, 1040–1041.

Sokal, A. (1989). *Monte Carlo Methods in Statistical Mechanics: Foundations and New Algorithms.* Lausanne: Course de Troisième Cycle de la Physique en Suisse Romande.

Smith, A.F.M. and Roberts, G.O. (1993). Bayesian computation via the Gibbs sampler and related Markov chain Monte Carlo methods. *J. R. Statist. Soc.* **B 55**, 3–23.

Stoyan, D., Kendall, W.S. and Mecke, J. (1995). *Stochastic Geometry and its Applications.* 2nd ed. New York: Wiley.

Strauss, D.J. (1975). A model for clustering. *Biometrika* **62**, 467–475.

Suomela, P. (1976). *Construction of Nearest Neighbour Systems.* Annales Academiae Scientiarum Fennicae Ser. A. I. Mathematica, Dissertationes 10. Department of Mathematics, University of Helsinki.

Swendsen, R.H. and Wang, J.-S. (1987). Nonuniversal critical dynamics in Monte Carlo simulations. *Phys. rev. Letters* **58**, 86–88.

Tierney, L. (1994). Markov chains for exploring posterior distributions. *Ann. Statist.* **22**, 1701–1728.

Widom, B. and Rowlinson, J.S. (1970). New model for the study of liquid-vapor phase transitions. *J. Chem. Physics* **52**, 1670–1684.

CHAPTER 5

Topics in Voronoi and Johnson–Mehl tessellations

J. Møller

Department of Mathematics and Computer Science, Aalborg University,
Fredrick Bajers Vej 7E, DK-9220 Aalborg, Denmark

5.1 Introduction

The study of random tessellations is one of the major subjects in stochastic geometry, see e.g. Stoyan, Kendall and Mecke (1995). A tessellation is roughly speaking a subdivision of the space into sets called cells, crystal, tiles etc. depending on the particular application. By 'space' is usually meant the d-dimensional Euclidean space $\mathbb{R}^d$, $d \geq 2$, and it is often assumed that the cells are bounded convex d-dimensional polytopes. Typically, the random mechanism is given by some stochastic process of simple geometrical objects which generate the tessellation in accordance to some rules.

Voronoi tessellations generated by point processes exemplify one such construction: Consider a point process $\Phi = \{x_i\} \subset \mathbb{R}^d$ of points called nuclei. Each nucleus x_i generates a convex Voronoi cell $C_i = C(x_i \mid \Phi)$ of all points in $\mathbb{R}^d$ which have x_i as nearest nucleus, i.e.

$$C(x_i \mid \Phi) = \{y \in \mathbb{R}^d \mid \forall x_j \in \Phi : \|x_i - y\| \leq \|x_j - y\|\}$$

where $\|\cdot\|$ denotes Euclidean distance. If Φ satifies certain consistency conditions (see Section 2) then $\{C_i\}$ is a Voronoi tessellation of $\mathbb{R}^d$. A part of a planar Voronoi tessellation generated by a binomial process is shown in Figure 5.1. Okabe *et al.* (1992) contains a comprehensive collection of results for Voronoi tessellations and related constructions but often without mathematical details and proofs; Møller (1994) provides a systematic study.

Alternative probabilistic models with non-convex cells are more scarce and complex. Johnson–Mehl tessellations form the perhaps most tractable and flexible class of such models: A Johnson–Mehl tessella-

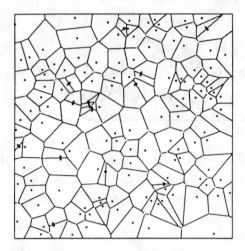

Figure 5.1 *Planar Voronoi tessellation generated by a binomial process of 150 independent and uniformly distributed points on the unit square. (Reprinted from Møller (1994) with permission of Springer-Verlag)*

tion results from a spatial birth process $\Phi = \{a_i\}$ of arrivals $a_i = (x_i, t_i) \in \mathbb{R}^d \times [0, \infty)$, where the nucleus x_i starts to grow at time t_i with a constant speed $v > 0$ in all directions so that we may imagine that a point $y \in \mathbb{R}^d$ is reached by x_i at time

$$T_i(y) = T(y, a_i) = t_i + \|x_i - y\|/v$$

If the set

$$C(a_i \mid \Phi) = \{y \in \mathbb{R}^d \mid \forall a_j \in \Phi : T_i(y) \le T_j(y)\} \qquad (5.1)$$

of points in $\mathbb{R}^d$ first reached by a_i is non-empty, then $C(a_i \mid \Phi)$ is said to be the crystal associated with a_i. In Section 2 we show that under certain consistency conditions these crystals constitute a random tessellation, the Johnson–Mehl tessellation generated by Φ. Illuminating figures of this construction can be found in Mahin *et al.* (1980) and Frost and Thompson (1987). Figure 5.2 shows a part of a planar Johnson–Mehl tessellation. Examples of 3-dimensional crystals are shown in Figures 5.3 and 5.4. Voronoi tessellations appear as the special case with all crystals being convex, i.e. when all birth times t_i are equal.

Notice that (5.1) agrees with the 'physical definition' of crystals where the growth of a crystal along a ray from its nucleus stops when it meets another growing crystal. The assumption of spherical growth

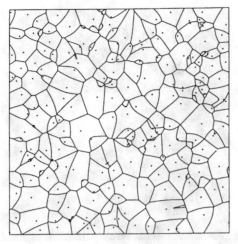

Figure 5.2 *Johnson–Mehl tessellation generated by 300 nuclei within the unit square. To each nucleus is attached an exponentially distributed waiting time for the start of growth (mean waiting time = 100). These waiting times are mutually independent and independent of the positions of the nuclei which are generated by a binomial process. Only those nuclei which generate non-empty crystals are shown (total = 222). (Reprinted from Møller (1994) with permission of Springer-Verlag)*

at a constant speed is essential at least in the non-convex case: Generalizations with non-spherical growth or a time dependent velocity seem not to be mathematically tractable when the physical definition is used. If the 'mathematical definition' (5.1) is used instead then rather strange phenomena are allowed to occur, cf. Evans (1945), Gilbert (1962) and Miles (1972): In the non-spherical case like e.g. when crystals grow as boxes, crystals given by (5.1) may be disconnected; and if e.g. the growth rate decreases as a function of the distance to the nucleus, then nuclei may lie outside the crystals (5.1) which they generate. Similar problems occur if nuclei grow with constant but different velocities.

The geometric structure of Johnson–Mehl tessellations is more complicated than for Voronoi tessellations. In the planar case the edges of Johnson–Mehl crystals are hyperbolic arcs and three neighbouring crystals may share one or two vertices, while the edges of Voronoi cells are line segments and three neighbouring cells meet at one vertex only, cf. Figures 5.1 and 5.2. In the spatial case and in higher dimensions Johnson–Mehl crystals violate topological relations which are satisfied

Figure 5.3 *Example of a spatial Johnson–Mehl crystal. The crystal consists of two faces and one edge, which is a closed curve; it has no vertices. (Reprinted from Møller (1995) with permission of the Applied Probability Trust)*

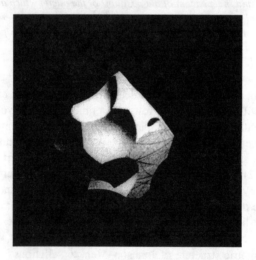

Figure 5.4 *Example of a spatial Johnson–Mehl crystal. (Reprinted from Møller (1995) with permission of the Applied Probability Trust)*

by Voronoi polytopes; e.g. the 'lense' shown in Figure 5.3 has two faces, one edge, and no vertices.

Johnson–Mehl tessellations have been introduced as (simplified) models for crystal growth as discussed in Kolmogoroff (1937), Johnson and Mehl (1939), Avrami (1939, 1940, 1941), and Evans (1945). Horálek (1990) argues that the so-called ASTM model, which is frequently used in metallurgy, can be described by a Johnson–Mehl tessellation.

The case $d = 1$ has been studied in detail by Meijering (1953), but the results do not lead to much insight for Johnson–Mehl tessellations in higher dimensions, cf. Gilbert (1962). Most authors consider birth processes Φ which are Poissonian with a constant intensity, i.e. the intensity measure is both time-homogeneous and invariant under translations in $\mathbb{R}^d$ (stationarity), but the time-inhomogeneous case may be included as well as demonstrated in Miles (1972), Horálek (1988, 1990), and Møller (1992, 1995).

Frost and Thompson (1987) study various models of planar John-son-Mehl tessellations by simulation. Mahin *et al.* (1980) present further simulated results for planar sections through spatial Poisson-Voronoi tessellations and spatial Johnson–Mehl tessellations generated by a time-homogeneous Poisson birth process. Møller (1994, 1995) shows how typical Poisson-Voronoi cells and Johnson–Mehl crystals in the case of time-inhomogeneous Poisson birth processes can be simulated. Heinrich and Schüle (1995) consider simulation procedures for some non-Poissonian cases.

In the following we shall review the current mathematical knowledge of Johnson–Mehl tessellations. The geometric structure of Voronoi and Johnson–Mehl tessellations is studied in Section 2. Stationary Voronoi and Johnson–Mehl tessellations are considered in Section 3. Poisson models are studied in Sections 4 and 5.

5.2 Geometric structure

Below we review some geometrical results established in Møller (1992, 1994).

Let $\Phi = \{a_i\} = \{(x_i, t_i)\}$ denote a simple point process on $\mathcal{A} = \mathbb{R}^d \times [0, \infty)$ which is almost surely non-empty and locally finite. Though most results in this section hold only with probability 1, this is for ease of representation not explicitly stated in the following.

We shall sometimes assume that Φ almost surely satisfies one or more of the following consistency conditions.

(C1) For any $(x, t) \in \Phi$ and any unit vector $u \in \mathbb{R}^d$ there exists

$(y, s) \in \Phi$ with $(y - x) \bullet u > v(s - t)$ where $\bullet$ is the usual inner product in $\mathbb{R}^d$.

(C2) No nucleus x_i is born at the time t_i at which another growing nucleus reaches x_i, i.e. $T_j(x_i) \neq t_i$ for all $i \neq j$.

(C3) No $m + 1$ nuclei lie on an $(m - 1)$-dimensional affine subspace of $\mathbb{R}^d$, $m = 1, \ldots, d$.

(C4) No $d + 2$ arrivals reach any point in $\mathbb{R}^d$ at the same instant.

Consider first the Voronoi case (all t_i equal). Then (C2) is clearly satisfied since the point process of nuclei is assumed to be simple. Further, (C1) states that any halfspace containing a nucleus in its topological boundary contains another nucleus in its topological interior. The conditions (C3)–(C4) mean that the nuclei are in 'general quadratic position' as (C4) reduces to that no $d + 2$ nuclei lie on the boundary of a sphere.

Now, consider the general case where the arrival times are not necessarily equal. Clearly, the crystal $C_i = C(a_i \mid \Phi)$ defined by (1.1) is closed and star-shaped with respect to its nucleus x_i. Moreover, (C1) ensures that C_i is bounded. Since Φ is a locally bounded subset of $\mathcal{A}$, the crystals are locally finite in the sense that

$$\forall B \subset \mathbb{R}^d \text{ bounded} : \#\{i \mid C_i \cap B \neq \emptyset\} < \infty. \tag{5.2}$$

In particular, (C1) implies that C_i has only a finite number of neighbouring crystals.

The crystals are also space-filling, i.e. $\mathbb{R}^d = \cup_i C_i$. This follows from the fact that Φ is locally finite and non-empty, so for an arbitrary point $y \in \mathbb{R}^d$ and $a_i \in \Phi$ there are only finitely many j with $T_j(y) \leq T_i(y)$, and hence $y \in C_j$ for the j which minimizes $T_j(y)$. Further, it is easily seen that the crystals have disjoint topological interiors.

Moreover, (C2) implies that the crystals are topological regular in the sense that $C_i = \text{cl} (\text{int } C_i)$, where int and cl denote topological interior and closure, respectively, cf. Proposition A.1 in Møller (1992).

Thus (C1)–(C2) ensure that the crystals constitute a tessellation of $\mathbb{R}^d$ where the crystals are star-shaped, bounded, locally finite, and topological regular. Recall that by definition crystals are the non-empty sets C_i given by (1.1). In the Voronoi case any C_i is a convex polytope of dimension d (Møller (1994), Proposition 2.1.4).

Next, consider a non-empty intersection $F_n = \cap_{i=0}^m C_i$ with $n = d - m$ and $m \geq 0$. This is called an n-facet. For $n = d$, F_d is just a crystal. Assuming (C4) we have $n \geq 0$ since $F_n = \emptyset$ otherwise. So let

$0 \leq n \leq d$ and define

$$G_n = G(a_0, \ldots, a_m) = \{y \in \mathbb{R}^d \mid T(y, a_0) = \cdots = T(y, a_m)\} \quad (5.3)$$

which is called a 'mathematical' n-facet. Note that F_n is of the form

$$F_n = F(a_0, \ldots, a_m \mid \Phi)$$
$$= \{y \in G_n \mid \forall a \in \Phi \setminus \{a_0, \ldots, a_m\} : T(y, a_0) \leq T(y, a)\}.$$
$$(5.4)$$

Especially, $G_n = F_n$ if $\Phi = \{a_0, \ldots, a_m\}$. From Lemma A.2 in Møller (1992) (or the lemma in the Appendix) it follows that G_n possesses the following properties if $G_n \neq \emptyset$ and (C2)–(C3) hold:

(i) For $n = 0$, G_0 consists of either one or two points.

(ii) For $0 < n < d - 1$, G_n is obtained by rotating a smooth curve in a $(n+1)$-dimensional affine subspace. The curve is either unbounded or closed.

(iii) For $n = d - 1$, G_{d-1} is a hyperplane if $t_0 = t_1$, whereas G_{d-1} is obtained by rotating one branch of a hyperbola about its axis of symmetry if $t_0 \neq t_1$.

In the Voronoi case, G_n is the n-dimensional affine subspace obtained by the intersection of the bisecting hyperplanes of the nuclei x_i and x_j, $0 \leq i < j \leq m$, cf. Lemma 2.1.1 in Møller (1994). Especially, G_0 becomes a vertex of the Voronoi tessellation.

In the general case, G_n with $n > 0$ may in extreme situations be degenerated to a single point (Møller (1992), Lemma A.2), but only if $n < d - 2$ (Møller (1992), Lemma A.3). Thus, in general the Hausdorff dimension of G_n is n; in particular, this is always the case when $d = 2$ or $d = 3$.

Now, suppose (C1)–(C4) hold and let $0 \leq n \leq d$. Suppose also that G_n is not degenerated to a single point if $0 < n < d - 2$. Then F_n is compact and n-dimensional, since $F_n = \mathrm{cl}(\mathrm{rel}\ \mathrm{int}\ F_n)$, where rel int denotes relative interior with respect to the induced topology on G_n, cf. Proposition A.4 in Møller (1992). Moreover, the boundary $\partial F_n = F_n \setminus \mathrm{rel}\ \mathrm{int}\ F_n$ is the union of all $(n - 1)$-facets contained in F_n for $0 < n \leq d$. The facets have also disjoint relative interiors. The connected components of F_n are called n-interfaces. Typically, a 0-interface is a vertex of $d + 1$ crystals, a 1-interface is an edge of d crystals, ... , and a $(d - 1)$-interface is a face of two crystals. In the Voronoi case the situation is again simpler as F_n is a n-dimensional convex polytope, cf. Proposition 2.1.4 in Møller (1994). In the general case

the collection of all interfaces of the Johnson–Mehl tessellation constitutes a d-dimensional PR-cell complex in the sense of Zähle (1988).
However, it should be noted that most of the results in Weiss and Zähle
(1988), Zähle (1988) and Leistritz and Zähle (1992) for 'general tesselations' do not immediately apply on Johnson–Mehl tessellations for
$d > 2$ and which are not of the Voronoi type because certain conditions in these papers are violated: n-interfaces are not always simply
connected when $1 < n < d$ and Euler's relation for polytopes does
not generalize to the interfaces contained in a crystal when $d > 2$.
For example, let $n = 2, d = 3$, and consider as in Figure 5.3 the case
where a crystal C_0 is surrounded by exactly two other crystals C_1 and
C_2. Then C_2 contains two faces and one edge but no vertices, which
clearly contradicts Euler's relation: # faces $-$ # edges $+$ # vertices $= 2$.
Notice also that the edge is a closed curve and the 2-interface $C_1 \cap C_2$
is not simply connected, since C_0 causes a 'gap' in the relative interior
of $C_1 \cap C_2$. Such gaps represent the possibility of forming lenses.

Finally, consider a non-empty intersection between F_n and a fixed
k-dimensional affine subspace L_k of $\mathbb{R}^d$ with $0 < k \leq d$. Conditions
must be imposed such that the Hausdorff dimension of $F_n \cap L_k$ is
$k - m$ for $n \leq d - k$. For instance, this is the case under the Poisson
model introduced in Section 4. Hence the connected components of the
sectional crystals constitute the cells of a tessellation of L_k. Therefore,
$F_n \cap L_k$ is called a $(k-m)$-facet in L_k and its connected components are
called $(k - m)$-interfaces in L_k. Note that $F_n \cap L_k$ may be disconnected
even though F_n is connected, and $F_n \cap L_k$ does not necessarily contain
p-interfaces in L_k for $p < k - m$. For example, the latter is not the
case for any $p < k - m$ if $L_k \cap F_n = L_k \cap$ rel int F_n. Notice also
that 'caps' appear whenever a k-facet in L_k is surrounded by another
k-facet in L_k; e.g. for $k = 2$ the boundary of such a cap consists only
of one edge which is a closed curve. The intersections between L_k and
mathematical facets are described in the Appendix.

In conclusion, these and other complications imply that the general
theory of random tessellations with non-convex cells is of limited use
for Johnson–Mehl tessellations (unless they are of the Voronoi type):
instead we shall utilize the properties discussed above.

5.3 Stationary Voronoi and Johnson–Mehl tessellations

In this section we define and study the concept of typical interfaces of
the intersection between a Voronoi or Johnson–Mehl tessellation and a
fixed k-dimensional affine subspace L_k of $\mathbb{R}^d$ with $0 < k \leq d$. We also

review some mean value relations of various characteristics for such interfaces. The definitions and results are taken from Møller (1989, 1992, 1994).

Henceforth we suppose for ease of presentation that the birth process $\Phi = \{a_i\} = \{(x_i, t_i)\}$ satisfies the following conditions which ensure that the consistency conditions (C1)–(C4) hold almost surely, and certain 'degenerate' cases of the geometric structure of facets and interfaces can be eliminated: We assume that Φ is almost surely non-empty, its distribution is invariant under translations in $\mathbb{R}^d$, and Φ restricted to any bounded Borel set $W \subset \mathbb{R}^d \times [0, \infty)$ is absolutely continuous with respect to a Poisson process as considered in Section 4.

Let $\Phi_p^{(k)}$ denote the set of p-interfaces in L_k, $p = 0, \ldots, k$. In order to define what is meant by the typical p-interface we first have to define the concept of a centroid of a p-interface $X \in \Phi_p^{(k)}$. Let $m = k - p$ and $n = d - m$ for $0 \le p \le k$. By definition of interfaces, X is a connected component of $F_n \cap L_k$ for some n-facet $F_n = F(a_0, \ldots, a_m \mid \Phi)$, cf. (5.4). Suppose we have chosen a point

$$c(X, a_0, \ldots, a_m \mid \Phi) \in L_k \tag{5.5}$$

called the centroid of X, which is equivariant under translations in L_k, i.e.

$$c(X + y, (x_m + y, t_0), \ldots, (x_m + y, t_m) \mid \{(x_i + y, t_i)\})$$
$$= c(X, (x_0, t_0), \ldots, (x_m, t_m) \mid \{(x_i, t_i)\}) + y \tag{5.6}$$

for all $y \in L_k$ (here $X + y = \{x + y \mid x \in X\}$). As in Møller (1989), Section 4, and Møller (1994), Section 3.2, it can be verified that the definitions and results presented in the present paper do not depend on the choice of the centroid (5.5) since (5.6) holds.

One possible choice of the centroid (5.5), which is convenient for purposes like sampling of interfaces, is the most 'extreme' point of either the interface X or the facet $F(a_0, \ldots, a_m \mid \Phi) \cap L_k$ in some fixed direction. Another possibility is first to define the centroids for $k = d$, and secondly, when $k < d$, to let $c(X, a_0, \ldots, a_m \mid \Phi)$ be equal to the orthogonal projection onto L_k of the centroid of the interface in $\mathbb{R}^d$ which contains X. This construction may be useful for proof-technical reasons when one wants to establish stereological relations between geometrical characteristics of interfaces in $\mathbb{R}^d$ and L_k, see e.g. Proposition 6.2 and Theorem 6.3 in Møller (1989). For example, when $k = d$ we can take $c(X, a_0, \ldots, a_m \mid \Phi) = x_0$ or $\frac{1}{m+1} \sum_0^m x_i$, the center of gravity of the nuclei $x_0, \ldots, x_m$. Then especially for $k = d = n$, $c(X, a_0 \mid \Phi)$ is the nucleus of the crystal $X = C(a_0 \mid \Phi)$.

Now, for $p = 0, \ldots, k$ and $m = k - p$ define the intensity of p-interfaces in L_k by

$$
I_p^{(k)} = \frac{1}{(m+1)!\lambda_k(B)}
$$

$$
\times E \sum_{a_0, \ldots, a_m \in \Phi}^{\neq} \sum_{X \in \Phi_p^{(k)}: X \subseteq F_n \cap L_k} 1[c(X, a_0, \ldots, a_m \mid \Phi) \in B]
$$

(5.7)

where $\neq$ in the first sum means that $a_0, \ldots, a_m$ are pairwise distinct, $B \subset L_k$ is an arbitrary Borel set with $0 < \lambda_k(B) < \infty$, and λ_k denotes Lebesgue measure in L_k (the right-hand side of (5.7) does not depend on B because of (5.6) and since Φ is stationary under translations in L_k). Note that the second sum in (5.7) is 0 if $F_n = F(a_0, \ldots, a_m \mid \Phi)$ does not intersect L_k, so the summation in (5.7) is just over all p-interfaces in L_k. In other words, $I_p^{(k)}$ is the mean number of p-interfaces per unit volume in L_k. We shall sometimes write $\lambda = I_d^{(d)}$ for the intensity of crystals.

We are now in a position to define 'typical interfaces' by means of Palm distributions (see also Baddeley's contribution in this volume). Suppose $0 < I_p^{(k)} < \infty$ for $0 \le p \le k \le d$. Let the space of closed subsets of $\mathbb{R}^d$ be equipped with a suitable σ-field $\mathcal{F}$ (by 'suitable' we mean of course that all mappings considered in the following should be measureable; see e.g. Matheron (1975)). Then the Palm distribution of the typical p-interface in L_k is the probability measure on $\mathcal{F}$ defined by

$$
P_p^{(k)}(A) = \frac{1}{(m+1)!I_p^{(k)}\lambda_k(B)}
$$

$$
\times E \sum_{a_0, \ldots, a_m \in \Phi}^{\neq} \sum_{X \in \Phi_p^{(k)}: X \subseteq F_n \cap L_k} 1[c(X, a_0, \ldots, a_m \mid \Phi) \in B,
$$

$$
X - c(X, a_0, \ldots, a_m \mid \Phi) \in A]
$$

(5.8)

with m, F_n and B as in (5.7) (again this definition depends not on B). The typical p-interface in L_k is defined as a random closed set $C_p^{(k)}$ having the distribution $P_p^{(k)}$.

In particular we let $C = C_d^{(d)}$ denote the typical crystal. The distribution of C is related to the distribution of the almost surely well-defined crystal $\mathcal{C}$ of the Johnson–Mehl tesselation which contains an arbitrary fixed point in $\mathbb{R}^d$:

$$P(\mathcal{C} \in F) = \lambda E(1[C \in F]\lambda_d(C))$$

for events F which are invariant under translations in $\mathbb{R}^d$. This relation implies that $\mathcal{C}$ is larger than C in the sense that $E\lambda_d(\mathcal{C}) \geq E\lambda_d(C)$.

By standard measure theoretical techniques we get immediately from (5.8) a so-called Campbell theorem:

$$(m + 1)! I_p^{(k)} \int_{L_k} Ef(C_p^{(k)}, y) \lambda_k(dy)$$

$$= E \sum_{a_0,\ldots,a_m \in \Phi}^{\neq} \sum_{X \in \Phi_p^{(k)} : X \subseteq F_n \cap L_k} f(X - c(X, a_0, \ldots, a_m \mid \Phi),$$

$$c(X, a_0, \ldots, a_m \mid \Phi)) \tag{5.9}$$

for non-negative functions f which are measurable with respect to the product σ-field of $\mathcal{F}$ and the Borel σ-field on L_k (the mean values may eventually be infinite).

We shall only consider statements which involve quantities like $I_p^{(k)}$ and $Ef(C_p^{(k)})$ where $f \geq 0$ is measurable and translation-invariant in the sense that $f(X + y) = f(X)$ for all $X \in \Phi_p^{(k)}$ and all $y \in L_k$. Then (5.9) gives that

$$(m + 1)! I_p^{(k)} Ef(C_p^{(k)}) \lambda_k(B)$$

$$= E \sum_{a_0,\ldots,a_m \in \Phi}^{\neq} \sum_{X \in \Phi_p^{(k)} : X \subseteq F_n \cap L_k} f(X) 1[c(X, a_0, \ldots, a_m \mid \Phi) \in B]$$

$$\tag{5.10}$$

and as mentioned it can be shown that $I_p^{(k)}$ and $Ef(C_p^{(k)})$ do not depend on the choice of the centroid.

We turn now to some mean value relations for the first order characteristics of interfaces. For $0 \leq q \leq p \leq k$, let

$$N_{pq}^{(k)} = E\#\{ q\text{-interfaces contained in } C_p^{(k)}\}$$

and

$$V_{pq}^{(k)} = E \sum_{X \subseteq C_p^{(k)} \; q\text{-interface}} \lambda_q(X)$$

where λ_q denotes q-dimensional Hausdorff measure. Further, consider the mean p-content of p-interfaces in L_k intersected with a Borel set $B \subseteq L_k$, i.e.

$$\mu_{kp}(B) = E \sum_{a_0,\ldots,a_m \in \Phi}^{\neq} \lambda_p(B \cap F(a_0,\ldots,a_m \mid \Phi))/(m+1)!$$

where as before $m = k - p$. Since $\mu_{kp}(\cdot)$ is a translation-invariant measure on L_k,

$$\mu_{kp}(B) = \mu_{kp}\lambda_k(B)$$

where the constant $\mu_{kp} \geq 0$ is the mean p-content of p-interfaces per unit volume in L_k. We call μ_{kp} the density of p-interfaces in L_k. Then by standard techniques (see e.g. Møller (1989), Section 5),

$$\mu_{kp} = E \sum_{a_0,\ldots,a_m \in \Phi}^{\neq} \lambda_p(B \cap F(a_0,\ldots,a_m \mid \Phi))/(m+1)!$$

$$= E \sum_{a_0,\ldots,a_m \in \Phi}^{\neq} 1[c(X,a_0,\ldots,a_m \mid \Phi) \in B]$$

$$\lambda_p(L_k \cap G(a_0,\ldots,a_m))/(m+1)! \quad (5.11)$$

with $\lambda_k(B) = 1$,

$$I_p^{(k)} N_{pq}^{(k)} = \binom{k-q+1}{p-q} I_q^{(k)} \quad (5.12)$$

and

$$I_p^{(k)} V_{pq}^{(k)} = \binom{k-q+1}{p-q} \mu_{kq}. \quad (5.13)$$

The following two theorems from Møller (1992) show how $I_p^{(k)}$, $N_{pq}^{(k)}$, and $V_{pq}^{(k)}$ are related to the intensity λ of crystals and the densities of interfaces in L_k; the latter quantities are ususally those which are estimated (in a straightforward manner) in statistical applications where the tessellation is observed within a bounded window $B \subset L_k$ only.

Theorem 5.1 *For all $d \geq k \geq 1$ and $0 \leq p \leq k$,*

$$I_d^{(d)} = \lambda, \quad N_{d0}^{(d)} = \frac{d+1}{\lambda}\mu_{d0}, \quad V_{dp}^{(d)} = \frac{d-p+1}{\lambda}\mu_{dp} \quad (5.14)$$

$$I_0^{(k)} = \mu_{k0}, \quad I_1^{(k)} \geq \frac{k+1}{2}\mu_{k0}, \quad N_{21}^{(k)} \geq N_{20}^{(k)}, \quad N_{10}^{(k)} \leq 2,$$

$$V_{11}^{(k)} \le \frac{2}{k+1} \frac{\mu_{k1}}{\mu_{k0}} \qquad (5.15)$$

$$I_2^{(2)} \ge \frac{1}{2}\mu_{20}, \quad N_{21}^{(2)} \le 6, \quad V_{21}^{(2)} \le \frac{4\mu_{21}}{\mu_{20}}, \quad V_{22}^{(2)} \le \frac{2}{\mu_{20}} \qquad (5.16)$$

$$I_1^{(1)} = I_0^{(1)} = \mu_{10}, \quad N_{10}^{(1)} = 2, \quad V_{11}^{(1)} = 1/\mu_{10}. \qquad (5.17)$$

The identity in any of these inequalities holds only in the Voronoi tessellation case or if $d \le 2$. Especially,

$$V_{dd}^{(d)} = E\lambda_d(C) = 1/\lambda = E\lambda_d(C)^{-1} \qquad (5.18)$$

Theorem 5.2 *For $0 < k \le d \le 3$,*

$$\sum_{i=0}^{k} (-1)^i I_i^{(k)} = 0. \qquad (5.19)$$

For $d = 3$,

$$I_3^{(3)} = \lambda, \quad I_2^{(3)} \ge \lambda + \mu_{30}, \quad I_1^{(3)} \ge 2\mu_{30}, \quad I_0^{(3)} = \mu_{30} \qquad (5.20)$$

$$I_2^{(2)} \ge \frac{1}{2}\mu_{20}, \quad I_1^{(2)} \ge \frac{3}{2}\mu_{20}, \quad I_0^{(2)} = \mu_{20}, \quad I_1^{(1)} = I_0^{(1)} = \mu_{10} \qquad (5.21)$$

$$N_{32}^{(3)} \ge 2 + 2\mu_{30}/\lambda, \quad N_{31}^{(3)} \ge 6\mu_{30}/\lambda, \quad N_{30}^{(3)} = 4\mu_{30}/\lambda \qquad (5.22)$$

$$N_{21}^{(3)} \ge N_{20}^{(3)}, \quad N_{20}^{(3)} \le \frac{6\mu_{30}}{\lambda + \mu_{30}}, \quad N_{10}^{(3)} \le 2 \qquad (5.23)$$

$$N_{20}^{(2)} \le N_{21}^{(2)} \le 6, \quad N_{10}^{(2)} \le 2, \quad N_{10}^{(1)} = 2 \qquad (5.24)$$

$$V_{33}^{(3)} = \lambda^{-1}, \quad V_{32}^{(3)} = 2\mu_{32}/\lambda, \quad V_{31}^{(3)} = 3\mu_{31}/\lambda \qquad (5.25)$$

$$V_{22}^{(3)} \le \frac{\mu_{32}}{\lambda + \mu_{30}}, \quad V_{21}^{(3)} \le \frac{3\mu_{31}}{\lambda + \mu_{30}}, \quad V_{11}^{(3)} \le \frac{\mu_{31}}{2\mu_{30}} \qquad (5.26)$$

$$V_{22}^{(2)} \le \frac{2}{\mu_{20}}, \quad V_{21}^{(2)} \le \frac{4\mu_{21}}{\mu_{20}}, \quad V_{11}^{(2)} \le \frac{2\mu_{21}}{3\mu_{20}}, \quad V_{11}^{(1)} = 1/\mu_{10} \qquad (5.27)$$

where equality in any of these inequalities holds only in the Voronoi tessellation case.

For $d = 2$,

$$I_2^{(2)} = \lambda, \quad I_1^{(2)} = 3\lambda, \quad I_0^{(2)} = 2\lambda, \quad I_1^{(1)} = I_0^{(1)} = \mu_{10} \qquad (5.28)$$

$$N_{21}^{(2)} = N_{20}^{(2)} = 6, \quad N_{10}^{(2)} = N_{10}^{(1)} = 2 \qquad (5.29)$$

$$V_{22}^{(2)} = \lambda^{-1}, \ V_{21}^{(2)} = 2\mu_{21}/\lambda, \ V_{11}^{(2)} = \mu_{21}/3\lambda, \ V_{11}^{(1)} = 1/\mu_{10}. \ (5.30)$$

The above results show that the actual distribution of Φ plays a more important role in the spatial case $d = 3$ than in the planar case $d = 2$. In the next sections we shall determine the intensity λ and the densities μ_{kp} when Φ is a Poisson process.

5.4 Poisson models

In the literature almost all probabilistic models of Johnson–Mehl tessellations assume the birth process Φ to be a Poisson process with an intensity measure which is invariant under translations in $\mathbb{R}^d$; non-Poissonian models seem in general to be analytically intractable. Most authors assume the Poisson process to be time-homogeneous too, but as remarked by Miles (1972) there is really no reason to do this, since the actual state of development is always inhomogeneous in time. In the rest of this paper Φ is also assumed to be a translation-invariant Poisson process, i.e. the intensity measure is of the form $M = \lambda_d \times \Lambda$, where as in the previous section λ_d is the Lebesgue measure in $\mathbb{R}^d$ and Λ is a locally finite measure on $[0, \infty)$ which satisfies the regularity conditions (5.33) and (5.35) below.

The usual definition of $\Phi = \{a_i\} = \{(x_i, t_i)\}$ being a Poisson process on $\mathcal{A} = \mathbb{R}^d \times [0, \infty)$ with intensity measure M is as follows: If $A \subset \mathcal{A}$ is a bounded Borel set with $M(A) < \infty$, then $\Phi(A)$ is Poisson distributed with mean $M(A)$, and for disjoint Borel sets $A_1, \dots, A_k \subset \mathcal{A}$ we have that $\Phi(A_1), \dots, \Phi(A_k)$ are mutually independent. Since $M = \lambda_d \times \Lambda$ we see that the distribution of Φ is actually invariant under motions in $\mathbb{R}^d$.

The distribution of Φ is in fact uniquely characterized by the following property, which can be considered as a form of Slivnyak's theorem (together with refined Campbell formula), see Slivnyak (1962) and Satz 3.1 in Mecke (1967):

$$E \sum_i f(\Phi \setminus \{(x_i, t_i)\}, (x_i, t_i)) = \int \int E f(\Phi, (x, t)) dx \Lambda(dt) \ (5.31)$$

for any measurable function $f \geq 0$ and where $dx = \lambda_d(dx)$. Using induction we obtain the following generalization of (5.31):

$$E \sum_{a_0,\ldots,a_m \in \Phi}^{\neq} f(\Phi \setminus \{a_0, \ldots, a_m\}, a_0, \ldots, a_m)$$

$$= \int \cdots \int \int \cdots \int Ef(\Phi, (x_0, t_0), \ldots, (x_m, t_m)) dx_0 \ldots dx_m$$
$$\Lambda(dt_0) \ldots \Lambda(dt_m) \qquad (5.32)$$

for non-negative measurable functions f. This formula has proven to be extremely useful for a number of applications in stochastic geometry, cf. Møller and Zuyev (1994) and the references therein. As shown in the following it is essentially (5.32) which makes Poissonian models for Johnson–Mehl tesselations so tractable.

We impose the following conditions on Λ: The condition that Φ is non-empty with probability one is just equivalent to that

$$\Lambda([0, \infty)) > 0. \qquad (5.33)$$

Further, from (5.7) and (5.14) we find that the intensity of crystals is given by

$$\lambda = I_d^{(d)} = E \sum_{(x_i, t_i) \in \Phi} 1[x_i \in [0, 1]^d, C(a_i \mid \Phi) \neq \emptyset]$$

$$= \int \int 1[x \in [0, 1]^d] P(C((x, t) \mid \Phi \cup \{(x, t)\}) \neq \emptyset] dx \Lambda(dt).$$

The translation-invariance of Φ implies that

$$P(C((x, t) \mid \Phi \cup \{(x, t)\}) \neq \emptyset) = \int p(t) \Lambda(dt)$$

where

$$p(t) = \exp\{-v^d \omega_d \int_0^t (t - s)^d \Lambda(ds)\} \qquad (5.34)$$

is the probability that an arbitrary fixed point in $\mathbb{R}^d$ is not reached by any growing nuclei at time t, and $\omega_d = \pi^{d/2}/\Gamma(\frac{1}{2}d + 1)$ is the volume of the unit ball in $\mathbb{R}^d$. Thus we require that

$$\lambda = \int p(t) \Lambda(dt) < \infty. \qquad (5.35)$$

From (5.10) and (5.32) with $m = 1, k = p = n = d$, and $c(X, a_0 \mid \Phi) = x_0$, we get immediately that the distribution of the typical crystals

$\mathcal{C}$ is given by

$$Eh(\mathcal{C}) = \int_{C((0,t)|\Phi\cup\{(0,t)\})\neq\emptyset} [Eh(C((0,t) \mid \Phi \cup \{(0,t)\})/\lambda]\Lambda(dt)$$

$$= \int Eh(C((0,t) \mid \Phi_t \cup \{(0,t)\}))\frac{p(t)}{\lambda}\Lambda(dt) \qquad (5.36)$$

where $\Phi_t = \{a \in \emptyset \mid T(0,a) > t\}$ so that $C((0,t) \mid \Phi_t \cup \{(0,t)\}) \neq \emptyset$. Hence, we can interpret $p(t)\Lambda(dt)/\lambda$ as the density of the birth time τ of $\mathcal{C}$, and $\mathcal{C}$ is stochastically equivalent to $C((0,\tau) \mid \Phi_\tau \cup \{(0,\tau)\})$. Notice also that if F is an event which is invariant under translations in $\mathbb{R}^d$, then $\mathcal{C} \in F$ almost surely if and only if all non-empty $C_i \in F$ almost surely. The latter identity in (5.36) is used in Møller (1995) for the simulation of typical crystals; see also Section 5.

The following proposition, taken from Møller (1992), ensures that all the desirable properties for crystals and facets discussed in Section 2 hold almost surely. Since these properties are translation invariant, they hold also with probability 1 for $\mathcal{C}$ and its facets and interfaces.

Proposition 5.1 *The consistency conditions (C1)–(C4) hold almost surely. Moreover, with probability 1, if $L_k \subseteq \mathbb{R}^d$ is a k-dimensional affine subspace with $0 < k \leq d$, any n-facet in L_k has Hausdorff dimension n for $n = 0, \dots, k$. Finally all crystals are almost surely convex if and only if Λ is concentrated at a singleton (the Voronoi tessellation case).*

As shown in the previous section numerous first-order characteristics are related to λ and the densities μ_{kp} of interfaces. Combining (5.11) and (5.32) we get in a similar way as the derivation of (5.35) above that for $0 < k \leq d, 0 \leq p \leq k, m = k - p, n = d - k + p$, and $\lambda_k(B) = 1$,

$$(m+1)!\mu_{kp}$$

$$= \int \dots \int \int \dots \int \int_{y\in B\cap G(a_0,\dots,a_m)} p(T(y,a_0))\lambda_p(dy)dx_0\dots dx_m$$

$$\Lambda(dt_0)\dots\Lambda(dt_m) \qquad (5.37)$$

with $a_i = (x_i, t_i), i = 0, \dots, m$. Because of the motion-invariance we can set $L_k = \mathbb{R}^k \times \{0^{d-k}\}$ and write vectors $x \in \mathbb{R}^d$ as $x = (x^k, x^{d-k})$ where $(x^k, 0^{d-k})$ is the orthogonal projection of x onto L_k (in the following we consider row vectors). Now, fix $t_0, \dots, t_m$ and make a change of variables $(x_0, \dots, x_m, y)$ with $y \in G(a_0, \dots, a_m) \cap L_k$ to $(t, y^k, q_0, u_0^k, u_0^{d-k}, \dots, q_m, u_m^k, u_m^{d-k})$ given by

$$x_j^k = y^k + vq_j^{1/2}s_ju_j^k, \quad x_j^{d-k} = v(1-q_j)^{1/2}s_ju_j^{d-k}, \quad j = 0, \dots, m$$

with $s_j = t - t_j$ and the following range:

$$t > \max_{0 \le i \le m} t_i, \ y^k \in \mathbb{R}^k, \ q_j \in (0, 1) \text{ if } k < d$$

$$q_j = 1 \text{ if } k = d, \ u_j^k \in S^{k-1}, u_j^{d-k} \in S^{d-k-1}$$

and so that (C2) is satisfied, i.e. $(u_i^k, u_i^{d-k}) \ne (u_j^k, u_j^{d-k})$ for all $i \ne j$; here $S^{-1} = \{0\}$ and S^{d-1} denotes the unitsphere in $\mathbb{R}^d$. Write this transformation as

$$(x_0, \ldots, x_m, y) = g(t, y^k, q_0, u_0^k, u_0^{d-k}, \ldots, q_m, u_m^k, u_m^{d-k})$$

and let

$$g' = \partial g / \partial(t, y^k, q_0, u_0^k, u_0^{d-k}, \ldots, q_m, u_m^k, u_m^{d-k})^*$$

where $*$ denotes transposition and differentiation with respect to u_j^k is more precisely differentation with respect to a $(k - 1)$-dimensional parametrization of unit vectors in $\mathbb{R}^k$ and similarly for u_j^{d-k} (if $k = d$ we omit all $q_j = 1$ and $u_j^0 = 0$ in the differentation). Then informally speaking

$$\lambda_p(dy)dx_0 \ldots dx_m = J_g dt dy^k \prod_{j=0}^m dq_j \lambda_{k-1}(du_j^k) \lambda_{d-k-1}(du_j^{d-k})$$

where the Jacobian

$$J_g = J_g(t, y, q_0, u_0^k, u_0^{d-k}, \ldots, q_m, u_m^k, u_m^{d-k}) = \sqrt{\det(g'g'^*)}$$

cf. e.g. Federer (1969) or Hoffmann-Jørgensen (1994). Observe that the rows/columns of the $k \times k$ matrix $[u_j^{k*} u_j^{'k*}]$ are orthonormal where $u_j^{'k} = \partial u_j^k / \partial u_j^{k*}$; similarly, for $k < d$, $[u_j^{d-k*}, u_j^{'d-k*}]$ has orthonormal rows/columns. Using this it is straightforwardly derived that J_g is of the form

$$J_g = v^{md+n} f(q_0, u_0^k, \ldots, q_m, u_m^k) \prod_{j=0}^m s_j^{d-1}$$

with

$$f(q_0, u_0^k, \ldots, q_m, u_m^k) =$$

$$(m + 1)^{1/2} 2^{-(m+1)} h(q_0, \ldots, q_m) \det(I_k + \sum_{j=0}^m (\omega_j - \overline{\omega})^* (\omega_j - \overline{\omega}))^{1/2}$$

where

$$h(q_0, \ldots, q_m) = \begin{cases} \prod_{j=0}^{m} q_j^{k-3/2}(1 - q_j)^{d-k-1} & \text{if } k < d \\ 1 & \text{if } k = d \end{cases}$$

and where

$$\omega_j = q_j u_j^k \text{ and } \overline{\omega} = \sum_0^m \omega_j / (m + 1).$$

Inserting all this into (5.37) gives

$$\mu_{kp} = v^{md+n} c_{kp} \int_0^\infty \{ \int_0^t (t - s)^{d-1} \Lambda(ds) \}^{m+1} p(t) dt$$

where c_{kp} is a constant which depends on the dimensions k, p, d, but not on the measure Λ. In the Voronoi tessellation case μ_{kp} can be calculated for every k, p, d using Møller (1989), Theorem 5.5, Theorem 6.3, and Theorem 7.2. Combining this with (5.38) we get an explicit expression of c_{kp} and hence the following theorem:

Theorem 5.3 *The densities μ_{kp} for $0 \leq p \leq k$ and $0 < k \leq d$ are given by*

$$\mu_{kp} = v^{md+n} c_{kp} \int_0^\infty \{ \int_0^t (t - s)^{d-1} \Lambda(ds) \}^{m+1} p(t) dt \qquad (5.38)$$

with

$$c_{kp} = \frac{2^{m+1} \pi^{(m+1)d/2} \Gamma\left(\frac{dm+n+1}{2}\right) \Gamma\left(\frac{k+1}{2}\right)}{(m + 1)! \Gamma\left(\frac{dm+n}{2}\right) \Gamma\left(\frac{d+1}{2}\right)^{m+1} \Gamma\left(\frac{p+1}{2}\right)}. \qquad (5.39)$$

To the best of my knowledge, this is the first general proof of (5.38)–(5.39) which so far has been published. Meijering (1953) derived the result in the time-homogeneous case $\Lambda(dt) = \alpha dt$ and with $k = d \leq 3$, treating each of the cases $p = 0, 1, 2$ separately. Miles (1972) stated the result for the inhomogeneous case when $k = d = 3$ but without including a proof. Møller (1992) verified (5.38)–(5.39) for each of the cases (i) $p = 0$ and $0 < k \leq d$, (ii) $p = 1$ and $0 < k \leq d$, (iii) $k = d$ and $p = 2$; this covers all cases of p and k when $d \leq 3$.

Notice that (5.38) can be written as

$$\mu_{kp} = \text{ constant } \int_0^\infty \text{haz}(t)^{m+1} p(t) dt \qquad (5.40)$$

where

$$\text{haz}(t) = -\frac{p'(t)}{p(t)} = v^d \sigma_d \int_0^t (t-s)^{d-1} \Lambda(ds)$$

is the 'hazard function' of the time at which an arbitrary point is crystallized and σ_d is the surface area of S^{d-1}. Formula (5.40) may be considered as a consequence of the strong independence properties of the Poisson process, but the perhaps surprising fact is that the constant in (5.40) depends only on the dimensions d, k, p and not on the measure Λ.

It should also be noted that if we in (5.38) replace ∞ by $T > 0$, then we get the density of those parts of p-interfaces in L_k which have been generated before time T.

Further first and higher-order moments can be derived for Voronoi and Johnson–Mehl tessellations generated by a Poisson process, but the expressions are often complicated and only suitable for numerical integration, see Møller (1989, 1992, 1994) and the references therein.

5.5 A special class of Poisson models

In this section we evaluate the expressions established in the previous section when Λ is of the particular form

$$\Lambda(dt) = \alpha t^{\beta-1} dt \tag{5.41}$$

with $\alpha > 0$ and $\beta > 0$. This seems to be the most tractable class of parametric models (Miles (1972), Horálek (1988, 1990), Møller (1992, 1995)). Further models are investigated in Møller (1992, 1995). Finally, we review how arrivals and typical crystals can be simulated under (5.41).

The distribution of the resulting Johnson–Mehl tessellation under (5.41) (as time tends to infinity) is parametrized by $\kappa = \alpha/v^\beta > 0$ and $\beta = 0$, and from (5.34) and (5.35) we find

$$p(t) = \exp\left(-\kappa v^{d+\beta} \omega_d B(d+1, \beta) t^{d+\beta}\right) \tag{5.42}$$

and

$$\lambda = \frac{\kappa^{d/(d+\beta)}}{\beta} \Gamma\left(1 + \frac{\beta}{d+\beta}\right) (\omega_d B(d+1, \beta))^{-\beta/(d+\beta)} \tag{5.43}$$

where $B(\cdot, \cdot)$ is the beta function. Thus another parametrization is provided by the scale parameter $\lambda^{-1/d} > 0$ and the shape parameter $\beta > 0$.

As verified in Møller (1992) the Johnson–Mehl tessellation tends to a Poisson-Voronoi tessellation with $\lambda = 1$ as $\kappa = \beta \to 0$. Letting

$$\eta = \kappa v^{d+\beta} \omega_d B(d + 1, \beta) = \alpha v^d \omega_d d! / \prod_{j=0}^{d} (\beta + j)$$

it follows from (5.42) that for the birth time τ of the typical crystal

$$\tau^{d+\beta} \sim \Gamma\left(\frac{\beta}{d + \beta}, \eta\right) \tag{5.44}$$

is Gamma-distributed, and so $\tau \to 0$ a.s. as $\kappa = \beta \to 0$.

Moreover, letting

$$\gamma_n^{(d)} = \frac{\Gamma\left(\frac{dm+n+1}{2}\right) \Gamma\left(\frac{d}{2}\right)^{m+1} (d! \pi^{d/2})^{m/d}}{m! \Gamma\left(\frac{n+1}{2}\right) \Gamma\left(\frac{dm+n}{2}\right) \Gamma\left(\frac{d+1}{2}\right)^{m} \Gamma\left(\frac{d}{2} + 1\right)^{m/d}}$$

and

$$g_n^{(d)}(\beta) = \frac{\Gamma\left(m - \frac{m}{d+\beta} + 1\right) (d + \beta)^m}{\left[\Gamma\left(1 + \frac{\beta}{d+\beta}\right) \prod_{i=1}^{d} (\beta + i)\right]^{m/d}} \gamma_n^{(d)}$$

formula (5.38) reduces to

$$\mu_{kp} = \frac{\Gamma\left(\frac{k+1}{2}\right) \Gamma\left(\frac{n+1}{2}\right)}{(m + 1) \Gamma\left(\frac{d+1}{2}\right) \Gamma\left(\frac{p+1}{2}\right)} g_n^{(d)}(\beta) \lambda^{m/d} \tag{5.45}$$

Hence $g_n^{(d)}(\beta)$ can be interpreted as the mean n-content of the n-interfaces contained in $\mathcal{C}$ when $E\lambda_d(\mathcal{C}) = 1$, cf. (5.18). Plots of the functions $g_n^{(d)}(\beta), 0 \le n < d \le 3$, are shown in Møller (1992). For every $n < d$ it turns out that $g_n^{(d)}(\beta)$ is a strictly decreasing function (unless $d = 2$ and $n = 0$, since $g_0^{(2)}(\cdot) = 6$ in accordance with (5.29)) with finite limits $g_n^{(d)}(\infty) = m! \gamma_n^{(d)}$ and

$$g_n^{(d)}(0) = \frac{\Gamma\left(m - \frac{m}{d} + 1\right) d^m}{(d!)^{m/d}} \gamma_n^{(d)}.$$

As expected, $g_n^{(d)}(0)$ is in agreement with the expression for the mean n-content of the n-interfaces contained in the typical cell of a Voronoi tessellation generated by a Poisson process with intensity 1. Inserting (5.43) and (5.45) into the formulae of Theorems 1 and 2, we find exact expressions or at least bounds of the characteristics $I_p^{(k)}$, $N_{pq}^{(k)}$, $V_{pq}^{(k)}$, as exhibited in Møller (1992).

Further first and second-order moments can be determined as shown in Møller (1992). For instance, consider the coefficient of variations $CV(\lambda_d(C))$ and $CV(\lambda_1(C^{(1)}))$ of the content of the typical crystal C and of the typical intercept $C^{(1)}$ obtained by sectioning the Johnson–Mehl tessellation with a line L_1 through 0. These are related to the content of the crystal $C = C^{(d)}$ of the Johnson–Mehl tessellation and the intercept $C^{(1)}$ in L_1 which both contain 0:

$$CV(\lambda_d(C)) = \{\lambda E\lambda_d(C^{(d)}) - 1\}^{1/2}$$

$$CV(\lambda_1(C^{(1)})) = \{\mu_{10} E\lambda_1(C^{(1)}) - 1\}^{1/2}.$$

Here $E\lambda_k(C^{(k)})$ can be expressed by a certain multivariate integral, which for $d = 3$ (which turns out to be the simplest case for $d \geq 2$) becomes a twofold integral,

$$E\lambda_k(C^{(k)}) = c_\beta(k)\lambda^{-k/3}$$

$$\times \int_{x>1} \int_{|y|<1} \frac{1}{2} W_\beta \left(\frac{x-y}{2}, \frac{x+y}{2}\right)^{-\frac{k}{3+\beta}-1}$$

$$\times \left[\frac{1}{\beta}\frac{x+y}{2}\frac{x-y}{2}\left(\frac{x-1}{2}\right)^{\beta} - \frac{x}{\beta+1}\left(\frac{x-1}{2}\right)^{\beta+1}\right.$$

$$\left. + \frac{1}{\beta+2}\left(\frac{x-1}{2}\right)^{\beta+2}\right] dx\, dy$$

with

$$W_\beta \left(\frac{x-y}{2}, \frac{x+y}{2}\right) = \frac{4}{(\beta+1)(\beta+2)(\beta+3)}$$

$$\times \left[\left(\frac{x-y}{2}\right)^{\beta+3} + \left(\frac{x+y}{2}\right)^{\beta+3} - \left(\frac{x-1}{2}\right)^{\beta+3}\right]$$

$$- \frac{1-y^2}{(\beta+1)(\beta+2)}\left(\frac{x-1}{2}\right)^{\beta+2}$$

and

$$c_\beta(k) = \frac{\sigma_k \beta \Gamma\left(\frac{k}{3+\beta}+1\right)\Gamma\left(\frac{\beta}{3+\beta}+1\right)^{3/k}}{(3+\beta)\sigma_2^{k/(3+\beta)}(\omega_3\beta B(4,\beta))^{k\beta/(3(3+\beta))}}$$

where $\sigma_k = 2\pi^{k/2}/\Gamma(k/2)$ is the surface area of the unit sphere in $\mathbb{R}^k$. This result was first established by Gilbert (1962) when $\beta = 1$. Plots of $CV(\lambda_d(C))$ and $CV(\lambda_d(C^{(1)}))$ versus β are shown in Møller (1992).

These are strictly increasing functions of β, which in the limit $\beta = 0$ agree with the CVs for a Poisson-Voronoi tessellation while the CVs tend to ∞ as $\beta \to \infty$.

The above result may be used for parameter estimation. For example, suppose $d = 3$ and we observe a sectional Johnson–Mehl tessellation of L_2 (i.e. $k = 2$). Using a systematic set of test lines (i.e. parallel line segments with a constant distance between successive line segments) we may obtain a non-parametric estimate of $CV(\lambda_1(C^{(1)}))$ and hence an estimate $\hat{\beta}$ from the plot of $CV(\lambda_1(C^{(1)}))$. Further, if L is the total length of the test lines and N the number of intersection points between the test lines and the edges of sectional tessellaiton, then

$$(EN)/L = \frac{\sqrt{\pi}}{2} g_2^{(3)}(\beta) \lambda^{1/3}.$$

This suggests to estimate λ by

$$\hat{\lambda} = \{2N/(L g_2^{(3)}(\hat{\beta}) \sqrt{\pi})\}^3.$$

In the remaining part of this section we take $v = 1$, so $\kappa = \alpha$. The following theorem from Møller (1995) shows how arrivals under (5.41) can be generated. For this we have ordered the arrivals $a_i = (x_i, t_i)$, $i = 1, 2, 3, \ldots$, such that $T_1 < T_2 < \cdots$ where $T_i = T_i(0)$ is the time the i'th arrival reaches the origin 0. This ordering is convenient for making simulated iid realisations of the typical crystal C as described below.

Theorem 5.4 *Let $0 < m_1 < m_2 < \cdots$ constitute a Poisson process on $(0, \infty)$ with intensity 1, let b_i be Beta-distributed with parameters β and d, and let u_i be a uniformly distributed unit vector in $\mathbb{R}^d$, $i = 1, 2, \ldots$, such that $\{m_i\}, b_1, b_2, \ldots, u_1, u_2, \ldots$ are mutually independent. Letting $x_i = (1 - b_i) T_i u_i$ and $t_i = b_i T_i$ with $T_i = (m_i/\eta)^{1/(d+\beta)}$, then $\Phi = \{(x_i, t_i)\}$ is a Poisson process on $\mathbb{R}^d \times [0, \infty)$ with intensity measure given by (5.41).*

Now, suppose we have generated $\tau = t_0$ in accordance with (5.44) and we have chosen the origin 0 as the nucleus of C. In order to generate C we need only to consider those arrivals $a_i = (x_i, t_i)$ with $\|x_i\| > |\, t_i - t_0\,|$. Letting

$$D_i = \{y \in \mathbb{R}^d \mid 1 \leq j \leq i, \|x_i\| > |\, t_i - t_0\,| \Rightarrow T_0(y) \leq T_j(y)\}$$

with $a_0 = (0, t_0)$ and

$$R_i = \sup\{\|y\| \mid y \in D_i\}$$

for $i = 1, 2, \ldots$, then

$$T_{i+1} \geq t_0 + 2R_i \Rightarrow C = D_i$$

Hence the following algorithm can be used for the simulation of those arrivals with $\|x_i\| > |t_i - t_0|$ which are needed to complete C: Set $m_0 = \eta t_0^{d+\beta}$. Then
for $i = 1, 2, \ldots$,

(a) generate a uniform random number w_i; set $m_i = m_{i-1} - \ln w_i$ and $T_i = (m_i/\eta)^{1/(d+\beta)}$;

(b) generate $g_{i1} \sim \Gamma(\beta, 1)$ and $g_{i2} \sim \Gamma(d, 1)$; set $b_i = g_{i1}/(g_{i1} + g_{i2})$;

(c) if $2b_i T_i < T_i + t_0$ then generate u_i and set $a_i = (x_i, t_i) = ((1 - b_i)T_i u_i, b_i T_i)$; else set $m_{i-1} = m_i$ and repeat (a)–(c);

until $T_i \geq t_0 + 2R_{i-1}$, where

$$R_{i-1} = \sup\{\|y\| \mid y \in \mathbb{R}^d, \ t_0 + \|y\| \leq t_j + \|y - x_j\|, \ j = 1, \ldots, i-1\}.$$

This algorithm is studied in detail in Møller (1995). In particular it is shown how R_{i-1} can be determined and how $C = D_{I-1}$ can be constructed where I is the i at which the algorithm is terminated. Note that clearly $I < \infty$ almost surely as Theorem 4 implies that with probability one $T_i \uparrow \infty$ and the R_i are decreasing and finite for all sufficiently large i. Simulated results in the planar ($d = 2$) and spatial ($d = 3$) cases are presented in Møller (1994, 1995).

Appendix A

Below we extend Lemma A.2 in Møller (1992) to the case of mathematical facets $G_n \cap L_k$ given by the intersection of a Johnson–Mehl tessellation with an affine subspace L_k of dimension $0 < k \leq d$.

For this we need a modification of (C3): Let as usual $\Phi = \{a_i\} = \{(x_i, t_i)\}$ denote the birth process of arrivals which generates the Johnson–Mehl tessellation. Consider any nuclei $x_0, \ldots, x_m$ corresponding to $m + 1$ arrivals of Φ. Suppose for ease of presentation that $L_k = \mathbb{R}^k \times \{0^{d-k}\}$ and write x_i^k and x_i^{d-k} for the k first and $d - k$ last coordinates of $x_i = (x_i^k, x_i^{d-k})$; here and elsewhere in the following vectors are row vectors and the superscript indicates the dimension of the space in which a vector is regarded to lie. Then if $0 < k \leq d$, $0 \leq p < k$, $m = k - p$, the condition is

(C3') $x_0^k, \ldots, x_m^k$ are affinely independent, i.e. $(x_1^k - x_0^k, 0^{d-k}), \ldots, (x_m^k - x_0^k, 0^{d-k})$ span a m-dimensional subspace L_m of L_k.

Let L_p denote the orthogonal complement to L_m within L_k,

$$L_p = \{x \in L_k \mid \forall y \in L_m : x \bullet y = 0\}$$

where $\bullet$ is the usual inner product. Write vectors $y \in L_k$ as $y = (y^m, y^p, 0^{d-k})$ where y^m and y^p are Cartesian coordinates of y with respect to L_m and L_p, respectively. Define the matrix

$$M = [(x_1^m - x_0^m)^* \dots (x_m^m - x_0^m)^*]^{-1}$$

where $*$ denotes transposition, and set

$$H = (t_1 - t_0, \dots, t_m - t_0)M$$

$$K = x_0^m - (1/2)(\|x_1^{d-p}\|^2 - \|x_0^{d-p}\|^2 + v^2(t_0^2 - t_1^2),$$
$$\dots, \|x_m^{d-p}\|^2 - \|x_0^{d-p}\|^2 + v^2(t_0^2 - t_m^2))M$$

$$A = v^2(1 - v^2\|H\|^2) \qquad B = 2v^2(H \bullet K - t_0)$$

$$C = v^2 t_0^2 - \|x_0^{d-k}\|^2 - \|K\|^2 \qquad D = B^2 - 4AC$$

where $x_j^{d-p} = (x_j^m, x_j^{d-k})$.

Lemma 5.1 *Let the situation be as described above and assume that (C2) and (C3') hold.*

Case $p > 0$: $G(x_0, \dots, x_m) \cap L_k \neq \emptyset$ if and only if one of the following three conditions holds:

$$A > 0 \text{ and } D \geq 0 \qquad (A.1)$$
$$A = 0 \text{ and } B > 0 \qquad (A.2)$$
$$A < 0 \text{ and } D \geq 0. \qquad (A.3)$$

If (A.1) holds, then

$$G(x_0, \dots, x_m) \cap L_k$$
$$= \{z + (\gamma(t), \sqrt{\delta(t)}u^m, 0^{d-k}) : \|u^m\| = 1, t \geq T\} \qquad (A.4)$$

where

$$T = (-b + \sqrt{D})/(2a)$$
$$z = (x_0^m - K + v^2 T H, x_0^P, 0^{d-k})$$
$$\gamma(t) = v^2(t - T)H$$
$$\delta(t) = At^2 + Bt + C.$$

If (A.2) holds then (A.4) remains true if T is now defined by $T = -C/B$.

If (A.3) holds then

$$G(s_0, \dots, x_m) \cap L_k$$
$$= \{z + (\gamma(t), \sqrt{\delta(t)}u^m, 0^{d-k}) : \|u^m\| = 1, T \le t \le T'\}$$

where $T, z, \gamma(t)$ and $\delta(t)$ are defined as in the case (A.1), and where $T' = (-B - \sqrt{D})/(2A)$.

Case $p = 0$: Define z as above and set

$$z' = (x_0^k - K + v^2 T' H, 0^{d-k}).$$

Then $G(x_0, \dots, x_k) \cap L_k \ne \emptyset$ implies that either (A.1) or (A.2) or (A.3) hold and $G(x_0, \dots, x_k) \cap L_k \subseteq \{z, z'\}$ with

$$z \in G(x_0, \dots, x_k) \cap L_k \Leftrightarrow T > \max_{0 \le i \le k} t_i$$

and

$$z' \in G(x_0, \dots, x_k) \cap L_k \Leftrightarrow T' > \max_{0 \le i \le k} t_i.$$

Hence, as in the case $k = d$, $G(x_0, \dots, x_k) \cap L_k$ consists of at most two points, while for $0 < p < k$, $G(x_0, \dots, x_m) \cap L_k$ is obtained by rotating a smooth curve in the orthogonal complement to L_p within L_k and around a line in $x_0 + L_m$, and this curve is either unbounded or closed. The proof of the lemma follows the same lines as the proof of Lemma A.2 in Møller (1992).

References

Avrami, M. (1939) Kinetics of phase change I. *J. Chem. Phys.*, **7**, 1103–1112.

Avrami, M. (1940) Kinetics of phase change II. *J. Chem. Phys.*, **8**, 212–224.

Avrami, M. (1941) Kinetics of phase change III. *J. Chem. Phys.*, **9**, 117–184.

Evans, U.R. (1945) The laws of expanding circles and spheres in relation to the lateral growth of surface films and the grain size of metals. *Trans. Faraday Soc.*, **41**, 365–374.

Federer, H. (1969) *Geometric Measure Theory*. Springer-Verlag, New York.

Frost, H.J. and Thompson, C.V. (1987) The effect of nucleation conditions on the topology and geometry of two-dimensional grain structures. *Acta Metall.*, **35**, 529–540.

Heinrich, L. and Schüle, E. (1995) Generation of the typical cell of a non-Poissonian Johnson–Mehl tessellation. *Commun. Statist. Stochastic Models*, **11**.

Hoffmann-Jørgensen, J. (1994) *Probability with a View toward Statistics*. Vol. II. Chapman and Hall, New York.

Johnson, W.A. and Mehl, R.F. (1939) Reaction kinetics in processes of nucleation and growth. *Trans. Amer. Inst. Min. Engrs.*, **135**, 416–458.

Kolmogoroff, A.N. (1937) Statistical theory of crystallization of metals. *Bull. Acad. Sci. USSR Mat. Ser.*, **1**, 355–359.

Leistritz, L. and Zähle, M. (1992) Topological mean value relations for random cell complexes. *Math. Nachr.*, **155**, 57–72.

Mahin, K.W. , Hanson, K. and Morris, J.W. (1980) Comparative analysis of the cellular and Johnson–Mehl microstructures through computer simulation. *Acta. Metall.*, **28**, 443–453.

Matheron, G. (1975) *Random Sets and Integral Geometry*. Wiley, New York.

Mecke, J. (1967) Stationäre zufällige Masse auf lokalkompakten Abelschen Gruppen. *Z. Wahrscheinlichkeitstheorie verw. Geb.*, **9**, 36–58.

Meijering, J.L. (1953) Interface area, edge length, and number of vertices in crystal aggregates with random nucleation. *Philips Res. Rep.*, **8**, 270–290.

Miles, R.E. (1972) The random division of space. *Suppl. Adv. Appl. Prob.*, 243–266.

Møller, J. (1989) Random tessellations in $\mathbb{R}^d$. *Adv. Appl. Prob.*, **21**, 37–73.

Møller, J. (1992) Random Johnson–Mehl tessellations. *Adv. Appl. Prob.*, **24**, 814–844.

Møller, J. (1994) *Lectures on Random Voronoi Tessellations*. Lecture Notes in Statistics. Springer-Verlag, New York.

Møller, J. (1995) Generation of Johnson–Mehl crystals and comparative analysis of models for random nucleation. *Adv. Appl. Prob. (SGSA)*, **27**, 367–383.

Møller, J. and Zuyev, S. (1994) Gamma-type results and other related properties of Poisson processes. Research Report 282, Dept. of Theoretical Statistics, University of Aarhus. To appear in *Adv. Appl. Prob. (SGSA)*.

Okabe, A. Boots, B. and Sugihara, K. (1992) *Spatial tessellations. Concepts and Applications of Voronoi Diagrams*. Wiley, Chichester.

Slivnyak, I.M. (1962) Some properties of stationary flows of homogeneous random events. *Teor. Veruyat. Primen.*, **7**, 347–352. (Translation in *Theory Probab. Appl.*, **7**, 336–341).

Stoyan, D. Kendall, W.S. and Mecke, J. (1995) *Stochastic Geometry and Its Applications*, 2nd ed. Wiley, New York.

Weiss, V. and Zähle, M. (1988) Geometric measures for random curved mosaics of $\mathbb{R}^d$. *Math. Nachr.*, **138**, 313–326.

Zähle, M. (1988) Random cell complexes and generalized sets. *Ann. Prob.*, **16**, 1742–1766.

CHAPTER 6

Current Topics in Applied Morphological Image Analysis

Luc Vincent

Xerox Scansoft, Inc.
3400 Hillview Avenue, Building 3
Palo Alto, CA 94304, U.S.A.
Phone: (1) 650 813 7863. Fax: (1) 650 813 6792
E-mail: lucv@adoc.xerox.com

6.1 Introduction

Morphology is an image processing methodology which was born in France in the late sixties and has been gaining increasing importance and popularity since then (Matheron 1975, Serra 1982, Serra 1988, Schmitt and Vincent 1997). Today, morphology is broadly accepted as one of the most useful 'toolboxes' for a wide variety of image analysis applications. A solid knowledge of this methodology enables the quick development of algorithms and products for solving complex detection, segmentation, and pattern recognition problems. Morphology has been successfully applied in many areas, including medicine (MR imagery, CT imagery, X-rays, angiographies, etc), biology, radar, sonar, infra-red, and remote sensing images, industrial inspection, robotics, material science, fingerprints, identification, document recognition, etc (See Fig. 6.1). It addresses binary and grayscale images, as well as color and 3-D imagery. A priori, any application involving images sharing a well-defined set of characteristics (e.g., size and shape of the objects to be detected in them) is a good candidate for a morphological solution.

The morphological approach to image analysis is natural and attractive: it begins by considering binary images as sets and grayscale images as functions or topographic reliefs. These sets and functions are then transformed – in the spatial domain – via morphological operators, whose definitions are usually based on *structuring elements*, i.e.

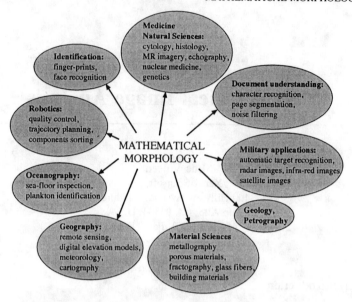

Figure 6.1 *Fields of application of morphological image analysis.*

particular shapes that are translated in the images and used as probes. From these basic operations, increasingly complex operators are derived, that are used towards more and more specific goals, like the detection of gaps and protrusions, valley and crest-line extraction in grayscale images, shape- and size-based feature extraction, etc. Despite this increasing algorithmic complexity, the operators remain extremely easy to use: they are intuitive, flexible, and powerful.

In the present chapter, we focus on two very very important aspects of mathematical morphology: image segmentation and granulometries. Segmentation is the task of partitioning an image into its 'meaningful' regions. This is often equivalent to detecting and extracting from an image the objects or zones of interest. These objects or zones can then be counted, measured, classified, etc. Accurate segmentation is therefore a critical step in a number of image analysis systems. On the other hand, granulometries are concerned with extracting size information from images *without* prior segmentation. This can be key when either the images under study are too difficult or too intricate to be segmented, or when processing speed is critical, or when size information is itself needed to guide the segmentation process. Beyond this, granulometries provide a set of general purpose feature extractors that turn out to be very powerful in a number of classification applications. Both morphological segmentation techniques and granulometries have seen significant advances in the past decade, and hundreds of papers have been published on these topics.

Our approach here is pragmatic and applied: indeed, in the field of image processing, the intuitive understanding of a few key concepts is usually more important than the knowledge of their theoretical foundations. Besides, a single chapter would not be sufficient to thoroughly cover both the theoretical and the applied aspects of segmentation and granulometries. For mathematical morphology theory, we invite the reader to refer to the monograph by Matheron (1975), the classic books by Serra (1982, 1988), or the more recent monograph by Heijmans (1994). French speaking readers can also consult the excellent new book by Schmitt and Mattioli (1994). Finally, it is worth mentioning that the topics covered in the present chapter will also be extensively discussed in a book to be published in 1997 (Schmitt and Vincent 1997).

This chapter is organized as follows: we begin in Section 6.2 with a brief review of mathematical morphology, introduce a few notations and conventions, and provide some reminders on morphological concepts that will be useful throughout the chapter. In Section 6.3, the emphasis is put on morphological segmentation, and such tools as *grayscale reconstruction, area openings*, and *watersheds* are described in detail. The effects of these operators on images are illustrated on numerous examples, and their use for image analysis problem-solving is emphasized. Finally, Section 6.4 deals with granulometries. In this section, while a number of applications are also presented, more emphasis is put on the *algorithmic* aspect. Indeed, regardless of how powerful a morphological operator is 'on paper', if no efficient implementation exists, this operator does not serve any practical purpose. Efficient binary and grayscale granulometry algorithms are described, and their use is illustrated in a variety of applications, ranging from size estimation, texture characterization, and classification. Problem-solving and algorithms are the two aspects of applied morphology that this chapter is mostly concerned with.

6.2 Brief review of mathematical morphology

For the sake of completeness of this chapter, the present section provides some background on mathematical morphology, its classic operators, and associated notations. Readers who are already very familiar with such concepts may skip directly to Section. 6.3; others will find here some reference material that will greatly help them understand Sections 6.3 and 6.4.

6.2.1 Basic definitions and notations

Definition 6.1 *A two-dimensional binary image is a subset X of the continuous plane* $\mathbb{R}^2$.

In the following, we exclusively deal with *bounded* sets. Besides, X is often considered equivalent to its *characteristic function*, i.e., the mapping f from $\mathbb{R}^2$ into $\{0, 1\}$ such that $f(p) = 1$ if $p \in X$ and $f(p) = 0$ otherwise.

Definition 6.2 *A two-dimensionsional grayscale image* I *is defined as a mapping from a bounded subset* D_I *of* $\mathbb{R}^2$ *(the domain of* I*) into* $\mathbb{R}$:

$$p \in D_I \longrightarrow I(p) \in \mathbb{R}.$$

Grayscale images are often called *functions* in the morphological literature. In the following, all the grayscale images we deal with are defined on a rectangular domain. From the two above definitions, it is clear that binary images are simply grayscale images that only take values 0 and 1. Therefore, any morphological operator defined for grayscale images is a perfectly valid binary operator as well.

The *operators* or *transformations* ψ we deal with typically transform binary images (i.e. sets) into binary images, or grayscale images into grayscale images. Some operator like the *threshold operation* transform a grayscale image into a binary image. Similarly, operators like the *distance function* turn a binary image into a grayscale image. Note that morphological operators can be defined in more general spaces such as *complete Boolean lattices* (Birkhoff 1983). Such theoretical aspects of morphology would however be beyond the scope of this chapter (See Heijmans 1994 for more detail).

Now, let ψ be a binary operator, i.e., a mapping from $\mathcal{P}(\mathbb{R}^2)$ into $\mathcal{P}(\mathbb{R}^2)$:

$$X \in \mathbb{R}^2 \longmapsto \psi(X) \in \mathbb{R}^2.$$

The following definitions will be useful throughout the present chapter:

Definition 6.3 ψ *is said to be an* extensive *transformation if and only if, for any set* X, $X \subseteq \psi(X)$. *Similarly,* ψ *is said to be* anti-extensive *if and only if, for any set* X, $\psi(X) \subseteq X$.

Definition 6.4 ψ is increasing *if and only if it preserves the inclusion relations between sets, i.e.:*

$$X \subseteq Y \Longrightarrow \psi(X) \subseteq \psi(Y). \tag{6.1}$$

Definition 6.5 ψ *is said to be* idempotent *when applying it several times in a row is equivalent to applying it only once:*

$$For\ any\ set\ X,\quad \psi(\psi(X)) = \psi(X). \tag{6.2}$$

Finally, with ψ' being another binary operator, we can give the following definition:

Definition 6.6 ψ *and* ψ' *are said to be* dual *of each other if applying one to a set X is equivalent to applying the other to the complement X^C of set X:*

$$\psi(X)^C = \psi'(X^C). \tag{6.3}$$

In the previous definition, the complement operator is denoted by superscript C: $X^C = \mathbb{R}^2 \setminus X$.

To extend the above definitions to grayscale images, we need to define a partial order between functions. The following is used:

$$f \geq g \Longleftrightarrow \forall x, f(x) \geq g(x). \tag{6.4}$$

Similarly, the 'complement' of a function $f : \mathbb{R}^2 \to \mathbb{R}$ is taken to be the function $(-f)$. Given this, definitions 6.3 to 6.6 easily extend to grayscale.

Most of the concepts this chapter deals with also extend to multidimensional images, or can be restricted to 1D signals. However, for simplicity, only the 2D case is considered here. Furthermore, although many of the concepts and methods discussed extend to color and multispectral images, due to the limitations of a single chapter, we must leave such topics aside.

In practice, the images we deal with are *discrete*, i.e., defined on a (rectangular) subset of the discrete plane $\mathbb{Z}^2$. Binary images take values 0 and 1 (the pixels with value 1 are sometimes called the ON pixels) whereas grayscale images take their values in the range $\{0, 1, \ldots, N\}$. For most applications, 8-bit per pixel is sufficient, therefore $N = 255$. In the following, new concepts are often defined in the continuous plane, and then applied to discrete images: indeed, the discrete version of a morphological operator is usually easy to derive from the 'continuous definition'. When it is more convenient to our purpose, we define the operator directly in the discrete plane $\mathbb{Z}^2$.

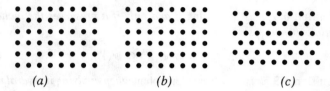

Figure 6.2 *The three classic grids used in discrete images. (a) 4-connected square grid; (b) 8-connected square grid; (c) hexagonal grid (c-connectivity).*

In the discrete plane, a *grid* G provides the connectivity, i.e., the neighborhood relationships between pixels. Commonly used grids are the square grid, for which a pixel p has either 4 (in 4-connectivity) or 8 neighbors (in 8-connectivity), as well as the hexagonal grid (6-connectivity). This is illustrated by Fig. 6.2. Two neighboring pixels p and q form an edge of G. The grid G induces a discrete distance in $\mathbb{Z}^2$, the distance between two pixels being the minimal number of edges required to join them. This distance metrics is denoted here by d_G, and in 4-, 6-, and 8-connectivity, we simply use d_4, d_6, and d_8. In a grid G, we denote by $N_G(p)$ the set of neighbors of a given pixel p. In 4-, 6- and 8-connectivity, we sometimes denote $N_G(p)$ by $N_4(p)$, $N_6(p)$ and $N_8(p)$ respectively.

In all our examples, binary images are printed 'black on white', that is, the ON pixels (pixels with value 1) are shown in black on a white background corresponding to the OFF pixels. The opposite convention is used for grayscale images: pixel values correspond to their brightness level. See Fig. 6.3 for examples of binary and grayscale images.

6.2.2 Basic morphological operators, binary case

As mentioned earlier, the basic morphological operators are defined based on the concept of structuring element. A structuring element is a particular set $B \subset \mathbb{R}^2$, usually small and of simple shape (disk, square, line segment, etc) that gets translated over an image X and whose relationships with this image are studied at each location. We denote by $\check{B}$ the *transposed* set:

$$\check{B} = \{-b \mid b \in B\}. \tag{6.5}$$

$\check{B}$ is simply obtained by rotating B by 180° about the origin O, which typically belongs to B. (The origin O is often called the *center* or *hot point* of structuring element B.) We also denote by B_x the translation

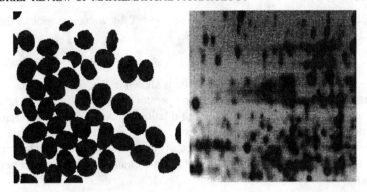

Figure 6.3 *(a) Binary image of coffee beans: the ON pixels (value 1) are shown in black; (b) grayscale image of two-dimensional electrophoresis gel: dark spots correspond to low pixel values.*

of B by x:

$$B_x = \{b + x \mid b \in B\}. \tag{6.6}$$

Definition 6.7 *The dilation of a set $X \subset \mathbb{R}^2$ by a structuring element B, denoted $\delta_B(X)$, is the set of points $x \in \mathbb{R}^2$ such that the translation of B by x has a non-empty intersection with set X:*

$$\delta_B(X) = \{x \in \mathbb{R}^2 \mid X \cap B_x \neq \emptyset\}. \tag{6.7}$$

One can show that the dilation of X by B, is equal to the *Minkowski Addition* (Minkowski 1897) of X and $\check{B}$, denoted $X \oplus \check{B}$:

$$\delta_B(X) = X \oplus \check{B} = \{x + b \mid x \in X, b \in \check{B}\}. \tag{6.8}$$

Note that some authors actually define the morphological dilation as a Minkowski addition (with no transposition of the structuring element). If we grossly generalize, we can say that the French school (Serra 1982) uses definition 6.7 whereas the American school (Sternberg 1986) defines dilation to be equal to the Minkowski addition. In practice, such distinctions are mostly academic, since useful structuring elements B are typically symmetric: $\check{B} = B$.

Definition 6.8 *The erosion of X by structuring element B, denoted $\varepsilon_B(X)$, is the set of points $x \in \mathbb{R}^2$ such that the translation of B by x is included in X:*

$$\varepsilon_B(X) = \{x \in \mathbb{R}^2 \mid B_x \subseteq X\}. \tag{6.9}$$

Here again, this definition of the erosion of X by B corresponds to the *Minkowski subtraction* of X and $\check{B}$:

$$\varepsilon_B(X) = X \ominus \check{B} = \{x \in \mathbb{R}^2 \mid \forall b \in \check{B}, x - b \in X\}. \qquad (6.10)$$

An example of erosion and dilation is shown in Fig. 6.4. From this illustration, it is clear that the dilation operator tends to 'grow' sets while the erosion 'shrinks' them. Through dilation, one can connect different components of a set X, fill some holes and gaps, whereas through erosion, holes become bigger, and objects (i.e., connected components) can vanish. More formally, these operations are equipped with a number of fundamental properties:

1. Erosion and dilation are *increasing* operators:

$$X \subseteq Y \implies \varepsilon_B(X) \subseteq \varepsilon_B(Y) \text{ and } \delta_B(X) \subseteq \delta_B(Y). \qquad (6.11)$$

2. Erosion and dilation are *dual* operations:

$$\varepsilon_B(X) = [\delta_B(X^C)]^C \text{ and } \delta_B(X) = [\varepsilon_B(X^C)]^C. \qquad (6.12)$$

3. When the structuring element 'contains its center', i.e., when $\vec{o} \in B$, then the erosion by B is anti-extensive and the dilation by B is extensive:

$$\vec{o} \in B \implies \varepsilon_B(X) \subseteq X \text{ and } X \subseteq \delta_B(X). \qquad (6.13)$$

4. Commutativity with union/intersection:

$$\delta_{B_1 \oplus B_2}(X) = \delta_{B_1}(\delta_{B_2}(X)) = \delta_{B_2}(\delta_{B_1}(X) \qquad (6.14)$$

$$\varepsilon_{B_1 \oplus B_2}(X) = \varepsilon_{B_1}(\varepsilon_{B_2}(X)) = \varepsilon_{B_2}(\varepsilon_{B_1}(X)). \qquad (6.15)$$

5. Structuring element combinations:

$$\delta_{B_1 \oplus B_2}(X) = \delta_{B_1}(\delta_{B_2}(X)) = \delta_{B_2}(\delta_{B_1}(X)) \qquad (6.16)$$

$$\varepsilon_{B_1 \oplus B_2}(X) = \varepsilon_{B_1}(\varepsilon_{B_2}(X)) = \varepsilon_{B_2}(\varepsilon_{B_1}(X)) \qquad (6.17)$$

$$\delta_{B_1 \cup B_2}(X) = \delta_{B_1}(X) \cup \delta_{B_2}(X) \qquad (6.18)$$

$$\varepsilon_{B_1 \cup B_2}(X) = \varepsilon_{B_1}(X) \cap \varepsilon_{B_2}(X). \qquad (6.19)$$

Property number 2 above is particularly interesting in practice: it means that eroding the foreground is equivalent to dilating the background. Stated differently, given a dilation operator, we can obtain the corresponding erosion by complementing, dilating, and complementing again. The same holds for a number of pairs of basic morphological operators.

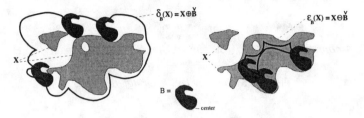

Figure 6.4 *(a) Dilation of set X by structuring element B. (b) Erosion of X by B. In each case, thick black outline represents contour of resulting set.*

Property 5 is also of practical interest: given a complex structuring element B, if we can find a decomposition of B as a union and/or Minkowski addition of simpler elements, then we can perform a dilation (resp. erosion) by B as a combination of dilations (resp. erosions) by simpler shapes. This can be extremely valuable from an algorithmic point of view. A specific instance of this property concerns the *homothetics* (i.e., scaled versions) of a basic structuring element. Denote by nB the n-fold Minkowski addition of B to itself:

$$nB = \underbrace{B \oplus B \oplus \cdots \oplus B}_{n \ times}, \qquad (6.20)$$

with $1B = B$. Then dilating a set X by nB is equivalent to iterating n dilations of X by B:

$$\delta_{nB}(X) = \underbrace{\delta_B(\delta_B(\ldots \delta_B(X)))}_{n \ times}. \qquad (6.21)$$

Similarly for erosions:

$$\varepsilon_{nB}(X) = \underbrace{\varepsilon_B(\varepsilon_B(\ldots \varepsilon_B(X)))}_{n \ times}. \qquad (6.22)$$

In practice, the most commonly used structuring elements are the homothetics of the *unit-size ball* B for the grid G being used. The unit-size ball centered at pixel p is equal to the union of p and its neighboring pixels for the grid. In other words, it is equal to the pixels (points of $\mathbb{Z}^2$) whose distance d_G to p is smaller or equal to 1. In 4-, 8-, and 6-connectivity, the unit-size ball B is respectively denoted by D ('diamond'), S (square), and H (hexagon), as illustrated by Fig. 6.5. In a given grid, a dilation (resp. erosion) by the unit-size ball B is often referred to as dilation (resp. erosion) of size 1. Similarly, a dilation (resp. erosion) by nB, which is equal to n consecutive dilations (resp.

4-connectivity: D   *8-connectivity: S*   *6-connectivity: H*

Figure 6.5 *Unit-size ball B for different grids*

erosions) by B, is called dilation (resp. erosion) of size n. This loose terminology will be used throughout the chapter.

From the erosion and dilation operators, two new extremely useful morphological transformations can be derived as follows:

Definition 6.9 *The opening of X by B, denoted by $\gamma_B(X)$, is given by:*

$$\gamma_B(X) = \delta_{\check{B}}(\varepsilon_B(X)). \tag{6.23}$$

The opening of X by B is often denoted by $(X)_B$, or $X \circ B$, notations that will be used in the present chapter when deemed more convenient.

Definition 6.10 *The closing of X by B, denoted by $\phi_B(X)$, is given by:*

$$\phi_B(X) = \varepsilon_{\check{B}}(\delta_B(X)). \tag{6.24}$$

In literature, the closing of X by B is sometimes denoted by $(X)^B$, or $X \odot B$.

Fig. 6.6 illustrates the effect of an opening and a closing by a disk. Clearly, both operators tend to smooth the boundary of the original set X, but each in a different manner: the opening tends to remove the small protrusions in which the structuring element cannot fit, whereas the closing fills in the gaps that cannot hold the structuring element. The opening also shrinks the original set whereas the closing makes it bigger. More formally, one can state the following properties:

1. Openings and closings are increasing operators.

2. Openings are anti-extensive, closings are extensive.

3. Openings and closings are duals of each other.

4. Openings and closings are *idempotent*.

5. The opening of X by B is the set of points of X that can be swept by structuring element B when B is translated inside X:

$$\gamma_B(X) = \bigcup_{B_x \subseteq X} B_x. \tag{6.25}$$

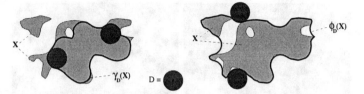

Figure 6.6 *(a) Opening of set X by disk D. (b) Closing of X by D. In each case, thick black outline represents contour of resulting set.*

A similar property holds for the closing:

$$\phi_B(X) = \left(\bigcup_{B_x \cap X = \emptyset} B_x \right)^C \tag{6.26}$$

While property 5 above provides an intuitive understanding of openings and closings, properties 1 through 4 are particularly important. For example, properties 1 and 4 together specify that openings and closings are particular types of *morphological filters* (Serra 1988, Serra and Vincent 1992). The theory of morphological filters is extensive and goes beyond the scope of this chapter. However, let us recall two important definitions:

Definition 6.11 *A* Morphological Filter *is an operator that is both increasing and idempotent.*

Definition 6.12 *An* algebraic opening *is an anti-extensive morphological filter, and an* algebraic closing *is an extensive morphological filter.*

Some of the morphological operators we deal with later in this chapter are *algebraic* openings according to Definition 6.12, but since they cannot be obtained by combining a dilation and an erosion, they are not *morphological* openings.

One can safely state that openings and closing are among the most useful operations in the morphological toolbox. They are directly useful for simple filtering tasks, and by combining them, one can design powerful filters that are tailored to specific applications. For example, Fig. 6.7 illustrates how *Alternating Sequential Filters (ASF)* based on grayscale openings and closings of increasing size (see Section 6.2.3) can be used for the restoration of noisy images (Sternberg 1986, Serra 1998, Serra and Vincent 1992). Furthermore, openings, closings, and the derived *top hat transformation* provide simple and powerful segmentation tools, as described in Section 6.2.3. Finally, openings and

Figure 6.7 *Image restoration using Alternating Sequential Filters.*

closings are at the foundation of *granulometries*, which are discussed in Section 6.4.

6.2.3 Extension to grayscale

All four operators described so far are *increasing*, and can therefore be extended to grayscale as follows: let I be a grayscale image, and denote by T_h the operator of *threshold at level h:*

$$T_h(I) = \{p \in D_I \mid I(p) \geq h\}. \tag{6.27}$$

Proposition 6.1 (Threshold Decomposition) *Let ψ be an* increasing *binary morphological operator, i.e., for all $X \subseteq Y$, we have $\psi(X) \subseteq \psi(Y)$. ψ can be extended to grayscale as follows:*

$$\psi(I)(p) = \sup\{h \mid p \in \psi(T_h(I))\}, \tag{6.28}$$

where I is a grayscale image and $p \in D_I$.

Indeed, the sets $(T_h(I))$ form a sequence of sets that characterize grayscale image I entirely, and are such that for any thresholds h and h',

$$h \geq h' \implies T_h(I) \subseteq T_{h'}(I),$$

as illustrated by Fig. 6.8. By definition, these inclusion relationships are preserved when applying increasing operator ψ, thus creating transformed grayscale image $\psi(I)$. For a more in-depth discussion, refer to Maragos and Ziff (1990).

In particular, using the above proposition, one can easily extend erosions, dilations, openings and closings, to the grayscale case. Intuitively,

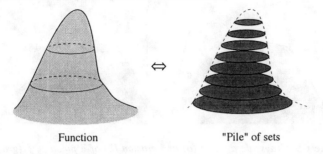

Function   "Pile" of sets

Figure 6.8 *Threshold decomposition of a graylevel image.*

the above proposition, often referred to as *threshold decomposition* or *threshold superposition* property (Maragos and Ziff 1990), states that to apply an increasing binary operator ψ to a grayscale image I, one can simply apply ψ to every 'slice' of I. When this principle is applied to erosions and dilations, one can prove the following:

Proposition 6.2 *The dilation and erosion of a grayscale image I with respect to structuring element B are given at pixel $x \in D_I$ by:*

$$\delta_B(I)(x) = \sup\{I(p) \mid p \in B_x\} \tag{6.29}$$

$$\varepsilon_B(I)(x) = \inf\{I(p) \mid p \in B_x\}. \tag{6.30}$$

This classic result, which can be easily proved (See for example Serra 1982), states that grayscale erosions and dilations can respectively be obtained by computing minima and maxima over a moving window – the structuring element. One can also prove that grayscale openings and closings are simply obtained as combinations of grayscale erosions and dilations:

$$\gamma_B(I) = \delta_{\check{B}}(\varepsilon_B(I)) \quad \text{and} \quad \phi_B(I) = \varepsilon_{\check{B}}(\delta_B(I)), \tag{6.31}$$

where I is a grayscale image.

In fact, the dilations (resp. erosions, openings and closings) defined above are often referred to as *flat* dilations, or dilations with *flat* structuring elements. More general dilations can be defined by considering the *umbra* or *subgraph* of grayscale image I:

$$U(I) = \{(x, h) \in \mathbb{R}^2 \times R \mid h \le I(x)\}, \tag{6.32}$$

and dilating this three-dimensional set by three-dimensional structuring elements using equation (6.7). The same holds for erosions, openings. and closings, but in general, it has little practical value. In the sequel, we focus on 'flat' structuring elements.

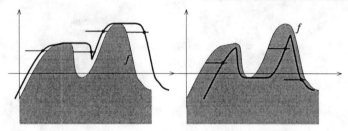

Figure 6.9 *Grayscale dilation (a) and erosion (b) of a function f (grayscale image).*

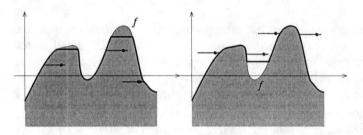

Figure 6.10 *(a) Grayscale opening. (b) Grayscale closing.*

Figure 6.9 illustrates the concepts of grayscale dilation and grayscale erosion. In this figure, a cross-section of the original grayscale image (or 'function' f) is shown: the horizontal axis corresponds to this cross-section whereas the vertical axis corresponds to grayscale values. On this type of figure, only a cross section of the (flat) structuring element can be shown, i.e., a horizontal line-segment. The (cross-section of the) resulting transformed functions are shown with a thick black stroke. One can observe that the dilation 'widens the peaks' and 'shrinks the valleys' whereas the erosion has the opposite effect.

Similarly, Fig. 6.10 illustrates the concepts of grayscale openings and closings. Here, the opening tends to clip the narrow peaks in which the structuring element cannot fit, whereas the closing fills-in the valleys that are too narrow for the structuring element.

Aside from a vast variety of morphological filters, a very useful basic morphological segmentation operator can be derived from gray openings and closings: the *top hat* transformation. Originally proposed by Meyer (1978), it provides an excellent tool for extracting light (resp. dark) objects from an uneven background. It relies on the fact that, by grayscale opening, one removes from an image the light structures

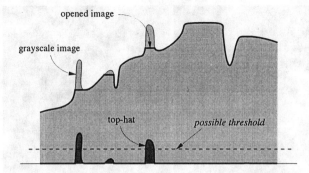

Figure 6.11 *Extraction of small bright structures ('peaks') from an uneven background using top hat followed by thresholding.*

that cannot hold the structuring element. Subtracting the opened image from the original one yields an image where the objects that have been removed by opening clearly stand out. This image can then easily be thresholded (see Fig. 6.11). Formally, the following definition can be proposed:

Definition 6.13 *The top hat $TH_\gamma(I)$ of a grayscale image I with respect to opening γ is given by:*

$$TH_\gamma(I) = I - \gamma(I), \tag{6.33}$$

where $-$ stands for the pixelwise subtraction operator.

Note that in the above definition, the opening γ does not have to be based on a structuring element, but could very well be an algebraic opening, as defined in Definition 6.12. The dual top hat TH^* can also be defined as the algebraic difference between a closing and the identity operator. It is used to extract small dark patterns from grayscale images.

An application is shown in Fig. 6.12: Fig. 6.12a is a scanning electron microscopy image where the balls in the lower right corner are to be extracted. These being compact and light compared to the background around them, they are removed by an opening of size 2 (see Fig. 6.12b). After subtraction of (b) from (a), i.e. top hat (see Fig. 6.12c), these small balls stand out and this image can be easily thresholded into Fig. 6.12d. The desired balls (right side) can now be extracted as the balls contained in the largest connected component of the dilation of Fig. 6.12d. The dilated image is shown in Fig. 6.12e, and the resulting segmentation is shown in Fig. 6.12f.

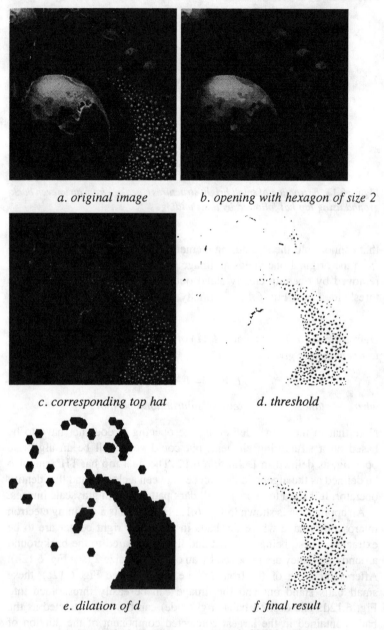

a. original image

b. opening with hexagon of size 2

c. corresponding top hat

d. threshold

e. dilation of d

f. final result

Figure 6.12 *Top-hat segmentation of SEM image.*

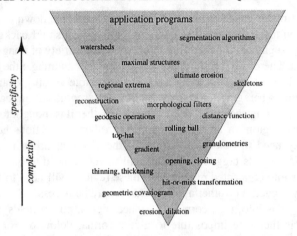

Figure 6.13 *Inverted pyramid of morphological operators.*

Let us conclude this first section by looking back at the approach that led to an operator such as the top hat: it is defined as the algebraic difference between the identity operator and a grayscale opening, itself obtained as the composition of a dilation and an erosion*. This is typical of the morphological approach to image analysis: given a few basic operations (erosions and dilations), one can compose them, iterate them, or combine them using Boolean and algebraic operators, thus creating increasingly complex and sophisticated transformations, that can be used towards increasingly specific tasks. This principle is illustrated by Fig. 6.13. A fair number of the operators listed in this illustration will be covered in the present chapter.

6.3 Advanced morphological segmentation techniques

6.3.1 Introduction, image segmentation

The present section is primarily concerned with the *segmentation* of binary and grayscale images. In the image analysis world, the word *segmentation* refers to the decomposition or partitioning of an image into its meaningful 'segments', i.e., zones, or objects. For example, a typical segmentation task in remote sensing, consists of decomposing an aerial (satellite) image into urban areas and rural areas. Similarly,

---

* Dilations and erosions can themselves be decomposed into maxima and minima of translations.

segmenting an image of biological material may come down to extracting individual cells and differentiating between cells and background.

Image segmentation therefore refers to a wide variety of image analysis tasks. There is never a unique way to segment an image: the process is heavily goal-driven. Going back to the remote sensing example in the previous paragraph, for some applications, segmentation may come down to tagging each pixel depending on whether it is thought to belong to a rural region, or an urban region. For other applications however, one may need to detect and count all the different urban regions of the image: pixels tagged as 'urban' will therefore need to be clustered together into connected regions, and these regions will need to be individually tagged. For other applications, it may be necessary to segment rural regions into, e.g., corn fields, wheat fields, potato fields, and the rest. It is therefore impossible to give a formal definition for 'image segmentation'.

The region-contour duality should also be noted. Partitioning an image into different regions is strictly equivalent to extracting the contours of these regions or objects: these contours simply form the separating lines between regions. *Contour extraction*, which has traditionally been seen as a separate image processing task, is therefore nothing more than a different perspective on segmentation, and is covered in the present section.

While segmentation can take so many different aspects, this section mainly deals with object segmentation, i.e., with images of objects or particles (such as cells, beans, etc), where segmenting is equivalent to extracting the objects of interest. Specifically, the goal of particle segmentation is to partition the image in as many connected components as there are objects or regions to extract, plus one or more *background* regions. We distinguish between binary and grayscale particle segmentation: in the binary case, i.e. when the images under study are binary, the segmentation task consists of separating the touching or overlapping particles (e.g., see the coffee beans example of Fig. 6.3a). In the grayscale case, the segmentation task is equivalent to a contour extraction problem (e.g., in Fig. 6.3b, the contours of the electrophoresis spots have to be extracted as accurately as possible).

This section is not intended as a review of all the existing image segmentation methodologies. Instead, we focus on morphological segmentation tools and techniques: they are extremely powerful, easy to use, and suitable for a wide variety of applications. We begin with *geodesic* operators and illustrate their use for object extraction from grayscale images. The *grayscale reconstruction* operator described is

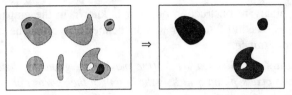

Figure 6.14 *Binary reconstruction from markers.*

the most versatile of these tools and is discussed at length. Grayscale *area openings and closings* are also described, and shown to provide a useful alternative to grayscale reconstruction in some segmentation applications. This progressively leads us to Section 6.3.4, where watershed segmentation techniques are described. We show that the watershed transformation is a compelling operation for the segmentation of both binary and grayscale images. Watersheds alone typically yield *oversegmented* results, but this issue is addressed using the grayscale reconstruction operator mentioned above, and this results in a consistent and widely applicable segmentation methodology. Finally, Section 6.3.5 describes the concept of *dynamics* and shows how, together with watersheds, it can be used for both automatic and interactive segmentation applications.

6.3.2 Grayscale reconstruction and applications to segmentation

Geodesy, binary reconstruction

Let I and J be two binary images defined on the same discrete domain D and such that $J \subseteq I$. In terms of mappings, this means that: $\forall p \in D$, $J(p) = 1 \implies I(p) = 1$. J is called the *marker* image and I is the *mask*. Let $I_1, I_2, \ldots, I_n$ denote the n connected components of I.

Definition 6.14 *The reconstruction $\rho_I(J)$ of mask I from marker J is the union of the connected components of I which contain at least a pixel of J:*

$$\rho_I(J) = \bigcup_{J \cap I_k \neq \emptyset} I_k.$$

This definition is illustrated by Fig. 6.14. It is extremely simple, but gives rise to several interesting applications and extensions, as we shall see in the following.

This binary reconstruction operator is known in the graphics community as 'seedfill'. Another way to define it involves the notion of *geodesic distance:*

Definition 6.15 *Given a set X (the mask), the geodesic distance between two pixels p and q in X is the length of the shortest paths joining p and q that are included in X.*

Note that the geodesic distance between two pixels within a mask highly depends on the type of connecticity being used: geodesic distances in 4-connectivity are always larger than geodesic distances in 8-connectivity. Besides, an object that is connected in 8-connectivity may be made of several connected components in 4-connectivity. The notion of geodesic distance is illustrated by Fig. 6.15. Geodesic distance was introduced in the framework of image analysis by Lantuéjoul and Beucher (1981) and is at the basis of numerous morphological operators (e.g., see Lantuéjoul and Maisonneuve 1984). In particular, one can define geodesic dilations (and similarly, geodesic erosions) as follows:

Definition 6.16 *Let $X \subset \mathbb{Z}^2$ be a discrete set and $Y \subseteq X$. The geodesic dilation of size $n \geq 0$ of Y within X is the set of the pixels of X whose geodesic distance to Y is smaller or equal to n:*

$$\delta_X^{(n)}(Y) = \{p \in X \mid d_X(p, Y) \leq n\}.$$

Note that the notations used in the above definition conflict somewhat with the definitions used in Section 6.2: in the present context, subscript X in δ_X refers to the *mask X* within which the geodesic dilation is being performed. The grid G (4-, 6-, or 8-connected) is supposed to be fixed.

From this definition, it is obvious that geodesic dilations are extensive transformations, i.e. $Y \subseteq \delta_X^{(n)}(Y)$. In addition, geodesic dilation of a given size n can be obtained by iterating n elementary geodesic dilations:

$$\delta_X^{(n)}(Y) = \underbrace{\delta_X^{(1)} \circ \delta_X^{(1)} \circ \cdots \circ \delta_X^{(1)}}_{n \text{ times}}(Y). \tag{6.34}$$

Fig. 6.16 illustrates successive geodesic dilations of a marker inside a mask, using 4- and 8-connectivity. The elementary geodesic dilation can itself be obtained via a standard dilation of size one followed by an intersection:

$$\delta_X^{(1)}(Y) = \delta_B(Y) \cap X. \tag{6.35}$$

This last statement only holds for *elementary* (i.e., unit-size) geodesic dilations.

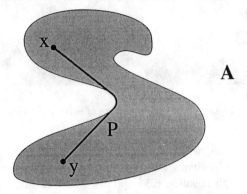

Figure 6.15 *Geodesic distance $d_G(x, y)$ within a set A.*

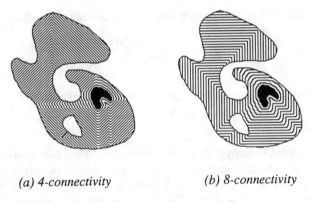

(a) *4-connectivity*   (b) *8-connectivity*

Figure 6.16 *Boundaries of the successive geodesic dilations of a set (black) within a mask.*

When performing successive elementary geodesic dilations of a set Y inside a mask X, the connected components of X whose intersection with Y is non empty are progressively flooded. The following proposition can thus be stated:

Proposition 6.3 *The reconstruction of X from $Y \subseteq X$ is obtained by iterating elementary geodesic dilations of Y inside X until stability. In other words:*

$$\rho_X(Y) = \bigcup_{n \geq 1} \delta_X^{(n)}(Y). \qquad (6.36)$$

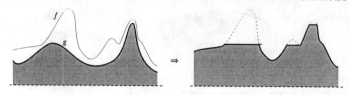

Figure 6.17 *Grayscale reconstruction of mask f from marker g.*

This proposition, which is easy to prove, forms the basis of one of the simplest algorithms for computing geodesic reconstructions. In addition, although equation 6.36 appears arcane at first, its extension to grayscale is straightforward.

Grayscale reconstruction

Binary geodesic reconstruction obviously is an increasing transformation in that it satisfies:

$$Y_1 \subseteq Y_2, X_1 \subseteq X_2, \quad Y_1 \subseteq X_1, Y_2 \subseteq X_2 \quad \Rightarrow \quad \rho_{X_1}(Y_1) \subseteq \rho_{X_2}(Y_2).$$
(6.37)

Therefore, following the threshold superposition principle of Prop. 6.1, we can define *grayscale reconstruction* as follows (Vincent 1992a, Vincent 1993a).

Definition 6.17 (Grayscale reconstruction, first definition) *Let J and I be two grayscale images defined on the same domain, taking their values in the discrete set $\{0, 1, \ldots, N\}$ and such that $J \leq I$ (i.e., for each pixel $p \in D_I$, $J(p) \leq I(p)$). The grayscale reconstruction $\rho_I(J)$ of I from J is given by:*

$$\forall p \in D_I, \quad \rho_I(J)(p) = \max\{k \in [0, N] \mid p \in \rho_{T_k(I)}(T_k(J))\}. \quad (6.38)$$

Fig. 6.17 illustrates this transformation. Just like binary reconstruction extracts those connected components of the mask which are marked, grayscale reconstruction extracts the *peaks* of the mask which are marked by the marker-image.

Definition 6.17 does not provide any interesting computational method to determine grayscale reconstruction in digital images. Indeed, even if a fully optimized binary reconstruction algorithm is used, one has to apply it 256 times to determine grayscale reconstruction for images on 8 bits! Therefore, it is most useful to introduce this transformation using the concept of geodesic dilations presented earlier.

Following the threshold decomposition principle, one can define the unit geodesic dilation $\delta_I^{(1)}(J)$ of grayscale image $J \leq I$ 'under' I:

$$\delta_I^{(1)}(J) = \delta_B(J) \wedge I, \qquad (6.39)$$

In this equation, $\wedge$ stands for the pointwise minimum operator, which is the direct extension to grayscale of the concept of Boolean intersection.

The grayscale geodesic dilation of size $n \geq 0$ is then given by:

$$\delta_I^{(n)}(J) = \underbrace{\delta_I^{(1)} \circ \delta_I^{(1)} \circ \cdots \circ \delta_I^{(1)}}_{n \text{ times}}(J). \qquad (6.40)$$

This leads to a second definition of grayscale reconstruction:

Definition 6.18 (Grayscale reconstruction, second definition)
The grayscale reconstruction $\rho_I(J)$ of I from J is obtained by iterating grayscale geodesic dilations of J 'under' I until stability, i.e.:

$$\rho_I(J) = \bigvee_{n \geq 0} \delta_I^{(n)}(J).$$

It is straightforward to verify that both this definition and definition 6.17 correspond to the same transformation.

Similarly, the elementary geodesic erosion $\varepsilon_I^{(1)}(J)$ of grayscale image $J \geq I$ 'above' I is given by

$$\varepsilon_I^{(1)}(J) = \varepsilon_B(J) \vee I, \qquad (6.41)$$

where $\vee$ stands for the pointwise maximum. The grayscale geodesic erosion of size $n \geq 0$ is then given by:

$$\varepsilon_I^{(n)}(J) = \underbrace{\varepsilon_I^{(1)} \circ \varepsilon_I^{(1)} \circ \cdots \circ \varepsilon_I^{(1)}}_{n \text{ times}}(J). \qquad (6.42)$$

We can now define the *dual* grayscale reconstruction in terms of geodesic erosions:

Definition 6.19 (Dual reconstruction) *Let I and J be two grayscale images defined on the same domain D_I and such that $I \leq J$. The dual grayscale reconstruction $\rho_I^*(J)$ of mask I from marker J is obtained by iterating grayscale geodesic erosions of J 'above' I until stability is reached:*

$$\rho_I^*(J) = \bigwedge_{n \geq 1} \varepsilon_I^{(n)}(J).$$

Application to segmentation

We will see in Section 6.3.4 that together with the watershed transformation, grayscale reconstruction forms one of the most powerful morphological tools for segmentation of binary and grayscale images. However, even without involving watersheds, grayscale reconstruction proves to be very useful for some segmentation applications.

For example, Fig. 6.18 represents an angiography of blood vessels in the eye, in which *microaneurisms* have to be detected. How should we approach this task? For every image analysis problem, a wise first step is to list all the pieces of information available on the collection of images to be processed. Indeed, as mentioned earlier, segmentation is a goal driven task, and completely general segmentation procedures simply do not exist. Here, our knowledge can be summarized as follows:

1. microaneurisms are small, compact, and brighter than the background,

2. blood vessels are thin, elongated, and brighter than the background,

3. microaneurisms are disconnected from the network of blood vessels,

4. microaneurisms are preferentially located in the dark areas of the image.

In general, such a list cannot be established from a single image, and a relatively large sample is required. However, subject again to the limitations of a single chapter, we content ourselves with the image shown in Fig. 6.18

Using element of knowledge number 1 above, one could think of extracting microaneurisms using a standard *top hat* transform (See Section 6.2.3). An opening of Fig. 6.18 by a small disk is shown in Fig. 6.19a, and the corresponding top hat is shown in Fig. 6.19b. This operation is clearly inappropriate here, because it extracts bright thin structures indiscriminately. The background of Fig. 6.19b does get normalized in the process, but the aneurisms have *not* been separated (segmented) from the blood vessels.

This preliminary experiment indicates that additional domain-specific knowledge should be used. The key pieces of information here are number 2 and 3: the aneurisms are not touching the blood vessels, and these vessels are – as one could expect – elongated. To take this into account, we use a series of morphological openings of Fig. 6.18 using line segments in different orientations. These segments are chosen to be longer than any possible aneurism, so that all the aneurisms are removed by

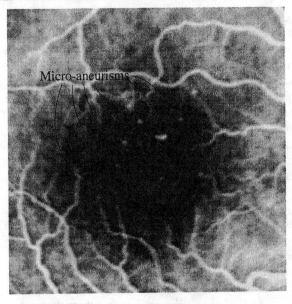

Figure 6.18 *Angiography showing eye blood vessels and micro-aneurisms.*

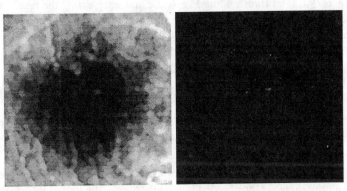

(a) opening with small disk   *(b) corresponding top hat*

Figure 6.19 *Inadequacy of the top hat for aneurism extraction.*

any such opening. On the other hand, the vessels being elongated, there is at least one orientation for which they are not completely removed by opening. After taking the supremum of these different openings, one obtains the result shown in Fig. 6.20a.

One can show that this maximum of openings using different structuring elements (specifically here: line segments in different orientations) is in fact an *algebraic opening* (see Section 6.2.2). It removes all the aneurisms, but the size of the line segments used is such that the network of blood vessels is not completely removed: plenty of *markers* remain, as can be seen in Fig. 6.20a. A grayscale reconstruction operation of Fig. 6.18 from Fig. 6.20a therefore enables us to reconnect (reconstruct) the blood vessels from their markers. But since the aneurisms are not connected to these vessels, they are not recovered by this operation. The result of this step is shown in Fig. 6.20b. Note that this operation of algebraic opening followed by a reconstruction is also an algebraic opening, sometimes refered to as *opening by reconstruction*. By subtracting this image from the original image, one obtains Fig. 6.20c, in which the aneurisms have been nicely extracted together with some extraneous small objects.

If we were to threshold the image of Fig. 6.20c, we would still obtain an imperfect result. At this point, we have to take into account piece of information number 4: aneurisms are primarily located in the darker areas of the image. We could extract these dark areas (here the central region of the image) by simply smoothing, then thresholding Fig. 6.19a; after thresholding Fig. 6.20c and intersecting the result with the extracted dark regions, we would be able to extract the desired aneurisms.

However, this technique, which involves *two* thresholding operation, would probably not be optimally robust. It may work adequately on this particular image, but is likely to fail when used on other images of the same type. As a general rule, one should:

- use as few fixed parameters as possible: fixed parameters limit the class of images on which the algorithm is likely to perform adequately,

- postpone thresholding operations as much as possible: indeed, once a gray image is thresholded into a binary one, the loss of information is huge and irreversible.

Our solution here is to perform a large opening of the original image (opening with a large isotropic element) to extract a background image, and to invert this image. The resulting image, shown in Fig. 6.20d, can be regarded as a fuzzy set: the brighter the pixel, the more likely it is that this pixel belongs to the dark image regions of the original image. By doing a pixelwise multiplication of this image with Fig. 6.20c, we enhance the candidate aneurisms of Fig. 6.20c in proportion to their likelihood of being part of a dark image region. The result is shown in Fig. 6.20e. A simple threshold of this image finally provides the result shown in Fig. 6.20f.

6.3.3 Grayscale area openings and closings

Background

In this section, we focus on the new concepts of area openings and closings, also called openings and closings *by area* (Vincent 1992b, Vincent 1993b). For a variety of segmentation applications, these operations provide a compelling alternative to openings followed by reconstruction.

Let us begin with a classic image analysis preprocessing problem: filtering out small light (respectively dark) particles from grayscale images without damaging the remaining structures. Often, simple morphological openings (respectively closings) with disks or approximations of disks, are adequate for this task. However, when the structures that need to be preserved are elongated objects, they can be either completely or partly removed by such operations.

Consider for example Fig. 6.21a, which represents a microscopic image of a metallic alloy. It is 'corrupted' by some black noise that one may wish to remove[†]. As shown in Fig. 6.21b, a closing of this image with respect to the elementary ball of the 8-connected metric, unit square S, severely damages most of the inter-grain lines, while still preserving some of the largests bits of noise (e.g., the blobs in the bottom right and left corners).

In this context, algebraic openings (resp. closings) defined as maxima (resp. minima) of openings (resp. closings) with line segments in different orientations often preserve elongated structures better. Unfortunately, such operations are rather expensive computationally, and as illustrated by Fig. 6.22, they may remain unsatisfactory.

The remedy to this last problem is to increase the number of orientations of the used line segments, but this in turn increases the computational complexity of the algorithm. In addition, even with a large number of orientations, very thin lines might still end up broken. We already described in Section 6.3.2 how this problem can be addressed using the grayscale reconstruction operator. In the present section, we describe an alternative solution, using the concepts of *area openings* and

---

[†] Note that part of what is called noise here is the intra-grain texture.

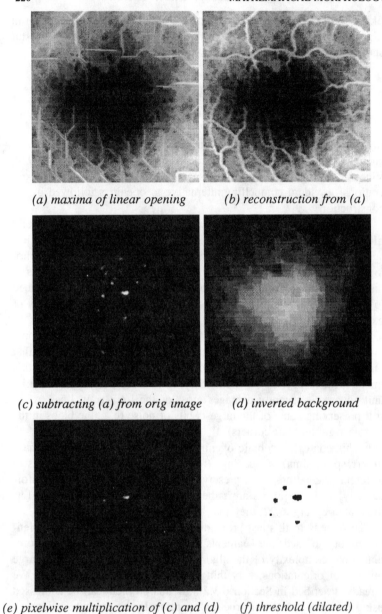

(a) maxima of linear opening   (b) reconstruction from (a)

(c) subtracting (a) from orig image   (d) inverted background

(e) pixelwise multiplication of (c) and (d)   (f) threshold (dilated)

Figure 6.20 *Extraction of micro-aneurisms using grayscale reconstruction*

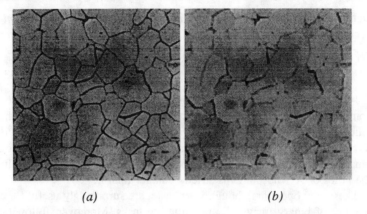

(a)   (b)

Figure 6.21 *(a) Microscopic image of a metallic alloy; (b) morphological closing of (a) by a unit-size square.*

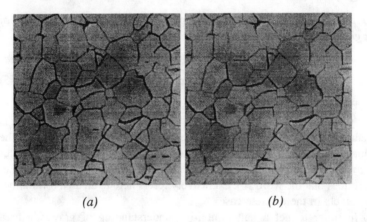

(a)   (b)

Figure 6.22 *Maxima of four linear openings of increasing size of Fig. 6.21a.
For small sizes, noise remains after opening, and for larger sizes, the inter-grain
boudaries start to get damaged.*

closings. We show that these operators can outperform reconstruction-
based algorithms and provide a new set of powerful segmentation tools.

Definition and intuitive interpretations

The easiest way *area openings* can be defined in the binary case is as
follows:

Definition 6.20 (Binary area opening) *Let X be a set made of n connected components $X_1, X_2, \ldots, X_n$. The area opening of parameter $\lambda > 0$ of X is given by:*

$$\gamma_\lambda^a(X) = \bigcup \{X_i \mid Area(X_i) \geq \lambda\}. \tag{6.43}$$

It follows that by area opening of size λ, we remove the connected components of X whose area is strictly smaller than λ, and preserve the others. Obviously, γ_λ^a is *increasing, idempotent*, and *anti-extensive*. It is therefore legitimate to call it an opening. By duality, the corresponding binary area closing ϕ_λ^a can easily be defined as the operator that fills the holes of X whose area is strictly smaller than λ.

These two operators, though very basic, are surprisingly useful for a number of binary image cleaning/filtering tasks. Moreover, following the threshold superposition principle of Prop. 6.1, they can be extended to grayscale, and the following definition can be proposed:

Definition 6.21 (Grayscale area opening) *The area opening $\gamma_\lambda^a(f)$ of a graylevel function f is given by:*

$$(\gamma_\lambda^a(f))(x) = \sup\{h \leq f(x) \mid Area(\gamma_x(T_h(f))) \geq \lambda\} \tag{6.44}$$

$$= \sup\{h \leq f(x) \mid x \in \gamma_\lambda^a(T_h(f))\}. \tag{6.45}$$

By duality, one similarly extends the concept of an area closing to grayscale. The interpretation of grayscale area openings and closings is relatively simple: a grayscale area opening basically removes from the image all the light structures which are 'smaller' than the size parameter λ, whereas the area closing has the same effect on dark structures. Let us stress that the word *size* exclusively refers here to an area (or number of pixels in the discrete case).

In order to get a better intuitive understanding of grayscale area openings and closings, the following theorem can be stated:

Theorem 6.22 *Denoting by A_λ the class of sets that are connected and whose area is greater than or equal to size parameter λ, the following equations hold:*

$$\gamma_\lambda^a = \bigvee_{B \in A_\lambda} \gamma_B, \tag{6.46}$$

$$\phi_\lambda^a = \bigwedge_{B \in A_\lambda} \phi_B. \tag{6.47}$$

In the discrete case, any connected set of area greater or equal to a positive integer λ contains a connected set of area equal to λ. We can therefore be more specific and state:

$$\gamma_\lambda^a(X) = \bigcup\{\gamma_B(X) \mid B \text{ connected }, Area(B) = \lambda\}, \quad (6.48)$$

$$\phi_\lambda^a(X) = \bigcap\{\phi_B(X) \mid B \text{ connected }, Area(B) = \lambda\}. \quad (6.49)$$

The above equations are valid in the grayscale case as well. This leads to a different understanding of area openings (resp. closings). As maxima of openings (resp. minima of closings) with all possible connected elements of a given area, they can be seen as *adaptive:* at every pixel location, the structuring element adapts its shape to the local image structure in order to 'remove as little as possible'. In particular, this means that bright elongated structures are usually very well preserved through area openings. Respectively, dark elongated structures, however fine, are typically unchanged by area closing.

Applications of area openings and closings

In spite of their apparent complexity, area openings and closings lend themselves very well to efficient implementations, without which they would remain mostly useless (Vincent 1993b). They have two main types of application: filtering and segmentation.

In section 6.3.3, we described a filtering problem that was hard to address with traditional openings and combinations thereof. Given how thin the grain boundaries in Fig. 6.21a are, it is clear that a good filtering solution must use operators involving connectivity. Experiments were conducted using two different techniques:

1. minima of linear closings followed by dual reconstruction,

2. direct area closing.

As shown in Fig. 6.23, the results provided by these two techniques look very similar. However, the area closing adapts to local image structure better than the reconstruction-based method, and therefore better preserves grain boundaries. This is illustrated by Fig. 6.23c, which is the thresholded algebraic difference between Fig. 6.23a and Fig. 6.23b: the ON pixels in this image show where the area closing outperforms the reconstruction-based method. In addition, technique number 2 above is a straightforward one-step operation, and is about one order of magnitude faster than technique number 1.

It is perhaps for segmentation that area openings and closings are most interesting. We saw in Section 6.2.3 that standard grayscale morphological openings led to a simple and powerful segmentation tool, the *top hat* transformation. In Section 6.3.2, we extended this concept

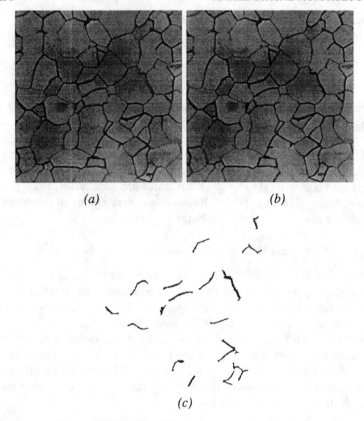

(a)   *(b)*

(c)

Figure 6.23 *(a) Result of reconstruction-based technique for filtering Fig. 6.21a from. (b) Area closing of Fig. 6.21a. (c) Zones where technique 2 outperforms technique 1, and preserves grain boundaries substantially better.*

to openings based on maxima of linear openings followed by reconstruction. Now, this approach extends to area openings and closings, and proves to be powerful and straightforward.

Let us revisit the problem of microaneurisms extraction from the angiography of Fig. 6.18. Instead of using linear openings followed by reconstruction, we can directly perform an area opening on this image. If we choose the size parameter in this operation to be larger than the area of any possible aneurism, we are guaranteed that all the aneurisms will be removed by area opening. However, except for occasional bright spikes inside them, the blood vessels will be perfectly preserved: indeed, they form a connected network whose overall area is much larger than the area of any aneurism. The result of this area opening operation is shown in Fig. 6.24a. A pixelwise subtraction of this image from the original image of Fig. 6.18 leads to Fig. 6.24b. This operation can be called *area top hat* or *top hat by area*.

The aneurisms are now clearly visible, but some other small structures not located on the dark image areas are also present. To get rid of them and extract the final aneurisms, we use the same technique used earlier: combine Fig. 6.24b with an image of the background in order to enhance the structures in proportion to the darkness of the background around them. This leads to Fig. 6.24c, which can now be thresholded into Fig. 6.24d, the final result of this segmentation.

6.3.4 Watershed segmentation

The segmentation problems addressed thus far in this section are in fact 'degenerate cases': what really matters with the angiography of Fig. 6.18 is that every aneurism in it gets detected. However, extracting the exact outline of each detected aneurism is not needed here. This is far from being always the case, and in the present section, we consider problems where extracting object or region boundaries as accurately as possible is critical.

Let us start with binary segmentation, and consider the image shown in Fig. 6.3a. This image shows a number of coffee beans, many of which are touching or slightly overlapping. In order to measure for example the area, length, and width of each bean, it will be necessary to extract the separating lines between the beans, i.e., to segment this image. Note however that full segmentation is not always required:

- If counting the beans is all that is needed, then extracting the outline of each bean is not necessary. Individually detecting each bean (e.g., its centroid) would be sufficient, and is a simpler problem. Another approach would be to use statistical models such as Boolean models (Serra 1988), from which a bean count estimate could be derived.

- If a rough estimate of dominant bean size is all that is needed, then segmentation may not be necessary either, and one could simply use *granulometries*, as described in Section 6.4.

In the present case, the segmentation problem comes down to separating touching beans. To address this, recall that the erosion operation not only makes object smaller, but tends to disconnect touching objects. Therefore, by performing erosions of increasing size on the coffee bean

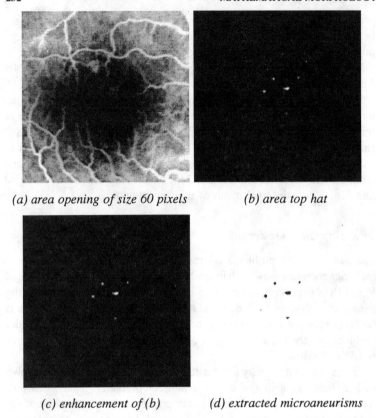

(a) area opening of size 60 pixels   *(b) area top hat*

(c) enhancement of (b)   *(d) extracted microaneurisms*

Figure 6.24 *Use of area opening in the detection of microaneurisms from angiographic imagery.*

image of Fig. 6.3a, we progressively disconnect the coffee beans as the 'bridges' between them are eroded away (See Fig. 6.25). Unfortunately, since the degree of overlap of these beans is not uniform throughout the image, they get disconnected for different erosion sizes. In other words, a single erosion size is not sufficient to disconnect the beans from each other, while not completely eroding away any bean.

To deal with this problem, a kind of 'multiscale erosion' operator is used, called the *distance function*. Recall that B denotes the unit-size ball for the grid being used (diamond D in 4-connectivity, hexagon H in 6-connectivity, square S in 8-connectivity). The distance is defined as follows:

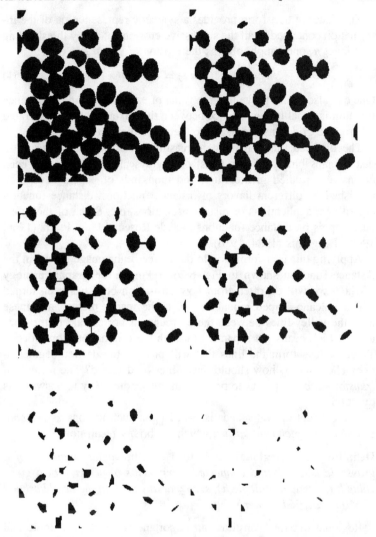

Figure 6.25 *Successive erosions of the coffee beans image*

Definition 6.23 *The distance function (or distance transform) $dist_X$ associated with a set X maps each pixel p of X to the smallest n such that p does not belong to the erosion of size n of X:*

$$dist_X(p) = \min\{n > 0 \mid p \notin \varepsilon_{nB}(X)\}. \qquad (6.50)$$

The distance transform provides a synthetic representation of the information contained in all the successive erosions of X: by thresholding it at level n, one simply obtains the erosion of size $n - 1$ of X:

$$\forall n > 0, \quad T_n(dist_X) = \varepsilon_{(n-1)B}(X). \tag{6.51}$$

One can also easily prove that the value of each pixel p in the distance function is equal to the distance between this pixel and the background X^C for the grid considered, hence the name *distance function*.

The distance function is an increasing operator that is unusual in that it maps binary images (sets) onto grayscale images. It has been extensively studied in literature, and thousands of papers have been published on different flavors of distance functions, distance function algorithms, applications of distance functions, etc. Some of the reference papers on distance functions include Rosenfeld and Pfaltz (1966, 1968), Borgefors (1984, 1986).

Applying this transformation to the coffee beans image results in the distance function shown in Fig. 6.26. Taking a topographic analogy – which we will do often from now on – each bean has been turned into a volcano-shaped mountain (or, if the reader prefers, a chinese hat); the valley-lines between these mountains correspond to the lines along which beans got separated through the original erosion process. These valleys form the lines that will provide the desired segmentation. The issue is: how should these lines be detected? The *watershed transformation* turns out to provide an outstanding way to answer this question.

The concept of watersheds is based upon such notions as *regional extrema* and *catchment basins*, which we briefly recall now.

Definition 6.24 (Regional maximum) *A regional maximum M of a grayscale image I is a connected components of pixels with a given value h (plateau at altitude h), such that any pixel in the neighborhood of M has a strictly lower value.*

Regional maxima constitute an important morphological topic and are extensively discussed in (Maisonneuve 1992). They should not be mistaken for *local* maxima: a pixel p of I is a local maximum for grid G if and only if its value $I(p)$ is greater than or equal to that of any of its neighbors. All the pixels belonging to a regional maximum are therefore themselves local maxima, but the converse is not true: for example, a pixel p belonging to the inside of a plateau is a local maximum, but the plateau may have neighboring pixels of higher altitude and thus not be a regional maximum.

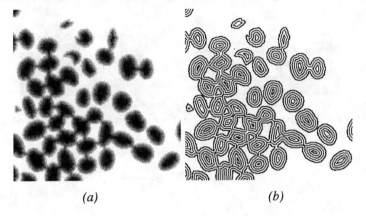

Figure 6.26 *(a) Distance function of the coffee beans image. Dark pixels corre-spond to large values of the distance function, which was here computed using the Euclidean metric for greater accuracy (Vincent 1991a). (b) Level lines of this function. (c) Representation of this distance function as a relief, using artificial shading and shadowing.*

Determining the regional maxima (resp. minima) of a grayscale im-age is relatively easy and several algorithms have been proposed in literature, some of which are reviewed in (Vincent 1990, 1993a). One of the most efficient methods makes use of grayscale reconstruction and is based on the following proposition:

Proposition 6.4 *The (binary) image $M(I)$ of the regional maxima of I is given by:*

$$M(I) = I - \rho_I(I - 1).$$

A proof for this result can be found in Vincent (1993b). It is il-lustrated by Fig. 6.27. By duality, a similar technique can be derived, enabling to extract regional minima through dual grayscale reconstruc-tion. In Section 6.3.5, we will see that this proposition forms the basis of the concept of *dynamics*.

Now, let us again regard grayscale image I as a topographic surface. If a drop of water falls at a pixel p of this relief, it will slide along the relief, following some steepest slope path, until it finally reaches one of the (regional) minima. We define the *catchment basin* $C(m)$ associated with a minimum m of I as follows:

Definition 6.25 (Catchment basin) *The catchment basin $C(m)$ asso-ciated with a (regional) minimum m of a grayscale image I, regarded*

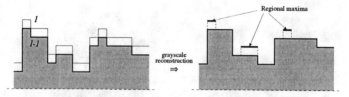

Figure 6.27 *Extracting the regional maxima of a graylevel image I by reconstruction of I from I − 1.*

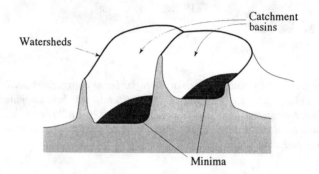

Figure 6.28 *Regional minima, catchment basins, and watershed lines.*

as a topographic relief, is the set of the points p such that a drop falling on p slides along the surface until it reaches m.

This definition, while not very formal, has the advantage of being intuitive. Obviously, catchment basins provide a tesselation of the space, and by definition, there is a one-to-one correspondence between minima and catchment basins. The crest-lines separating different basins are called *watersheds lines* or simply *watersheds*.

Definition 6.26 (Watersheds) *The watersheds (lines) of a grayscale image I are the lines that separate the different catchment basins of I.*

These notions are illustrated by Fig. 6.28. Going back to our binary segmentation problem, the distance function $dist_X$ exhibits regional maxima roughly located near the centroid of each bean. These maxima of the distance function provide *markers* of each bean, and form what is known as the *ultimate erosion* of the original binary image. They are separated by the valley lines that we are trying to detect. If we invert this relief, by duality, regional maxima are turned into re-

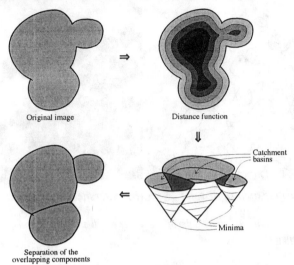

Figure 6.29 *Binary segmentation using catchment basins of the opposite of the distance function.*

gional minima and valley lines are turned into crest lines. By extracting the catchment basins of $-dist_X$, we associate with each minimum (i.e., bean marker) its catchment basin. The catchment basins are separated by watershed lines located on the valley lines of the original distance function $dist_X$. In summary, binary segmentation is achieved using the watersheds of the inverted distance function, as illustrated by Fig. 6.29.

In practice, things are almost never that easy, and even our academic coffee bean segmentation problem cannot be segmented that straightforwardly. As illustrated by Fig. 6.30a, there is not one-to-one correspondence between coffee beans and regional maxima of the distance function: the centroid of each bean is marked by at least one such maximum, but some beans are multiply marked, and some maxima are even located in-between beans (an example of this can be seen in the lower left corner of the image). As a result, watershedding the inverted distance function produces an oversegmented result, as shown in Fig. 6.30b.

This problem is not insurmountable: the maximal regions of the distance function are rather obvious structures, and the reason why they sometimes produce more than one regional maximum each is due to the fact that the workspace is discrete. The distance value that gets

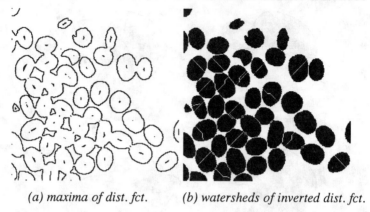

(a) maxima of dist. fct.   (b) watersheds of inverted dist. fct.

Figure 6.30 *Direct watersheds of the inverted distance function produce an oversegmented result.*

assigned to each pixel may sometimes get rounded up or down, which typically results in unit gray level oscillations in the vincinity of the top of the maximal structures. Specifically, if a structure (i.e., a bean) produces multiple maxima, these maxima are all at the same altitude h; in addition, for each pair of such maxima, there exists a path connecting them in such a way that the distance values along the path are all equal to h or $h - 1$. In other words, these maxima may be relatively far from each other in terms of distance (along the grid), but are very close in terms of gray-level.

In order or connect these maxima, the versatile grayscale reconstruction operator is used: the principle is to reconstruct $dist_X$ from $dist_X - 1$. In doing so, maxima marking the same structure are connected into one single blob. In addition, irrelevant maxima, such as the one in the bottom left corner of Fig. 6.30a, are removed in the process. This technique is illustrated in Fig. 6.31.

Going back to the coffee beans example, we can apply this method directly. Fig. 6.32a shows the modified distance function obtained after grayscale reconstruction. Extracting the maxima of this function results in Fig. 6.32b: each bean is now marked once, and there is no extraneous marker left. Therefore, the watersheds of the inverted modified distance function provide the desired segmentation, as shown in Fig. 6.32c.

Watersheds stand out as a powerful morphological crest-line extractor. It is therefore most interesting to apply the watershed transformation to gradient images: indeed, the contours of a grayscale image

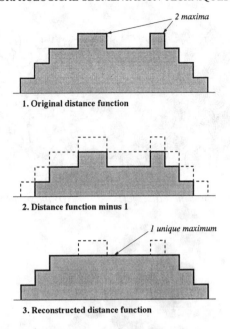

Figure 6.31 *Connection of regional maxima using grayscale reconstruction of the distance function dist$_X$ from dist$_X$ − 1.*

can be viewed as the regions where the gray levels exhibit the fastest variations, i.e., the regions of maximal gradient. These regions form the *crest-lines of the gradient*. This principle is illustrated by Fig. 6.33 and is at the basis of the use of watersheds for grayscale segmentation, as described and illustrated in (Beucher and Lantuéjoul 1979, Beucher 1982, Vincent and Beucher 1989, Vincent and Soille 1991).

Note that in morphology, the word *gradient* refers to an operation associating with each image pixel the *modulus* of its gradient − in the classical sense of the word. Most of the time, the gradient known as *Beucher's gradient* (Serra 1982) is used. It is obtained as the algebraic difference of a unit-size dilation and a unit-size erosion of image I:

$$grad(I) = (I \oplus B) - (I \ominus B).$$

However, depending on the type of image contours to be extracted, other gradients may be of interest: directional gradients, asymmetric gradients, regularized gradients, etc (Rivest *et al.* 1992).

The watershed transformation always provides *closed* contours and

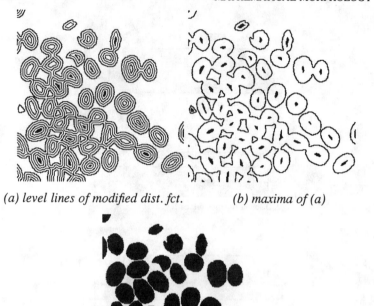

(a) level lines of modified dist. fct.   *(b) maxima of (a)*

(c) watersheds of inverted modified distance: final result

Figure 6.32 *Watershed segmentation of coffee beans image.*

constitutes a very general approach to contour detection. However, rarely can it be used directly on gradient images without resulting in dramatic *over-segmentations:* the image gets partitioned in far too many regions, i.e., the correct contours are lost in a large number of irrelevant ones. This problem is mainly due to noise in the data: noise in the original image results in noise in its morphological gradient, this in turn causing it to exhibit far too many regional minima. This directly translates into far too many catchment basins, i.e., over-segmentation.

In the rest of this section, we use the classic image of Fig. 6.3b to illustrate the methodology exposed[‡]. This image represents a two-

---

[‡] This image is classic because it has been used in at least a dozen papers dealing with watershed segmentation.

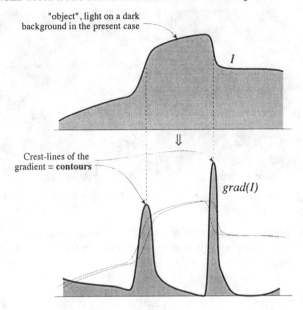

Figure 6.33 *Principle of grayscale segmentation via watersheds of the gradient.*

dimensional electrophoretic gel, obtained through the migration of proteins in an electric field. Such gels are becoming increasingly useful in criminal cases, as they provide a means to attach a 'genetic signature' with any individual. Here, our purpose is not only to detect all the dark spots in this image, but also to extract their outline as accurately as possible.

The standard morphological gradient of this image is shown in Fig. 6.34a. As mentioned above, if we simply compute the watersheds of Fig. 6.34a, the result is clearly disappointing, as proved by Fig. 6.34b. Indeed, the gradient exhibits a large number of minima, mainly due to the presence of noise in the original data. Note however that *all* the correct spot contours are present: they are just lost among hundreds of 'illegal' contour elements.

Several approaches have been proposed in literature to overcome this over-segmentation: for example, some techniques remove arcs of the watersheds based on an integration of the gradient's gray values along them. Others, known as *region-growing* techniques, take the dual point of view and merge adjacent regions (i.e., catchment basins here) when the gray level of the original image over them is comparable. None of

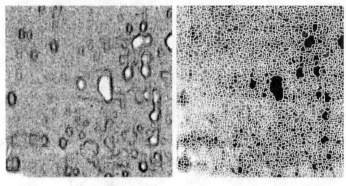

(a) gradient of gel image   (b) watersheds of gradient

Figure 6.34 *Using the watershed operator directly on the original image's gradient typically results in extreme oversegmentation. In (b), the watershed lines are superimposed on the original gel image.*

these techniques is satisfactory in that it is very difficult to incorporate to them knowledge specific to the collection of images under study.

An approach developed in the mathematical morphology community has proved to be a lot more robust and widely applicable to a range of image segmentation problems. This approach is as follows:

- Make use of image-specific knowledge (e.g., size, shape, location or brightness of the objects to extract) to design robust object marking procedures (Vincent and Beucher 1989, Vincent and Soille 1991, Beucher and Meyer 1992). This step of the segmentation can be completely different from one problem to another. Not only must each object be uniquely marked, the *background* – if any – is also a region, and needs to be marked as well. The idea behind this marker extraction step is that markers are generally easier to extract than actual object contours. By extracting markers first, we reduce the complexity of the problem and make it easier to incorporate domain-specific information into the segmentation.

- Use these markers to *modify the gradient image* on which watersheds are computed. Let us stress that the markers are not used to help in the postprocessing of the oversegmented watershed image, but instead, to alter the (gradient) image on which the watershed transformation is computed.

Let us describe this second step in more detail: let I denote the original grayscale image, let $J = grad(I)$ be its morphological gradient (or rather, a gradient chosen to best enhance the desired edges), and let M denote the binary image of markers previously extracted. The 'modification' of J should result in a grayscale image J' with the following characteristics:

— its only regional minima are exactly located on the connected components of M (M is the set of 'imposed' minima);

— its only crest-lines are the highest crest-lines of J that are located between the imposed minima.

The watersheds of J' would then be the highest crest-lines of $grad(I)$ that separate our markers, that is, the optimal contours with respect to our markers M and our gradient J.

The actual computation from J and M of an image J' with these characteristics has been classically achieved using a three-step process (Vincent and Beucher 1989):

1. Set to h_{min} any pixel of J that is located on a marker, h_{min} being chosen such that $\forall p, h_{min} < J(p)$. This results in a new image J^*:

$$\forall p, \quad J^*(p) = \begin{cases} h_{min} & \text{if } M(p) = 1 \\ J(p) & \text{otherwise.} \end{cases}$$

2. Create the following grayscale image M^*

$$\forall p, \quad M^*(p) = \begin{cases} h_{min} & \text{if } M(p) = 1 \\ h_{max} & \text{otherwise,} \end{cases}$$

where h_{max} is chosen such that $\forall p, J(p) < h_{max}$.

3. Use M^* to remove all the unwanted minima of J^* while preserving its highest crest-lines between markers. This is done using the dual grayscale reconstruction operation ρ^*:

$$\forall p, \quad J'(p) = \rho^*_{J^*}(M^*). \tag{6.52}$$

This process is illustrated by Fig. 6.35. The watersheds of the resulting image J' then provide the desired segmentation.

The whole procedure presented above is often referred to as *marker-driven watershed segmentation*. It is extremely powerful in a number of complex segmentation cases, where it essentially reduces the segmentation task to (1) the choice of a gradient and (2) the extraction of object markers (this latter task being itself very complex in some cases).

In practice, the two steps of gradient modification followed by watershed extraction can be folded into one. For example, the watershed

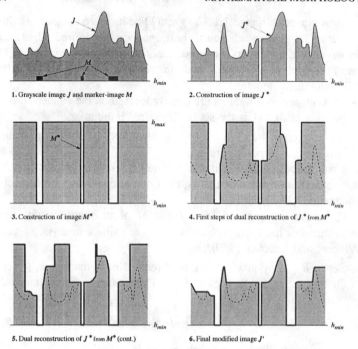

1. Grayscale image J and marker-image M

2. Construction of image J^*

3. Construction of image M^*

4. First steps of dual reconstruction of J^* from M^*

5. Dual reconstruction of J^* from M^* (cont.)

6. Final modified image J'

Figure 6.35 *Use of dual grayscale reconstruction to 'impose' a set M of minima to a graycale image J.*

algorithm introduced by Vincent and Soille (1991) can easily be modified to take a gradient image *and* a marker image as input, and directly produce the set of watershed lines of the modified gradient.

On our electrophoresis gel example, avoiding over-segmentation requires the prior extraction of correct spot markers. Since the spots constitute the dark part of the image, they can be interpreted as the image minima. However, once again, the noise in this data would make a direct minima extraction completely useless. Fortunately, filtering the original gel image using a standard morphological filter called an *alternating sequential filter (ASF)* (Serra 1988, Serra and Vincent 1992) is sufficient to produce an image whose minima correctly mark the spots, as shown in Fig. 6.36a.

In fact, this marker extraction step is followed by binary watershed segmentation in order to cut markers like the upper-right corner one, which clearly should mark two different spots. Arguably, even after this, our set of markers is still not perfect, but it is good enough for our present purpose. Note that the quality of the final result is directly related to the quality of the initial marker extraction: poor markers automatically result in poor segmentation. (On the other hand, good markers do not absolutely guarantee that the final segmentation will be perfect.)

The background marker of the gel image is easily extracted as the set of the highest crest-lines of the original image that separate the spot markers. This is the best way to ensure that this marker is located on the brightest areas of the image and separates all the spot markers. Its determination is done in a similar way as gradient modification (see Eq. 6.52 and Fig. 6.35). It is shown in Fig. 6.36b.

Both sets of markers are then combined in a final marker image, shown in Fig. 6.36c. It is used to modify the gradient of Fig. 6.34a, this resulting in Fig. 6.36d. The watersheds of the latter image provide the desired segmentation, as shown in Figs 6.36e, f.

The result is in accordance with our expectations: each spot has a unique contour which is located on the inflexion points of the initial luminance function – i.e. the original image. Given the extracted set of markers, we found the optimal gradient crest-lines, i.e. the best possible segmentation for these markers and, to a lesser extent, this gradient. Notice that in a few places, the contours shown in Figs 6.36e, f leave to be desired: this means that the gradient should have been more care-fully designed for this particular segmentation problem (the standard morphological gradient was used here). Similarly, some spots have not been detected, and some appear to have been merged: this is directly related to the less-than-perfect marking method that was crafted here.

Numerous applications illustrating the power of this marker-driven watershed segmentation paradigm could be shown, but then again, a single chapter would not be sufficient. We refer the reader to the litera-ture, and in particular to the following articles and chapters: Talbot and Vincent 1992, Vincent and Beucher 1989, Vincent and Masters 1992, Vincent 1993a, Vincent and Dougherty 1994.

6.3.5 Dynamics and hierarchical watershed segmentation

This section on state-of-the-art morphological segmentation techniques would not be complete without a brief summary on the concept of *dynamics*, its use for segmentation, and some of the derived trans-formations. Earlier in this section, the usefulness of regional *minima* and *maxima* was illustrated. Not only do maxima and minima provide

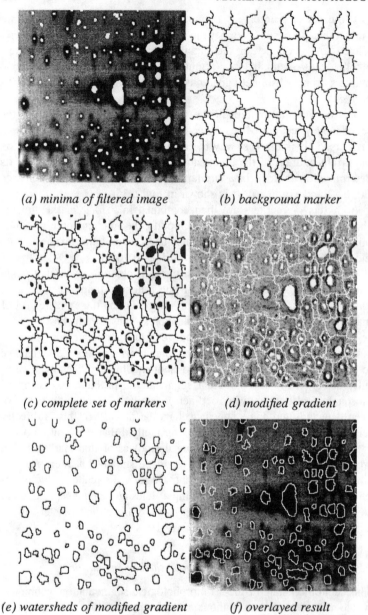

(a) minima of filtered image   *(b) background marker*

(c) complete set of markers   *(d) modified gradient*

(e) watersheds of modified gradient   *(f) overlayed result*

Figure 6.36 *Watershed segmentation of electrophoresis gel image.*

means to extract object markers for binary and grayscale segmentation, they are also the basis for such concepts as catchment basins and watersheds.

In Section 6.3.4, we also showed how irrelevant maxima could be eliminated using grayscale reconstruction (see Prop. 6.4): the technique proposed there, based on reconstructing the image I from $I-1$, enabled us to eliminate maxima based on their relative height (brightness) with respect to other maxima. Specifically, maxima that are such that one can 'walk' from one to the other on a path that never dips down more than one gray level got merged (see Fig. 6.27). In addition, if these maxima were such that one could reach a higher maximum by traveling along a path that dips down by at most one gray level, they were eliminated altogether.

Now, instead of reconstructing I from $I-1$, one could instead reconstruct from $I-2$. This would eliminate even more maxima. Utlimately, by reconstructing I from all the $I-n$, with $n=0,1,2\ldots$, one can tag each maximum of an image by the first value n for which this maximum disappears. This is called the *dynamics* of this maximum. Alternatively, we can also state the following definition, originally proposed by Grimaud (1992):

Definition 6.27 *The dynamics of a regional maximum m of a grayscale image I is defined as:*

$$Dyn(m) = \min_{P}\{I(m) - I_{\min}(P)\} \qquad (6.53)$$

where P is a path (connected set of pixel) joining m and a maximum higher than m, and where $I_{\min}(P)$ is the minimal value of image I along P.

In simpler terms, and using again a topographic analogy, the dynamics of m is equal to the minimal altitude loss that has to occur when traveling between m and a maximum of higher altitude. This is illustrated by Fig. 6.37. The dynamics of image *minima* is defined in a similar fashion.

This definition raises a couple of questions: first, what is the dynamics of the highest maximum in the image? According to the above definition, it is undefined. For consistency, we just need to make sure that it is higher than the dynamics of any other maximum in the image. Second: what happens with maxima of the same altitude? Indeed, the definition did not specify whether the maxima higher than m were (a) strictly higher, or (b) higher or equal. Neither is the correct answer, and explaining why would be beyond the scope of this chapter (this issue

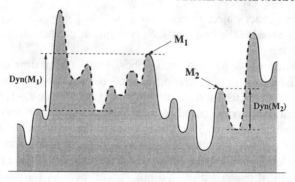

Figure 6.37 *Dynamics of maxima.*

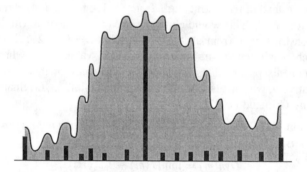

Figure 6.38 *Dynamics of the maxima inside a white blob: only* one *maximum has a significant dynamics value.*

is discussed at length in Grimaud 1991, 1992). The important thing to remember is that Definition 6.27 is not complete, and choices will have to be made in the implementation.

The dynamics provides a new and robust way to select maxima/minima based purely on their contrast – size and shape are irrelevant. For example, the task of extracting markers of white blobs in an image can be tackled using regional maxima. However, even if the image is only slighly noisy, it is more than likely that the blobs will exhibit multiple maxima. This can be addressed via dynamics: indeed, as illustrated by Fig. 6.38, only *one* of these maxima will have a significant dynamics value, whereas all the others will have low dynamics.

This principle can be put to good use with our electrophoresis gel image: in order to extract blob markers, the method used in Section 6.3.4

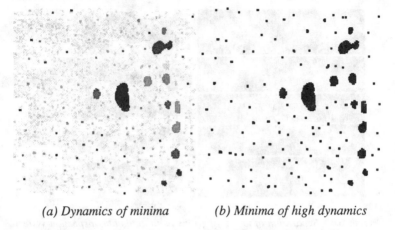

(a) Dynamics of minima   (b) Minima of high dynamics

Figure 6.39 *Using the dynamics of minima to extract robust spot markers.*

consisted of smoothing the image by alternating sequential filtering, and then extracting minima. An alternative method is simply to extract the minima whose dynamics is greater than or equal to a given threshold. This is ilustrated in Fig. 6.39, and provide a set of markers that is at least as good as the one used previously.

The dynamics is one of the most useful tools recently added to the morphological toolbox. Just like the distance function can be seen as a multiscale erosion, and the granulometry function (see Section 6.4) can be seen as a multiscale opening, the dynamics is the multiscale version of the operation consisting of subtracting a constant h from a gray image I, and reconstructing I from $I - h$. As such, it contains a lot of information that is invaluable in numerous segmentation application.

Fig. 6.40 illustrates this last point: the dynamics is the key element of the method that was designed to robustly separate touching grains in an automated inspection system. The actual segmentation algorithm designed to solve this particular application is a combination of both binary and grayscale segmentation techniques presented in Section 6.3.4.

To conclude this section, let us mention that the notion of dynamics extends to watershed lines in the following way: given an image I, and given h and h', $h < h'$, we can extract the set S_h of minima of dynamics greater than or equal to h, and the set $S_{h'}$ of minima of

Figure 6.40 *Separation of touching grains in an automatic inspection system.*

dynamics greater than or equal to h'. Obviously:

$$S_{h'} \subseteq S_h.$$

Now, using S_h and $S_{h'}$ as sets of markers, we can use the method proposed in the Section 6.3.4 and extract the corresponding *marker-constrained* watershed lines of I, denoted by W_h and $W_{h'}$ respectively. Here again:

$$W_{h'} \subseteq W_h.$$

Indeed, a watershed lines L located between a minimum m such that $m \in S_h$, $m \notin S_{h'}$, and any other minimum of S_h exists in W_h, but not in $W_{h'}$. From this, a method for tagging each watershed line with its dynamics is derived: the dynamics of a watershed line L is equal to the dynamics value h such that L exists in W_h, but not in W_{h+1}. This was originally proposed by Najman and Schmitt (1994).

As an illustration, consider the metallic alloy image of Fig. 6.21a. Using as markers the *maxima* of increasing dynamics, and extracting marker-driven watersheds on the *inverted* original image (to turn grain boudaries into crest-lines), a series of watershed images is derived, each image containing the next one. The information contained in this series of images can be condensed into one image, namely the image of the dynamics-tagged watershed lines, also known as hierarchical watershed image. This is illustrated by Fig. 6.41.

Hierarchical watersheds constitute a very new concept, which the morphological image analysis community is just beginning to explore.

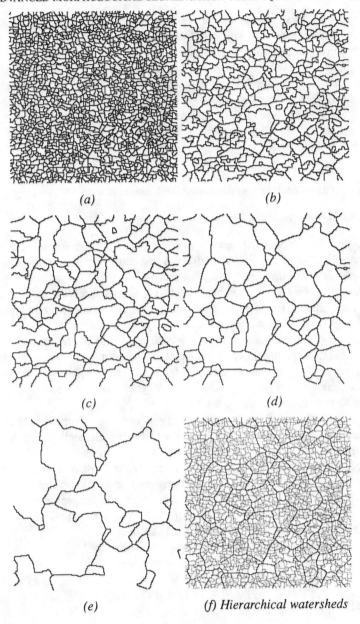

Figure 6.41 (a)–(e): watershed lines of Fig. 6.21a (inverted) when the set of markers used are maxima of increasing dynamics. (f): corresponding image of dynamics-tagged watershed lines, or hierarchical watersheds.

Their usefulness for interactive segmentation (i.e., where the dynamics threshold is manually adjusted by a user) is already very clear. Hierarchical watersheds are also potentially very powerful for complex segmentation tasks. Furthermore, beyond dynamics, other criteria can be used to tag watershed edges, as described by Vachier and Vincent (1995). This leads to a large class of new watershed-based tools, and we can expect to read more and more about them, as image analysts begin to use them on practical applications and realize how powerful these techniques are.

6.4 Granulometries: applications and algorithms

6.4.1 Introduction, background on granulometries

In the present section, another major topic in morphological image analysis is covered: *granulometries*. They constitute one of the most useful and versatile sets of tools of morphological image analysis, and can be applied to a wide range of tasks, from feature extraction, to texture characterization, to size estimation, to image segmentation, etc. Though these operations are nearly 30 years old, they have become more popular and more useful in recent years, due to the general increase in computing power available at one's desktop, and due to recent advances in algorithms. These two factors combined make it possible to efficiently compute a whole range of granulometries, and this has led to new uses for these old tools.

Even today, traditional granulometry algorithms, which involve sequences of openings or closings with structuring elements of increasing size, are often prohibitively costly on non-specialized hardware. The primary goal of this section is to cover the most recent algorithmic advances. The proposed algorithms are often orders of magnitude faster than previously available techniques, thereby opening a range of new applications for granulometries. Several such applications of granulometries will be described in this section.

The concept of granulometries was introduced by Matheron (1967) as a new tool for studying porous media. The size of the pores in such media was characterized using series of openings with structuring elements of increasing size (Serra 1982). The theoretical study of these operations led Matheron to propose the following definition:

Definition 6.28 *Let* $\Phi = (\phi_\lambda)_{\lambda \geq 0}$ *be a family of image transformations depending on a unique parameter* λ. *This family constitutes a granu-*

lometry if and only if the following properties are satisfied:

$$\forall \lambda \geq 0, \phi_\lambda \text{ is increasing,} \tag{6.54}$$

$$\forall \lambda \geq 0, \phi_\lambda \text{ is anti-extensive,} \tag{6.55}$$

$$\forall \lambda \geq 0, \mu \geq 0, \phi_\lambda \phi_\mu = \phi_\mu \phi_\lambda = \phi_{\max(\lambda,\mu)}. \tag{6.56}$$

Property (6.56) implies that for every $\lambda \geq 0$, ϕ_λ is an idempotent transformation. Therefore, $(\phi_\lambda)_{\lambda \geq 0}$ is nothing but a decreasing family of algebraic openings (Serra 1982). Conversely, one can prove that for any convex set B, the family of the openings with respect to $\lambda B = \{\lambda b \mid b \in B\}$, $\lambda \geq 0$, constitutes a granulometry (Matheron 1975).

More intuitively, suppose now that the transformations considered are acting on discrete binary images, or sets. In this context, a granulometry is a sequence of openings ϕ_n, indexed on an integer $n \geq 0$. Each opening is smaller than the previous one:

$$\forall X, \quad \forall n \geq m \geq 0, \quad \phi_n(X) \subseteq \phi_m(X). \tag{6.57}$$

The granulometric analysis of X with family of openings $(\phi_n)_{n \geq 0}$ is often compared to a *sifting* process: X is sifted through a series of sieves with increasing mesh size. Each opening (corresponding to one mesh size) removes more than the previous one, until the empty set is finally reached. The rate at which X is sifted is characteristic of this set and provides a 'signature' of X with respect to the granulometry used. Denote by $m(A)$ the measure of a set A (area or number of pixels in 2-D, volume in 3-D, etc):

Definition 6.29 *The granulometric curve or pattern spectrum (Maragos 1989) of a set X with respect to a granulometry* $\Phi = (\phi_n)_{n \geq 0}$ *is the mapping* $PS_\Phi(X)$ *given by:*

$$PS_\Phi(X)(n) = m(\phi_n(X)) - m(\phi_{n-1}(X)). \tag{6.58}$$

Since $(\phi_n(X))_{n \geq 0}$ is a decreasing sequence of sets ($\phi_0(X) \supseteq \phi_1(X) \subseteq \phi_2(X) \supseteq \cdots$), it is possible to condense its representation by introducing the concept of granulometry function (Laÿ 1987, Vincent and Beucher 1989, Haralick, Chen and Kanungo 1992):

Definition 6.30 *The granulometry function or opening function* $G_\Phi(X)$ *of a binary image X for granulometry* $\Phi = (\phi_n)_{n \geq 0}$ *maps each pixel* $x \in X$ *to the size of the first n such that* $x \notin \phi_n(X)$:

$$G_\Phi(X)(x) = \min\{n > 0 \mid x \notin \phi_n(X)\}. \tag{6.59}$$

For any $n > 0$, the threshold of $G_\Phi(X)$ above a value n is equal to

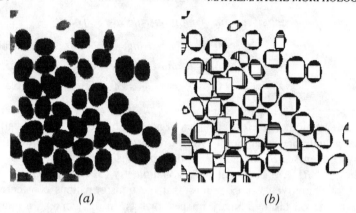

(a)   *(b)*

Figure 6.42 *(a) Square granulometry function of coffee beans image of Fig. 6.3a, in which dark regions correspond to higher pixel value; (b) level lines of this granulometry function.*

$\phi_n(X)$:

$$\phi_n(X) = \{p \in X \mid G_\Phi(X)(p) > n\}.$$

The following property follows immediately and states that the pattern spectrum can be obtained as histogram of the granulometry function:

Proposition 6.5 *The pattern spectrum $PS_\Phi(X)$ of X for granulometry $\Phi = (\phi_n)_{n \geq 0}$ can be derived from the granulometry function $G_\Phi(X)$ as follows:*

$$PS_\Phi(X)(n) = \mathrm{card}\{p \mid G_\Phi(X)(p) = n\}, \qquad (6.60)$$

where card *stands for the cardinal (number of pixels) in a set.*

This result is straightforward to prove. The concept of granulometry function is central to the algorithms described in section 6.4.4. An example of square granulometry function is shown in Fig. 6.42.

The granulometries that have been described so far are often referred to as granulometries *by openings*. By duality, granulometries *by closing* can also be defined; the granulometric analysis of a set X with respect to a family of closings is strictly equivalent to the granulometric analysis of X^C (complement of X) with the family of dual openings. Therefore, from now on, only granulometries by openings are considered. Similarly, these notions can be directly extended to grayscale images; in this context, the measure m chosen is the 'volume' of the image processed, i.e. the sum of all its pixel values.

The granulometric analysis of Fig. 6.3a with respect to a family of openings with squares (as was used for Fig. 6.42) is shown in Fig. 6.43. From the resulting pattern spectrum, the dominant size of the beans ('size' being defined as size of the largest square a bean can contain) in this image can easily be derived. Granulometries therefore allow one to extract size information without any need for prior segmentation: the beans in this image are highly overlapping, yet their size can be estimated without individually identifying each bean.

Granulometries have been used for a variety of other image analysis tasks, including shape characterization and feature extraction (see for example Yang and Maragos 1993), texture classification (Chen and Dougherty 1992), and even segmentation (Dougherty *et al.* 1992). Nonetheless, until recently, granulometric analysis involved performing a series of openings and/or closings of increasing size, which is prohibitively expensive for most applications, unless dedicated hardware is used.

In section 6.4.2, the literature on granulometry algorithms is briefly reviewed. A comprehensive set of fast algorithms for computing granulometries in binary images is then proposed: linear granulometries (i.e., granulometries based on openings with line segments) form the easiest case, and are computed using image 'run-length'. They constitute the topic of Section 6.4.3. The 2-D case (granulometries with square or 'diamond'-shaped structuring elements, or granulometries with unions of line-segments at different orientations) is covered in Section 6.4.4. It involves the determination of opening functions, and following Proposition 6.5, pattern spectra are then derived by simple histogramming. The grayscale case is then addressed in Section 6.4.5, and a very efficient algorithm for grayscale linear granulometries is described. Finally, the concept of an *opening tree* is proposed in Section 6.4.6, and used to efficiently compute granulometries based on maxima of linear openings, minima of linear closings, and even 'pseudo-granulometries' by minima of linear openings or maxima of linear closings. The efficiency of these latter algorithms enables one to use granulometries where previously unthinkable, a point which we illustrate on a number of examples.

6.4.2 Background on granulometry algorithms

The literature on mathematical morphology is not short of algorithms for computing erosions and dilations, openings and closings, with various structuring elements, in binary and in grayscale images. Reviewing them would be beyond the scope of this chapter. But no matter how

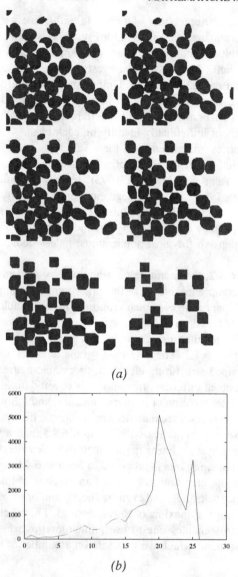

(a)

(b)

Figure 6.43 *(a) Successive openings of a Fig. 6.3a using squares of increasing size as structuring elements. (b) Corresponding granulometric curve, or pattern spectrum: the peak at size 20 indicates the dominant size of the beans in the original binary image.*

efficient an opening algorithm is used, determining a pattern spectrum using a sequence of openings is a very time-consuming task given the number of operations involved. Furthermore, since the size of the structuring element increases with n, so does the computation time of the corresponding opening. Even if we assume that the computation time of $\phi_n(X)$ (n-th opening in the series) can be done in constant time (which is not always true depending on the structuring element and on the opening algorithm used), determining the pattern spectrum up to size n using openings is still an $O(n)$ algorithm.

The few granulometry algorithms found in literature only deal with the binary case, and have in common the use of granulometry functions as an intermediate step. The algorithm proposed by Yuan (1991) for determining binary square granulometries consists of first determining the *quench function* of the original set X. The quench function maps each pixel p of the skeleton (medial axis) S_X of X to the size (radius) $S_X(p)$ of the corresponding maximal square. (See Vincent (1991b) for more details on these concepts.) In a second step, each pixel p of the skeleton is replaced by a square centered at this pixel, with size $S_X(p)$, and gray-level $S_X(p)+1$. The pixelwise maximum of all these squares provides granulometry function of X. This algorithm is faster than the brute force method described in the previous paragraph, but still requires a significant amount of image scans. In addition, the more complicated the image or the larger the objects in it, the longer this method takes.

Suprisingly, a better algorithm can be found in an earlier paper by Laÿ (1987), in which the author devotes a few lines to the description of a sequential algorithm (Rosenfeld and Pfaltz 1966, Vincent 1991a) based on the distance function (Rosenfeld and Pfaltz 1968, Borgefors 1986), and also using the granulometry function as an intermediate step. This algorithm still provides one of the most efficient implementations to date for binary granulometries with structuring elements such as squares and hexagons. In section 6.4.4, this technique is described in detail, and is extended to other types of binary granulometries.

The algorithm proposed by Haralick, Chen and Kanungo (1992) is interesting in that it allows in principle to compute granulometry functions with respect to any family of homothetic elements[§]. However, for simple structuring elements such as squares, this technique is not as efficient as the one mentioned in the previous paragraph, because

[§] The base element does not even have to be convex!

its elementary steps (propagation and merging of lists of 'propagators')
are rather computationally intensive, therefore slow.

6.4.3 Linear granulometries in binary images

Linear granulometries in binary images constitute the simplest possible
case of granulometries. Let us for example consider the horizontal granulometry, i.e., the granulometry by openings with the $(L_n)_{n \geq 0}$ family
of structuring elements, where:

$$L_n = \underbrace{\bullet \ \bullet \ \bullet \ \cdots \ \bullet \ \bullet}_{n+1 \text{ pixels}} \tag{6.61}$$

From now on, we use the convention that the center of a structuring
element is marked using a thicker dot than is used for the other pixels.
Note that the location of the center of the structuring elements used has
no influence on the resulting granulometry.

   Let us analyze the effect of an opening by L_n, $n \geq 0$ on a discrete set X (binary image). The following notations are used from now
on: the neighbors of a given pixel p in the square grid are denoted
$N_0(p), N_1(p), \ldots , N_7(p)$, and the eight elementary directions are en-
coded in the following way:

$$\begin{matrix} 3 & 2 & 1 \\ 4 & \bullet & 0 \\ 5 & 6 & 7 \end{matrix}$$

For a direction $d \in \{0, 1, \ldots , 7\}$ and $k \geq 0$, we denote by $N_d^{(k)}(p)$ the
k-th order neighbor of pixel p in direction d:

$$N_d^{(0)}(p) = p, \quad \text{and} \quad k > 0 \Longrightarrow$$
$$N_d^{(k)}(p) = N_d(N_d^{(k-1)}(p)). \tag{6.62}$$

The opposite of direction d is denoted $\check{d}$. For example, if $d = 3$, then
$\check{d} = 7$.

Definition 6.31 *The* ray *in direction d at pixel p in set X is given by:*

$$r_{X,d}(p) = \{N_d^{(k)}(p) \mid k \geq 0 \text{ and}$$
$$\forall 0 \leq j \leq k, N_d^{(j)}(p) \in X\}. \tag{6.63}$$

With each pixel $p \in X$, we also associate a *run* in direction d, defined
as the union of the rays in direction d and in direction $\check{d}$.

Definition 6.32 *The* run *in direction d at pixel p in set X is given by:*

$$R_{X,d}(p) = r_{X,d}(p) \cup r_{X,\check{d}}(p). \qquad (6.64)$$

The number of pixels in a run R is called the *length* of this run and denoted $l(R)$.

The following proposition is immediate:

Proposition 6.6 *The opening of X by L_n, denoted here $X \circ L_n$, is the union of the horizontal runs $R_{X,0}(p)$ whose length is strictly greater than n:*

$$X \circ L_n = \bigcup_{p \in X} \{R_{X,0}(p) \mid l(R_{X,0}(p)) > n\}. \qquad (6.65)$$

Therefore, any horizontal run of length n is left unchanged by all the openings with L_k, $k < n$, and is removed by any opening with L_k, $k \geq n$. Hence, the corresponding pattern spectrum PS_0 satisfies:

$$PS_0(X)(n) = \text{card}\{p \in X \mid l(R_{X,0}(p)) = n\}. \qquad (6.66)$$

An extremely efficient 1-scan horizontal granulometry algorithm is easily derived from this formula:

Algorithm: horizontal binary granulometry

- Initialize pattern spectrum: for each $n > 0$, PS[n] $\leftarrow 0$
- Scan each line of image from left to right.
- In this process, each time a run R is discovered, do:
  PS[$l(R)$] $\leftarrow$ PS[$l(R)$] $+ l(R)$;

In applications where directional information is of interest, this algorithm provides a very useful and efficient way to extract size information characterizing the image under study. Consider for example Fig. 6.44a, which is a binary image of lamellar eutectics; Schmitt (1991) proposed a variety of methods for extracting the defect lines present in this image. Different methods used different kind of information about this image, and some required a knowledge of the typical width of the lamellae. This width can be accurately estimated by adapting the previous algorithm to the computation of linear granulometries at +45 degree orientation (direction perpendicular to the lamellae). The resulting pattern spectrum is shown in Fig. 6.44b, and its peak at index 3 indicates that the typical width of the lamellae is of 3 pixels.

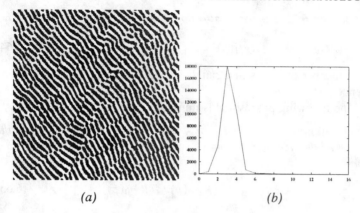

Figure 6.44 *Binary image of lamellar eutectics (a) and its granulometric curve using line segments at +45 degrees orientation (b).*

6.4.4 Granulometry functions on binary images

For non 1-D granulometries, the direct approach described in the previous section becomes intractable. Consider for example the case of a granulometry $(\phi_n)_{n \geq 0}$ where ϕ_n is a maximum of openings with the horizontal segment L_n and its vertical counterpart $L_n^{\perp}$. For each pixel, it becomes necessary to know the size of the horizontal run as well as the vertical run it belongs to.

Linear granulometry functions are therefore the required step. Given the horizontal and the vertical granulometry functions of X, the granulometry function of X corresponding to the $(\phi_n)_{n \geq 0}$ of previous paragraph is simply obtained as a pixelwise maximum. More generally, the same is true for any two granulometry functions, and the following proposition can be stated:

Proposition 6.7 *Let* $\Phi = (\phi_n)_{n \geq 0}$ *and* $\Psi = (\psi_n)_{n \geq 0}$ *be two granulometries. Then,* $\max(\Phi, \Psi) = (\max(\phi_n, \psi_n))_{n \geq 0}$ *is also a granulometry and for any set* X:

$$G_{\max(\Phi, \Psi)}(X) = \max(G_\phi(X), G_\psi(X)). \tag{6.67}$$

Indeed, for any n, $\max(\phi_n, \psi_n)$ is an opening, and for any $n, m, n \leq$

Figure 6.45 *Linear granulometry function of image 6.44a, whose histogram directly provides the curve of Fig. 6.44b.*

m, $\max(\phi_n, \psi_n) \geq \max(\phi_m, \psi_m)$. Moreover, for any $x \in X$:

$$
\begin{aligned}
G_{\max(\Phi, \Psi)}(X)(x) &= \min\{n > 0 \mid x \notin (\max(\phi_n, \psi_n))(X)\} \\
&= 1 + \max\{n > 0 \mid x \in (\max(\phi_n, \psi_n))(X)\} \\
&= 1 + \max\{n > 0 \mid x \in \phi_n(X) \cup \psi_n(X)\} \\
&= \max\{1 + \max\{n > 0 \mid x \in \phi_n(X)\}, \\
&\qquad\quad 1 + \max\{n > 0 \mid x \in \phi_n(X)\}\} \\
&= \max\{G_\phi(X)(x), G_\psi(X)(x)\}
\end{aligned}
$$

which completes the proof. Determining the linear granulometry function of a binary image is a relatively straightforward task. Take for example the horizontal case: like in the previous section, the principle of the granulometry function algorithm is to locate each horizontal run. But now, in addition, each run R gets also tagged with its length $l(R)$. This involves scanning the black (ON) pixels of the image twice, and the white (OFF) pixels only once. The resulting algorithm is straightforward and is hardly more time consuming than the one described in Section 6.4.3. Examples of linear granulometry functions are shown in Fig. 6.45 and in Fig. 6.46.

The case of truly 2-D binary granulometry functions is the next level up in complexity. In the rest of this section, we focus on granulometry functions $G_S(X)$ based on openings with the homothetics of elementary square S. The algorithm described extends to granulometry functions

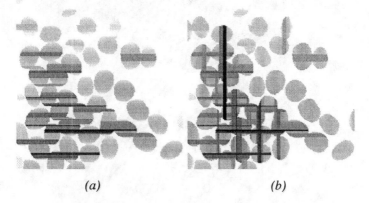

$$(a) \qquad\qquad\qquad (b)$$

Figure 6.46 *(a) Horizontal granulometry function. (b) Pixelwise maximum of horizontal and vertical granulometry functions provides the granulometry function corresponding to maxima of openings with vertical and horizontal line segments.*

$G_D(X)$, based on openings with the homothetics of 'diamond' shape D, as described by Vincent (1994a).

$$S \ = \ \begin{matrix} \bullet & \bullet \\ \bullet & \bullet \end{matrix} \quad ; \quad D \ = \ \begin{matrix} & & \bullet \\ \bullet & \bullet & \bullet \\ & & \bullet \end{matrix} \qquad (6.68)$$

Together with the linear case, these granulometries cover 99% of all practical needs.

Like in Haralick's algorithm (Haralick, Chen and Kanungo 1992), the first step of the proposed technique consists of computing what some authors have called a *generalized distance function* (Bertrand and Wang 1988, Haralick and Shapiro 1991). Let B be an arbitrary structuring element containing its center. Let

$$nB = \underbrace{B \oplus B \oplus \cdots \oplus B}_{n\,\text{times}}$$

denote the structuring element 'of size n'. Let also ε_B denote the erosion by structuring element B (Serra 1982):

$$\varepsilon_B(X) = X \ominus \check{B}.$$

Definition 6.33 *The generalized distance function $d_B(X)$ with respect to the family of structuring elements $(nB)_{n>0}$ assigns to each pixel $p \in$*

X the smallest $k > 0$ such that $p \notin \varepsilon_{kB}(X)$:

$$d_B(X)(p) = \min\{k > 0 \mid p \notin \varepsilon_{kB}(X). \tag{6.69}$$

Generalized distance functions are determined using sequential algorithms that are straightforwardly derived from the original algorithm proposed by Rosenfeld and Pfaltz (1966, 1968). When the center of the structuring element is in the bottom-right corner of element B (last pixel met in a raster-order scan of this element), the distance function $d_B(X)$ can be computed in one single raster scan.

In the case where $B = S$ (see Eq. (6.68)), the following algorithm can be proposed:

Algorithm: Generalized dist. func. with square S

- Input: binary image I of set X
- Scan I in raster order;
    - Let p be the current pixel;
    - if $I(p) = 1$ (p is in X):
        $I(p) \leftarrow \min\{I(N_4(p)), I(N_3(p)), I(N_2(p))\} + 1$;

An example of generalized distance function resulting from this algorithm is shown in Figs 6.47a–b. A way to interpret the result is to say that, for each pixel p, if one was to translate structuring element $[d_S(X)(p)]S$ so that its center coincides with p, this translated element – denoted $p + [d_S(X)(p)]S$ – would be entirely included in X. However, $p + [d_S(X)(p) + 1]S \notin X$. We can therefore state the following proposition:

Proposition 6.8 *The granulometry function $G_S(X)$ is obtained from $d_S(X)$ as follows:*

$$\forall p \in X, \quad G_S(X)(p) = \max\{d_S(X)(q) \mid$$
$$p \in (q + d_S(X)(q)S)\}. \tag{6.70}$$

This result is intuitively obvious and can be proved easily. In algorithmic terms, it means that we can compute $G_S(X)$ by propagating the value $d_S(X)(p)$ of each pixel p over the square $p + d_S(X)(p)S$, and then by taking the pixelwise maximum of the values propagated at each pixel.

In the technique proposed by Haralick, Chen and Kanungo (1992), this propagation step is achieved via an anti-raster scan of the distance function image, in which, at each pixel, a list of propagated values is maintained. In the particular case of square granulometry function $G_S(X)$, computing the value at pixel p as well as the list of propagated values at p, requires a merging of the lists of propagated values at pixels $N_0(p)$, $N_6(p)$, and $N_7(p)$.

This merging step turns out to be expensive, and in the case of square granulometry function $G_S(X)$, a less general, but much more efficient technique can be proposed. This technique takes advantage of the fact that square S can be decomposed into the Minkowski addition of the two elementary line segments E_1 and E_2:

$$S = \quad\begin{matrix}\bullet & \bullet \\ \bullet & \bullet\end{matrix}\quad = \quad\begin{matrix}\bullet \\ \bullet\end{matrix}\ \oplus\ \begin{matrix}\bullet & \bullet\end{matrix}\quad = E_1 \oplus E_2. \qquad (6.71)$$

Therefore, the complex propagation step of the granulometry function algorithm described by Haralick, Chen and Kanungo (1992) can in fact be decomposed into two much simpler propagations, with substantial speed gain. The distance function extraction step is followed by two linear propagation steps that are identical, except that one propagates distance values leftward in each line, whereas the other one propagates values upward in each column.

The algorithm for right-to-left propagation of distance values is given below. Its principle is to propagate each pixel value $I(p)$ to the left $I(p) - 1$ times, or until a larger value v is found, in which case the list of propagated values is reset to this value ... The algorithm maintains an array propag containing the number of times each value remains to be propagated.

Algorithm: Left propag. of dist. values of $d_S(X)$

- Input: image I of the generalized distance function $d_S(X)$;
- For each line of the image, do:
    - Initializations: maxval ← 0 (current maximal value propagated);
    - Scan line from right to left:
        Let p be the current pixel;
        If $I(p) \neq 0$:
            If $I(p) >$ maxval:
                maxval ← $I(p)$;
            propag[$I(p)$] ← $I(p)$;
            ∀0 $< i <=$ maxval, propag[i] ← propag[i] − 1;
            maxval ← largest $i \leq$ maxval with propag[i] >= 0;
            $I(p)$ ← maxval

A few implementation tricks can speed up computation by substantially reducing the number of times the entire array propag is scanned per scanline. Their description would be beyond the scope of this chapter (see Vincent 1994a) for more a more in-depth discussion). The resulting algorithm is quasi-linear with respect to the number of pixels in the image, and is almost independent of object size (see Table. 6.1). Using again the coffee bean image as running example, the result of this propagation step is shown in Figs 6.47c–d, and the final granulometry function obtained after upward propagation in each column is shown in Figs 6.47e, f.

This algorithm can be adapted for granulometry functions with any structuring element that can be decomposed as a Minkowski addition of the elementary line segments E_1, E_2, E_3, and E_4 (See Eqs. (6.71) and (6.72)). It also extends to the computation of hexagonal opening functions in the hexagonal grid (Laÿ 1987) . Furthermore, although diamond D cannot be decomposed as the Minkowski addition of two line segments:

$$D = \begin{smallmatrix} & \bullet & \\ \bullet & \bullet & \bullet \\ & \bullet & \end{smallmatrix} \neq \begin{smallmatrix} \bullet & \oplus & \bullet \\ & & \bullet \end{smallmatrix} = E_3 \oplus E_4, \qquad (6.72)$$

the algorithm can be adapted to the computation of granulometry functions based on openings with the homothetics of D, as described by Vincent (1994a).

Table 6.1 summarizes the speed of these granulometry functions on the 256×256 coffee bean image used as running example. We chose not to compare these timings with those of traditional opening-based algorithms. The speed of the latter algorithms can indeed vary tremendously depending on the quality of the implementation. Note however that for this coffee bean image, which has approximately 30000 black pixels, Haralick, Chen, and Kanungo's algorithm (1992) takes between 0.5 s and 0.6 s to compute the square granulometry function shown in Fig. 6.47e, on a Sparc Station 2. This workstation being between two and three times slower than a Sparc Station 10, we can conclude that the algorithm described here is between three and four times faster.

6.4.5 Linear grayscale granulometries

Grayscale granulometries are potentially even more useful than binary ones, because they enable the extraction of information directly from grayscale images. A number of theoretical results have been published on them (see e.g. Kraus et al. 1992); however, since until recently, no efficient technique was available to compute grayscale granulome-

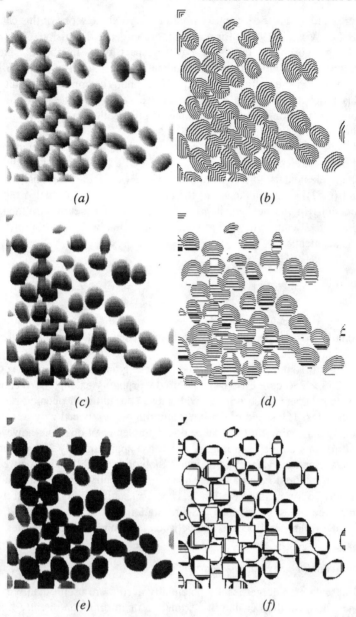

Figure 6.47 *Computation of granulometry function using square structuring elements. (a) generalized distance function; (b) level lines; (c) propagation of values from right to left; (d) level lines; (e) final granulometry function; (f) level lines of granulometry function.*

Table 6.1 *Execution time of various granulometry function algorithms on the* 256×256 *coffee bean image, measured on a Sun Sparc Station 10 workstation.*

Type of granulometry function	Execution time
horizontal	0.018s
max in 4 directions	0.207s
square	0.085s
'diamond'	0.094s

tries, they have not been used very much in practice. In this section, we remedy this situation and describe a new algorithm for computing linear grayscale granulometries. In the next section, we extend it to granulometries based on maxima of linear openings, as well as pseudo square granulometries. For more details, refer to (Vincent 1994b).

Without loss of generality, let us consider the horizontal case. The structuring elements considered now are the L_n's of equation (6.61). Let I be a discrete grayscale image.

Definition 6.34 *A horizontal maximum M of length $l(M) = n$ in grayscale image I is a set of pixels* $\{p, N_0^{(1)}(p), N_0^{(2)}(p), \ldots, N_0^{(n-1)}(p)\}$ *such that*

$$\forall i, \ 0 < i < n, I(N_0^{(i)}(p)) = I(p) \quad and$$

$$I(N_4(p)) < I(p), \quad I(N_0^{(n)}(p)) < I(p). \tag{6.73}$$

This notion is the 1-D equivalent of the classic *regional maximum* concept described in Section 6.3. The study of how such maxima are altered through horizontal openings is at the basis of the algorithm described here. The following proposition holds:

Proposition 6.9 *Let M be a horizontal maximum of I. Let $p_l \in M$ and $p_r \in M$ respectively denote the extreme left pixel and the extreme right pixel of M. Let $n = l(M)$ be the length of this maximum. Then, for any $p \in M$:*

$$\forall k < n, (I \circ L_k)(p) = I(p), \tag{6.74}$$

$$for \ k = n, (I \circ L_n)(p) = \max\{I(N_4(p_l)),$$

$$I(N_0(p_r))\} < I(p), \tag{6.75}$$

$$\forall k > n, (I \circ L_k)(p) < I(p). \tag{6.76}$$

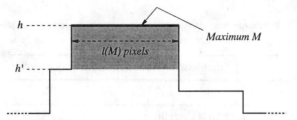

Figure 6.48 *Horizontal cross section of I with a maximum M. The shaded area, of volume $(h - h') \times l(M)$ shows the local contribution of this maximum to the $l(M)$-th bin of the horizontal pattern spectrum.*

The proof for this proposition is straightforward. Phrased differently, this result states that any opening of I by a line segment L_k such that $k < n$ leaves this maximum unchanged, whereas for any $k \geq n$, all the pixels of M have a lower value in $I \circ L_k$ than in I. Furthermore, we can *quantify* the effect of an opening of size n on the pixels of this maximum: the value of each pixel $p \in M$ is decreased from $I(p)$ to $\max\{I(N_l(p_0)), I(N_r(p_{n-1}))\}$. In granulometric terms, the contribution of maximum M to the n-th value of the horizontal pattern spectrum $PS_h(I)$ is:

$$n \times [I(p) - \max\{I(N_l(p_0)), I(N_r(p_{n-1}))\}]. \qquad (6.77)$$

This is illustrated by Fig. 6.48.

Additionally, the local effect of a horizontal opening of size n on M is that a new 'plateau' of pixels is created at altitude $\max\{I(N_l(p_0)), I(N_r(p_{n-1}))\}$. This plateau P contains M, and may be itself a maximum of $I \circ L_n$. If it is, we say that P is part of the *maximal region* $R(M)$ surrounding maximum M, and we can now compute the contribution of P to the $l(P)$-th bin of the pattern spectrum, etc.

Following these remarks, the principle of the present grayscale granulometry algorithm is to scan the lines of I one after the other. Each horizontal maximum M of the current line is identified, and its contribution to $PS_0(I)(l(M))$ is determined. If it turns out that after opening of size $l(M)$, the new plateau formed is still a maximum, the contribution of this maximum to the pattern spectrum is computed as well. The process is iterated until the plateau formed by opening is no longer a maximum, or until it becomes equal to the entire scanline considered. The next maximum of the current line is then considered, etc. This process is illustrated by Fig. 6.49. Specifically, the algorithm works as follows on each image line:

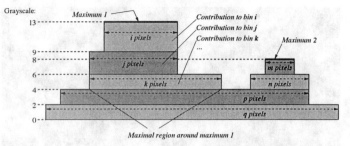

Figure 6.49 *Illustration of horizontal grayscale granulometry algorithm for a line with two maxima. In this case, maxima were scanned from left to right. Processing maximum 1 results in bins i, j, and k of the pattern spectrum being incremented. While processing maximum 2, bins m and n are first incremented, then the algorithm skips over the already processed maximal region, and bins p and q are incremented. The maximal region around maximum 2 is the entire line.*

Algorithm: horizontal granul. of an image line

- for each maximum M of this line (in any order) do:
  - Adds contribution of this maximum to $l(M)$-th bin of pattern spectrum;
  - Let P be the plateau of pixels formed by opening of size $l(M)$ of M; if P is itself a maximum of $I \circ L_{l(M)}$, computes its contribution to bin $l(P)$ of pattern spectrum;
  - Iterate previous step until the new plateau formed is no longer a maximum;
  - 'Mark' the maximal region around M as already processed;

Note that this algorithm is inherently recursive: once a maximal region R has been processed, all of its pixels are regarded as having the gray-level they were given by the last opening considered for R. In practice though, there is no need to physically modify the values of all the pixels in R: keeping track of the first and last pixels of R is sufficient, allowing the algorithm to efficiently skip over already processed maximal regions (see Fig. 6.49). Thanks to this trick, the algorithm only considers each image pixel *twice* in the worst case.

This algorithm was compared to the traditional opening-based technique. For the latter, a fast opening algorithm was used, whose speed is proportional to the number of pixels in the image and (almost) inde-

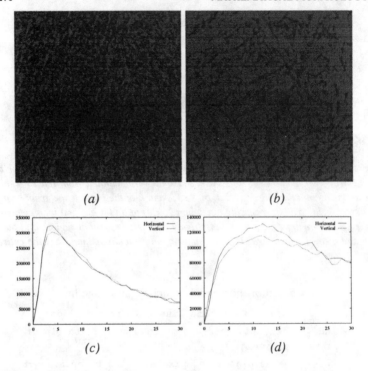

Figure 6.50 *Using linear grayscale granulometries to estimate object size without prior segmentation. Curve (c) clearly indicates that the typical width and height of the white patterns in image (a) is 4 pixels. Similarly, curve (d) shows that the typical width/height of the patterns in (b) is 12.*

pendent of the length of the line segment used as structuring element. As illustrated by Table 6.2, the new algorithm described in this section is *three orders of magnitude faster*.

We compared the speed of this algorithm to the traditional opening-based technique. For this latter, a highly optimized opening algorithm was used, which is linear with respect to the number of pixels in the image, and whose speed is (almost) independent of the length of the line segment used as structuring element. Both original 512×512 weld images of Fig. 6.50 were used for this comparison. As illustrated by Table 6.2, the algorithm described in this section is up to three orders of magnitude faster.

The speed of this new algorithm opens a range of new applications

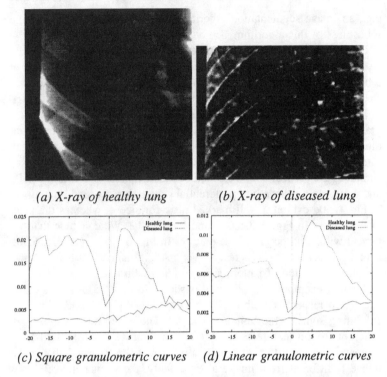

(a) X-ray of healthy lung   (b) X-ray of diseased lung

(c) Square granulometric curves   (d) Linear granulometric curves

Figure 6.51 *Example of texture differenciation using grayscale granulometries by openings and closings. The pattern spectra of (a) and (b) are completely different, whether linear or square granulometries are used (original study due to M. Grimaud).*

for grayscale granulometries. Traditionally, the practical problems granulometries have been used to address dealt with either texture discrimination, or feature extraction for object recognition. In the first case, either computation time was not an issue, or the discrimination task could be performed off-line. In the second case, granulometries were computed on very small images (e.g. characters), so that computation time could remain reasonable.

With this algorithm, it becomes possible to use grayscale granulometries more systematically: these tools indeed provide an efficient and accurate way to extract *global* size information directly from a grayscale image. Extracting this information is sometimes a goal in itself; but this size estimation can also be essential to calibrate the parameters of, e.g., an image segmentation algorithm, thereby greatly enhancing the robustness of this algorithm.

Figs 6.50a–b are used to illustrate how grayscale granulometries can be used to estimate size information[J]. These figures represent welds at a high magnification. The quality of these welds is related to the size, shape, and organization of the light patterns observed in Figs 6.50a, b. To estimate the size of the typical patterns in each image, linear granulometries were used, both in the vertical and in the horizontal direction. The resulting pattern spectra are shown in Figs 6.50c, d. First, one can observe that the horizontal granulometric curve is very similar to the vertical one; we conclude that the patterns in images 6.50a and 6.50b do not have any preferential orientation. Second, the curves in Fig. 6.50c exhibit a well-marked peak for size 4, whereas the peak of the curves in Fig. 6.50d is found for size 12. We conclude that the typical width and height of the patterns in Fig. 6.50a and in Fig. 6.50b is of 4 pixels and 12 pixels respectively. Additionally, Fig. 6.50d shows the the distribution of pattern sizes in Fig. 6.50b is rather wide.

Another example of application is shown in Fig. 6.51. Images 6.51a and 6.51b respectively show X-rays of a healthy lung and of a lung exhibiting signs of the 'miner's disease'. These two images look very different from a texture point of view, thought it would be hard for a segmentation algorithm to specifically extract the 'objects' that make image b different from image a. Essentially, the former image shows a very smooth texture while the latter is scattered with white nodules, which makes the texture rougher. Grayscale granulometries are the instrument of choice to discriminate between these two types of textures. To that end, M. Grimaud, author of the original study, had used grayscale granulometries by openings and closings with squares. The resulting pattern spectra for both X-ray images, shown in Fig. 6.51a–b, are extremely different from one another. However, in order to effectively discriminate between these two images, expensive grayscale granulometries with squares are not necessary: using the algorithms presented here to extract *horizontal* grayscale granulometric curves is just as effective, as shown in Fig. 6.51d, and is orders of magnitude faster.

6.4.6 Granulometries with maxima of linear openings

In order to deal with more complicated cases, we now describe a technique that can be seen as a generalization of the concept of opening

---

[J] Images graciously provided by DMS, CSIRO, Australia.

functions for the grayscale case. When performing openings of increasing size of a binary image, each 'on' pixel p is turned 'off' for an opening size given by the value of the opening function at pixel p. In other words, the opening function encodes for each pixel the successive values it takes for increasing opening sizes (namely, a series of 1's followed by a series of 0's). Similarly, in the grayscale case, as the size of the opening increases, the value of each pixel decreases monotonically. If each pixel was assigned the list of values it takes for every opening size, then the corresponding grayscale granulometry could be extracted straightforwardly.

Unfortunately, even if it were possible to compute such lists of values quickly, assigning one to each image pixel would require far too much memory. A more compact representation needs to be designed, that takes into account the intrinsic 'redundancy' of opened images, characterized by their large plateaus of pixels. If we again consider the linear case, an elegant solution can be proposed to both the problem of computing these lists of values, and the problem of storing them compactly:

Let M be a horizontal maximum of image I, with altitude (grayscale) h. We pointed out in the previous section that a horizontal opening of size $l(M)$ of M takes all of its pixels down to a new value h'. Beyond this, the following proposition can easily be proved:

Proposition 6.10 *Let $n > 0$, I a grayscale image such that $I = I \circ L_{n-1}$. Then, for every pixel p in I:*

$$(I \circ L_n)(p) < I(p) \iff$$
$$\exists M \text{ horiz maximum}, l(M) = n \text{ and } p \in M. \tag{6.78}$$

Therefore, at opening n in the sequence, the only pixels affected are those which belong to maxima of length n. Furthermore, all the pixels belonging to the same maximum M, $l(M) = n$, will be affected in the same way for any opening of size greater than or equal to n. The list of decreasing values we wish to associate with each pixel in M can therefore 'converge' into one single list for size n. For larger opening sizes, this list may itself be merged with other lists, etc.

Based on this principle, the present algorithm represents each image line as a tree T, which we call its *opening tree*. The leaves of T are the image pixels, and the nodes are made of pairs (h, n), where h is a grayscale value and n is an opening size. Every pixel that can be reached by going upwards in the tree starting from node (h, n) is such that its value for the opening of size n is h. Conversely, starting from a pixel

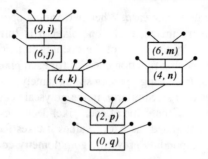

Figure 6.52 *Opening tree representation of the image line of Fig. 6.49. The leaves of this tree (•) correspond to the image pixels.*

p, successive pairs $(h_1, n_1), (h_2, v_2), \ldots, (h_1, v_i), \ldots$, are reached by going down towards the root of the tree. By convention, for this pixel, $(h_0, n_0) = (I(p), 0)$. These pairs satisfy:

$$\forall i > 0, \quad h_i > h_{i+1} \text{ and } n_i < n_{i+1}. \tag{6.79}$$

For $n \geq 0$, the value of the opening of size n of I at pixel p is given by:

$$(I \circ L_n)(p) = h_j,$$
$$\text{where } j \text{ is such that } n_j \leq n < n_{j+1}. \tag{6.80}$$

Opening trees can be computed using an algorithm very similar to the one described in the previous section. An example of opening tree is shown in Fig. 6.52.

Opening trees provide a hierarchical description that can be used to compactly represent *all* the horizontal openings of a grayscale image. In this respect, this notion is a grayscale equivalent of the opening function mentioned earlier (see Fig. 6.42). One can prove that, in the worst case, the opening tree has one node per image pixel. In practice though, only between 0.3 and 0.9 nodes per pixel are needed depending on the complexity of the image processed.

Any horizontal opening of I can be straightforwardly derived from its opening tree. In addition, the horizontal pattern spectrum of I, $\mathrm{PS}_h(I)$, can be computed from T as follows:

Algorithm: horiz granul of I from its opening tree

- initialize each bin of pattern spectrum $\mathrm{PS}_h(I)$ to 0;
- for each pixel p of I do:
  - $v \leftarrow I(p)$; $(h, n) \leftarrow$ node pointed at by p;

```
− while (h, n) exists, do:
    PS_h(I)(n) ← PS_h(I)(n) + (v − h);
    v ← h;    (h, n) ← next node down in tree;
```

This algorithm is obviously less efficient for horizontal granulometries than the one described in the previous section. However, it easily generalizes to the computation of granulometries using maxima of linear openings in several orientations. For example, to determine the granulometric curve corresponding to maxima of horizontal and vertical openings, one first extracts the horizontal opening tree T_1 and the vertical opening tree T_2. Then, for each pixel p, the technique consists of descending the corresponding branches of T_1 and T_2 simultaneously as follows:

Algorithm: granul of I from trees T_1 and T_2

• initialize each bin of pattern spectrum PS(I) to 0;
• for each pixel p of I do:

```
− v ← I(p);
− (h₁, n₁) ← node of T₁ pointed at by p;
− (h₂, n₂) ← node of T₂ pointed at by p;
− while (h₁, n₁) and (h₂, n₂) exist, do:
    size ← max(n₁, n₂);
    while n₁ ≤ size and (h₁, n₁) exists do:
        (h₁, n₁) ← next node down in T₁;
    while n₂ ≤ size and (h₂, n₂) exists do:
        (h₂, n₂) ← next node down in T₂;
    PS(I)(size) ← PS(I)(size) + (v − max(h₁, h₂));
    v ← max(h₁, h₂);
```

The same technique extends to any number of opening trees. The whole granulometry algorithm (extraction of trees followed by computation of the pattern spectrum from these trees) is once again orders of magnitude faster than traditional techniques, as illustrated by Table 6.2. In this table, a granulometry by maxima of linear openings at 4 orientations was computed. In terms of memory, the computation of this particular granulometry requires, in the worst case, 1 pointer (4 bytes) and 1 node (8 bytes) per pixel, for each orientation. This comes to a total of 48 bytes/pixel, i.e., a worst case scenario of 12 Megabytes for a 512×512 image. This is a reasonable tradeoff given the speed of the algorithm, and is not a strain on modern systems.

The algorithm given above can be easily adjusted to the computation of *pseudo-granulometries* by *minima* of linear openings. Although

Table 6.2 *Execution time of traditional opening-based techniques and of present algorithms for the computation of a horizontal grayscale granulometry (left) and a granulometry by maxima of linear openings in 4 orientations (right). A complex 512×512 image was used for this comparison, done on a Sun Sparc Station 10. Granulometries were computed for opening sizes 1 to 512.*

horizontal		
traditional	new	improvement factor
204s	0.206s	**990**
in 4 orientations		
traditional	new	improvement factor
824s	2.78s	**296**

minima of openings are not themselves openings (Matheron 1975), minima of openings with line segments of increasing length constitute a decreasing family of image operators. The resulting pseudo-granulometric curves often characterize the same image features as square granulometries.

Together with the linear granulometries described in this section, these pseudo square granulometries have been very successful as one of the feature sets used to characterize plankton in towed video microscopy images[||] (Davis *et al.* 1992, Tang *et al.* 1996). As an illustration, Fig. 6.53 shows four different images of copepod oithona and their corresponding pattern spectra, for a granulometry using openings with squares. These curves are all relatively flat, exhibiting a single well-marked maximum, between size 5 and 10, which corresponds to the size of the body of the organisms. By contrast, the same granulometric curves for the pteropods shown in Fig. 6.54 are entirely different, climbing sharply until a strong maximum, reached for a size between 10 and 20. Together with other structural and shape-based features, these pattern spectra are used as input to a sophisticated classifier, which currently discriminates between 5 different types of organisms with over 90% accuracy. The ultimate goal of the project is to be able to clas-

[||] The support of the Office of Naval Research, through grant N00014–93–1–0606, is gratefully acknowledged for the plankton study.

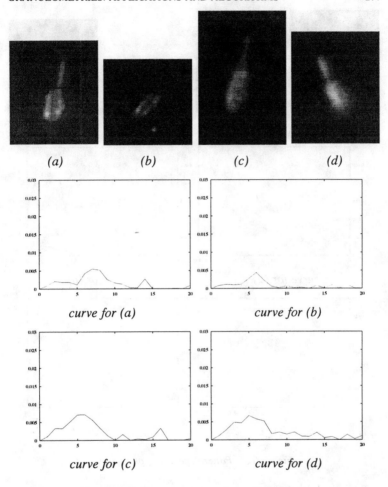

Figure 6.53 *Pattern spectra of copepod oithona.*

sify more than a dozen different kinds of organisms with over 90% accuracy. For more details, refer to (Tang *et al*. 1996).

6.4.7 *Final notes on granulometries*

Even though the concept of granulometries was introduced close to three decades ago, the computation time required to extract granulometric curves has long made it impossible to use them for most practical

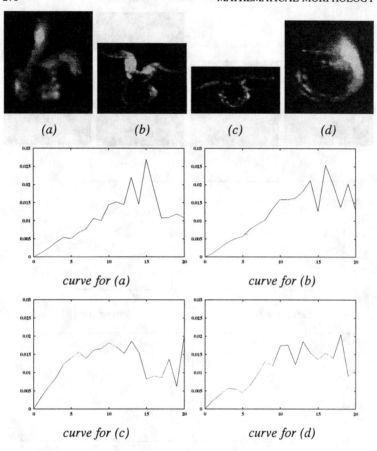

Figure 6.54 *Pattern spectra of pteropods.*

applications. To address this issue, a comprehensive set of fast algorithms for computing granulometries in binary and grayscale images
was proposed (Vincent 1994a, 1994b). These algorithms constituted
the topic of the present section. Some of them are several orders of
magnitudes faster than previously available techniques, so it now becomes possible to use granulometries on a 'routine' basis. A number
of examples of applications illustrated the use of these tools for the
extraction of global size information directly from grayscale or binary
images, the extraction of granulometric feature vectors, and the characterization of textures.

One of the key new concepts that enabled the efficient computation of grayscale granulometries is that of *opening trees:* such structures are shown to provide a compact representation for the successive openings of a grayscale image by line segments of increasing size. They are at the heart of the algorithm proposed for grayscale granulometries with maxima of linear openings. In addition, while the efficient computation of *exact* grayscale granulometries with square openings remains an open problem, opening trees provide a way to extract pseudo-granulometries by minima of openings with line segments at different orientations, which can be used to approximate square granulometries. These techniques even extend to the fast computation of granulometries by area openings and closings (Vincent 1992b).

Beyond simply turning granulometries into a useful and computationally efficient set of tools, the algorithms described in the present section are expected to significantly contribute to expanding the range of problems that can be addressed using granulometries. For example, they can easily be extended to *local granulometries*, where they prove useful for texture segmentation and related problems, as described in (Vincent 1996). In addition, the concept of opening trees proves useful not only for granulometries, but for a variety of new operations such as grayscale *size transforms* (Vincent 1996). This new set of tools should therefore help popularize the use of grayscale granulometries and related techniques in the image and signal analysis community.

6.5 Conclusion

After some reminders on classic morphological image analysis concepts, operators, and notations, this chapter focused on two very important topics of morphology: segmentation and granulometries. Because of the introduction of new notions such as the dynamics and the opening trees, and because of advances in morphological algorithms (e.g., watersheds and granulometry algorithms), these two areas have boomed recently: an increasing number of papers is being published on watershed segmentation and derived techniques, and the same is true with granulometries. Moreover, the range of image analysis applications that are being approached using these techniques is growing steadily.

Two fundamental aspects of applied mathematical morphology were covered here: problem-solving (i.e., how to use the available tools to solve a particular image analysis problem), and algorithms (i.e., given a useful tool, how to implement it efficiently). The section on segmentation (Section 6.3) focused on the problem-solving aspect, whereas the section on granulometries (Section 6.4) focused on the algorithmic aspect. To further tie these two sections together, let us stress that the granulometry algorithms described enable the efficient and robust automatic extraction of size information directly from grayscale images. This information is often invaluable for segmentation: automatic extraction of structure size as part of a segmentation algorithm eliminates the need for 'hard' size parameters or thresholds, thereby greatly enhancing robustness.

For more information on Mathematical Morphology in general, and in particular on the theoretical aspects of morphology, the reader is invited to refer to Serra (1982, 1988) and to Heijmans (1994). The reference section below also lists a number of papers and chapters that can be helpful to further probe any of the topics covered in this chapter. For a complete in-depth handbook of applied mathematical morphology, the reader may have to wait until 1997 or 1998 (Schmitt and Vincent 1997).

References

G. Bertrand and X. Wang. An algorithm for a generalized distance transformation based on Minkowski operations. In *9th International Conference on Pattern recognition*, pages 1163–1167, Rome, Nov. 1988.

S. Beucher. Watersheds of functions and picture segmentation. In *IEEE Int. Conf. on Acoustics, Speech and Signal Processing*, pages 1928–1931, Paris, May 1982.

S. Beucher and C. Lantuéjoul. Use of watersheds in contour detection. In *International Workshop on Image Processing, Real-Time Edge and Motion Detection/Estimation*, Rennes, France, 1979.

S. Beucher and F. Meyer. The morphological approach to segmentation: the watershed transformation. In E. R. Dougherty, editor, *Mathematical Morphology in Image Processing*, pages 433–481. Marcel-Dekker, Sept. 1992.

G. Birkhoff. *Lattice Theory*, volume 25. A.M.S. Colloq., third edition, 1983.

G. Borgefors. Distance transformations in arbitrary dimensions. *Comp. Vis., Graphics and Image Processing*, 27:321–345, 1984.

G. Borgefors. Distance transformations in digital images. *Comp. Vis., Graphics and Image Processing*, 34:334–371, 1986.

L. Calabi and W. Harnett. Shape recognition, prairie fires, convex deficiencies and skeletons. Technical Report 1, Parke Math. Lab. Inc., One River Road, Carlisle MA, 1966.

Y. Chen and E. Dougherty. Texture classification by gray-scale morphological granulometries. In *SPIE Vol. 1818, Visual Communications and Image Processing*, Boston MA, Nov. 1992.

P. Danielsson. Euclidean distance mapping. *Comp. Graphics and Image Processing*, 14:227–248, 1980.

C.S. Davis, S.M. Gallager, and A.R. Solow. Microaggregations of oceanic plankton observed by towed video microscopy. *Science*, 257:230–232, Jan. 1992.

E. Dougherty, J. Pelz, F. Sand, and A. Lent. Morphological image segmentation by local granulometric size distributions. *Journal of Electronic Imaging*, 1(1), Jan. 1992.

M. Grimaud. *La Géodésie Numérique en Morphologie Mathématique: Application à la Détection Automatique de Microcalcifications en Mammographie Numérique.* PhD thesis, Ecole des Mines, Paris, Dec. 1991.

M. Grimaud. A new measure of contrast: Dynamics. In *SPIE Vol. 1769, Image Algebra and Morphological Image Processing III*, pages 292–305, San Diego CA, July 1992.

R.M. Haralick, S. Chen, and T. Kanungo. Recursive opening transform. In *IEEE Int. Computer Vision and Pattern Recog. Conference*, pages 560–565, Champaign IL, June 1992.

R.M. Haralick and L.G. Shapiro. *Computer and Robot Vision.* Addison-Wesley, 1991.

H. Heijmans. *Morphological Image Operators.* Academic Press, Boston, 1994.

E.J. Kraus, H. Heijmans, and E.R. Dougherty. Gray-scale granulometries compatible with spatial scalings. *Signal Processing*, 34:1–17, 1993.

C. Lantuéjoul. Issues of digital image processing. In R. M. Haralick and J.-C. Simon, editors, *Skeletonization in Quantitative Metallography.* Sijthoff and Noordhoff, Groningen, The Netherlands, 1980.

C. Lantuéjoul and S. Beucher. On the use of the geodesic metric in image analysis. *Journal of Microscopy*, 121:39–49, Jan. 1981.

C. Lantuéjoul and F. Maisonneuve. Geodesic methods in quantitative image analysis. *Pattern Recognition*, 17(2):177–187, 1984.

B. Laÿ. Recursive algorithms in mathematical morphology. In *Acta Stereologica Vol. 6/III*, pages 691–696, Caen, France, Sept. 1987. 7th International Congress For Stereology.

F. Maisonneuve. Extrema régionaux: Algorithme parallèle. Technical Report 781, Ecole des Mines, CGMM, Paris, 1982.

P. Maragos. Pattern spectrum and multiscale shape representation. *IEEE Trans. Pattern Anal. Machine Intell.*, 11(7):701–716, July 1989.

P. Maragos and R. Ziff. Threshold superposition in morphological image analysis. *IEEE Trans. Pattern Anal. Machine Intell.*, 12(5), May 1990.

G. Matheron. *Eléments pour une Théorie des Milieux Poreux.* Masson, Paris, 1967.

G. Matheron. *Random Sets and Integral Geometry.* John Wiley and Sons, New York, 1975.

F. Meyer. Contrast feature extraction. In J.-L. Chermant, editor, *Quantitative Analysis of Microstructures in Material Sciences, Biology and Medicine*,

Stuttgart, FRG, 1978. Riederer Verlag. Special issue of Practical Metallography.

H. Minkowski. Allgemein lehrsätze über konvexe polyeder. *Nach. Ges. Wiss. Göttingen*, pages 198–219, 1897.

L. Najman and M. Schmitt. A dynamic hierarchical segmentation algorithm. In J. Serra and P. Soille, editors, *EURASIP Workshop ISMM'94, Mathematical Morphology and its Applications to Image Processing*, Fontainebleau, France, Sept. 1994. Kluwer Academic Publishers.

J.-F. Rivest, P. Soille, and S. Beucher. Morphological gradients. In *SPIE/SPSE Vol. 1658, Nonlinear Image Processing III*, pages 139–150, Feb. 1992.

A. Rosenfeld and J. Pfaltz. Sequential operations in digital picture processing. *J. Assoc. Comp. Mach.*, 13(4):471–494, 1966.

A. Rosenfeld and J. Pfaltz. Distance functions on digital pictures. *Pattern Recognition*, 1:33–61, 1968.

M. Schmitt. Variations on a theme in binary mathematical morphology. *Journal of Visual Communication and Image Representation*, 2(3):244–258, Sept. 1991.

M. Schmitt and J. Mattioli. *Morphologie Mathématique*. Masson, Paris, 1994.

M. Schmitt and L. Vincent. *Morphological Image Analysis: a Practical and Algorithmic Handbook*. Cambridge University Press, (to appear in 1997).

J. Serra. *Image Analysis and Mathematical Morphology*. Academic Press, London, 1982.

J. Serra, editor. *Image Analysis and Mathematical Morphology, Volume 2: Theoretical Advances*. Academic Press, London, 1988.

J. Serra and L. Vincent. An overview of morphological filtering. *Circuits, Systems and Signal Processing*, 11(1):47–108, Jan. 1992.

S.R. Sternberg. Grayscale morphology. *Comp. Vis., Graphics and Image Processing*, 35:333–355, 1986.

H. Talbot and L. Vincent. Euclidean skeletons and conditional bisectors. In *SPIE Vol. 1818, Visual Communications and Image Processing*, Boston, MA, Nov. 1992.

X. Tang, K. Stewart, L. Vincent, H. Huang, M. Marra, S. Gallager, and C. Davis. Automatic plankton image recognition. *International Artificial Intelligence Review Journal*, 1996.

C. Vachier and L. Vincent. Valuation of image extrema using alternating filters by reconstruction. In *SPIE Vol. 2568, Neural, Morphological, and Stochastic Methods in Image and Signal Processing*, pages 94–103, San Diego, CA, July 1995.

L. Vincent. *Algorithmes Morphologiques à Base de Files d'Attente et de Lacets: Extension aux Graphes*. PhD thesis, Ecole des Mines, Paris, May 1990.

L. Vincent. Efficient computation of various types of skeletons. In *SPIE Medical Imaging V*, San Jose, CA, 1991.

L. Vincent. Exact euclidean distance function by chain propagations. In *IEEE Int. Computer Vision and Pattern Recog. Conference*, pages 520–525, Maui,

HI, June 1991.

L. Vincent. New trends in morphological algorithms. In *SPIE/SPSE Vol. 1451, Nonlinear Image Processing II*, pages 158–169, San Jose, CA, Feb. 1991.

L. Vincent. Morphological area openings and closings for grayscale images. In *NATO Shape in Picture Workshop*, pages 197–208, Driebergen, The Netherlands, Sept. 1992.

L. Vincent. Morphological grayscale reconstruction: Definition, efficient algorithms and applications in image analysis. In *IEEE Int. Computer Vision and Pattern Recog. Conference*, pages 633–635, Champaign IL, June 1992.

L. Vincent. Grayscale area openings and closings: their efficient implementation and applications. In *EURASIP Workshop on Mathematical Morphology and its Applications to Signal Processing*, pages 22–27, Barcelona, May 1993.

L. Vincent. Morphological grayscale reconstruction in image analysis: Applications and efficient algorithms. *IEEE Transactions on Image Processing*, 2:176–201, Apr. 1993.

L. Vincent. Fast grayscale granulometry algorithms. In J. Serra and P. Soille, editors, *EURASIP Workshop ISMM'94, Mathematical Morphology and its Applications to Image Processing*, pages 265–272, Fontainebleau, France, Sept. 1994. Kluwer Academic Publishers.

L. Vincent. Fast opening functions and morphological granulometries. In *SPIE Vol. 2300, Image Algebra and Morphological Image Processing V*, pages 253–267, San Diego, CA, July 1994.

L. Vincent. Local grayscale granulometries based on opening trees. In *ISMM'96, International Symposium on Mathematical Morphology*, Atlanta, GA, May 1996.

L. Vincent and S. Beucher. The morphological approach to segmentation: an introduction. Technical report, Ecole des Mines, CMM, Paris, 1989.

L. Vincent and E.R. Dougherty. Morphological segmentation for textures and particles. In E. R. Dougherty, editor, *Digital Image Processing Methods*, pages 43–102. Marcel-Dekker, New York, 1994.

L. Vincent and B. Masters. Morphological image processing and network analysis of corneal endothelial cell images. In *SPIE Vol. 1769, Image Algebra and Morphological Image Processing III*, pages 212–226, San Diego, CA, July 1992.

L. Vincent and P. Soille. Watersheds in digital spaces: an efficient algorithm based on immersion simulations. *IEEE Trans. Pattern Anal. Machine Intell.*, 13(6):583–598, June 1991.

P.-F. Yang and P. Maragos. Morphological systems for character image processing and recognition. In *IEEE International Conference on Acoustics, Speech, and Signal Processing*, pages V.97–100, Minneapolis, MN, Apr. 1993.

L.-P. Yuan. A fast algorithm for size analysis of irregular pore areas. In *SPIE/SPSE Vol. 1451, Nonlinear Image Processing II*, pages 125–136, San Jose, CA, Feb. 1991.

CHAPTER 7

Random closed sets: results and problems

Ilya Molchanov

Department of Statistics,

University of Glasgow, G12 8QW Glasgow, Scotland, U.K.

7.1 Introduction

As the word 'set' has a broad sense ranging from everyday meaning to a mathematical and philosophical concept, it is possible to give several possible 'definitions' of a random set.

Biometrical 'definition'. Close your eyes, then go one mile in a random direction in a forest. Open your eyes. What you will see is a random set.

Business-manager 'definition'. Take all sets, make a list and then pick one set from this list at random.

Housewife 'definition'. Take a glass of red wine and pour it out onto a beige carpet. The stain you will see is a random set.

Image analysis 'definition'. Take a grid of pixels. If some of them are randomly coloured in black or white, then the resulting picture is a random set.

Statistical 'definition'. A random set is a confidence region for an estimated parameter.

Mathematical definition. Consider the family of all sets of interest (usually closed sets). Equip it with a σ-algebra and then define a random set as a measurable map from a given probability space to this space of sets.

The mathematical theory of random closed sets was essentially laid by Matheron (1975). Starting from Choquet's theory of positively defined functions on cones, he characterised distributions of random sets and established many natural links to integral geometry and mathematical morphology. More recently, also connections with lattice theory, harmonic analysis on semigroups, set-valued analysis and optimisation became clearer.

We will definitely stick to the mathematical definition in order to discuss probabilistic/statistical results on random sets and outline possible generalisations, open problems, and interdisciplinary connections of the theory of random sets. This survey starts with distributions and examples of random sets, then discusses convergence of random set distributions, limit theorems for Minkowski sums and unions of random sets, and finally outlines some problems in statistical inference. There are several exercises scattered through the text and a number of (open) problems. The principal difference between them is that the author knows answers to the exercises but can only conjecture about the problems.

7.2 Distributions of random sets

7.2.1 Definition and examples

We will deal only with random closed sets in the Euclidean space $\mathbb{R}^d$, although many results can be easily generalised for sets in locally compact spaces that have a countable topological base. Moreover, some results hold for random sets in Banach spaces. The corresponding record space is the space $\mathcal{F}$ of all (topologically) closed subsets of $\mathbb{R}^d$. If $(\Omega, \Sigma, \mathsf{P})$ is a probability space, then a measurable map $X : \Omega \to \mathcal{F}$ is said to be a random set. The measurability is guaranteed by the fact that

$$\{\omega : \ X(\omega) \cap K \neq \emptyset\} \in \Sigma \tag{7.1}$$

for all compact sets K, see Matheron (1975). This means that observing a realisation of X we can always say if X hits or misses a given compact set K. Note that X is a function whose values are closed sets. Such functions are a usual object of set-valued analysis. For instance, the measurability condition (7.1) coincides with the so-called weak measurability of set-valued functions, see Aubin and Frankowska (1990) and Wagner (1977).

This definition is quite flexible and includes interesting random sets. At the same time, the definition ensures that many functionals $f(X)$ are measurable, i.e. they become random variables. For instance, $\mu_d(X)$ (d-dimensional Lebesgue measure of X) is a measurable functional. The same is true for the surface area (if it is well-defined) and many other functionals known from convex geometry, see Schneider (1993).

Fubini's theorem implies that, for a general σ-finite measure μ, $\mu(X)$ is a random variable, and

$$\mathsf{E}\mu(X) = \int_{\mathbb{R}^d} \mathsf{P}\{x \in X\}\mu(dx). \qquad (7.2)$$

This result goes back to Robbins (1944). A further generalisation is suggested by Baddeley and Molchanov (1997).

Measurability of some set-theoretic operations can be deduced from the topological properties (the so-called semicontinuity) of related maps. For instance, if X and Y are random closed sets, then $X \cup Y$, $X \cap Y$, $\overline{X^c}$ (the closure of the complement $X^c = \mathbb{R}^d \setminus X$), ∂X (the boundary of X), $\overline{\text{conv}(X)}$ (the closed convex hull of X), $\overline{X \oplus Y}$ (the closure of the Minkowski sum $X \oplus Y = \{x + y : x \in X, y \in Y\}$) are random closed sets. (If X is compact almost surely, then $X \oplus Y$ and $\text{conv}(X)$ are closed.)

Many interesting examples of random closed sets are related to random functions. If $\xi(x)$, $x \in \mathbb{R}^d$, is a continuous random function, then its level (homecoming) $\{x : \xi(x) = t\}$ is a random closed set. Such sets have been throughly investigated for the case when $\xi(t)$ is a Wiener process, see Itô and McKean (1965), Lévy (1992) and references therein. It is well-known that the set of zeroes of the Wiener process is of fractal type (Falconer, 1990) and has the Hausdorff dimension $1/2$. The structure of level sets have been investigated also for stationary random processes and random fields, see Adler (1981) and Cramér and Leadbetter (1967). If $\xi(x)$, $x \in \mathbb{R}$, is a strong Markov stochastic process, then X_t is said to be a (strong) Markov (or regenerative) set, see Krylov and Yushkevitch (1964), Hoffman-Jørgensen (1969), Kingman (1973), Maisonneuve (1974). Recent related results can be found in Fristedt (1996) and Peres (1996).

Problem 7.1 It is interesting to characterise classes of random processes by *intrinsic* properties of their level sets. Which conditions on $X \subset \mathbb{R}^1$ must be imposed in order to ensure existence of a Gaussian continuous random process $\xi(x)$ such that $X = \{x : \xi(x) = t\}$ for some t?

Function f is said to be *lower semicontinuous* if $f(a) \leq \liminf_{x \to a} f(x)$ for all $a \in \mathbb{R}^d$ (then $(-f)$ is said to be upper semicontinuous). If $\xi(x)$ is a random lower semicontinuous function, then its *epi-graph*

$$\text{epi}\,\xi = \{(x, t) \in \mathbb{R}^d \times \mathbb{R} : \xi(x) \leq t\}.$$

is a random closed subset of $\mathbb{R}^d \times \mathbb{R}$. Epi-graphs of lower semicontinuous random functions are extensively studied in stochastic optimisation, see Salinetti (1987) and Salinetti and Wets (1986), and in the studies of extreme values for random functions, see Norberg (1987).

7.2.2 Capacity functionals

A random set X gives rise to the hitting probabilities

$$T(K) = T_X(K) = \mathsf{P}\{X \cap K \neq \emptyset\} \tag{7.3}$$

defined for K running through the family $\mathcal{K}$ of compact sets in $\mathbb{R}^d$. This functional $T : \mathcal{K} \to [0, 1]$ is said to be the *capacity functional* (or hitting functional) of X.

The following properties of the capacity functional are easy to derive.

(T1) $T(\emptyset) = 0$.

(T2) $T(K_1) \leq T(K_2)$ if $K_1 \subseteq K_2$, so that T is a monotone functional.

(T3) T is upper semicontinuous on $\mathcal{K}$, i.e. $T(K_n) \downarrow T(K)$ as $K_n \downarrow K$.

(T4) The functionals given by

$$S_1(K_0; K) = T(K_0 \cup K) - T(K_0)$$
$$\cdots \quad \cdots$$
$$S_n(K_0; K_1, \dots, K_n) = S_{n-1}(K_0; K_1, \dots, K_{n-1})$$
$$- S_{n-1}(K_0 \cup K_n; K_1, \dots, K_{n-1})$$

are non-negative for all $n \geq 0$ and $K_0, K_1, \dots, K_n$ from $\mathcal{K}$.

Note that **(T3)** follows from continuity of P. Property **(T4)** strengthens the monotonicity property **(T2)** and becomes trivial if we notice that the value of $S_n(K_0; K_1, \dots, K_n)$ is equal to the probability that X misses K_0 but hits each of the sets $K_1, \dots, K_n$. The properties of T resemble those of the distribution function. However, in contrast to measures, functional T is not additive, but only *subadditive*, i.e.

$$T(K_1 \cup K_2) \leq T(K_1) + T(K_2)$$

for all compact sets K_1 and K_2.

Example 7.1 Let $X = (-\infty, \xi]$ be a random set in $\mathbb{R}^1$, where ξ is a random variable. Then $T(K) = \mathsf{P}\{\xi > \inf K\}$ for all $K \in \mathcal{K}$.

Example 7.2 Let $X = \{\xi\}$ be a random singleton in $\mathbb{R}^d$. Then $T(K) = \mathsf{P}\{\xi \in K\}$ is the probability distribution of ξ. It can be proven that the capacity functional T is additive if and only if X is a random singleton.

Example 7.3 If $X = B_1(\xi)$ is the unit ball with the random centre ξ, then $T_X(K) = P\{\xi \in K \oplus B_1(0)\}$, which is *not* a measure.

Example 7.4 Let w_t, $t \geq 0$, be a Wiener process and let $X_0 = \{t \geq 0 : w_t = 0\}$ be its set of zeroes. Then, for all $t, s \geq 0$,

$$T_{X_0}([t, t+s]) = P\{X_0 \cap [t, t+s] \neq \emptyset\} = 1 - \frac{2}{\pi} \arcsin \sqrt{\frac{t}{t+s}}.$$
(7.4)

Example 7.5 Let ξ be a lower semicontinuous random function. Then, for $K = K_0 \times (-\infty, a]$ with $K_0 \subset \mathbb{R}^d$, $T_{\text{epi}\,\xi}(K) = P\{\inf_{x \in K_0} \xi(x) \leq a\}$.

7.2.3 Capacities

A map $\phi : \mathcal{P} \to [-\infty, +\infty]$ is called a *capacity*, see Choquet (1953/54), Dynkin (1961), and Landkof (1972), if $M \subseteq M'$ implies $\phi(M) \leq \phi(M')$, $M_n \uparrow M$ implies $\phi(M_n) \uparrow \phi(M)$, and $K_n \downarrow K$ for compact sets K_n, K implies $\phi(K_n) \downarrow \phi(K)$. Property **(T4)** singles out those capacities which are called *completely alternating* or *alternating of infinite order*. If only $S_1(\cdot; \cdot)$ and $S_2(\cdot; \cdot, \cdot)$ are non-negative, then the corresponding capacity is called 2-alternating.

Exercise (see Matheron, 1975) Let ϕ be a completely alternating non-negative capacity. Prove that $T(K) = 1 - \exp\{-\phi(K)\}$ satisfies **(T4)**, so that T is also a completely alternating capacity. More generally, are the alternation orders of T and ϕ equal (the latter problem is open)?

Many capacities appear as upper envelopes of measures. For instance, a family M of σ-finite measures on $\mathbb{R}^d$ gives rise to the capacity

$$\phi(K) = \sup\{\mu(K) : \mu \in M\}.$$
(7.5)

Such capacities appear in robust statistics, in particular, their 2-alternating property is important to derive simple minimax tests, see Huber (1981). (Then M is the family of all distributions in the 'neighbourhood' of a given distribution.) Usually, it is very difficult to prove the alternating property of infinite order for the capacity given by (7.5), so that it is not easy to establish direct relationships between random sets theory and robust statistics.

Problem 7.2 Is it possible to develop 'robust' statistical theory for the case when the family of possible measures is determined through the distributions of random elements ξ satisfying $\xi \in X$ for some random set X?

The capacity functional T is said to be *maxitive* if

$$T(K_1 \cup K_2) = \max(T(K_1), T(K_2))$$

for all compacts K_1, K_2. Such capacities arise naturally in the theory of extremal processes, see Norberg (1986, 1987), and large deviations theory, see O'Brien and Vervaat (1995). It is also possible to define *random* capacities that include both random closed sets and random semicontinuous functions as particular cases, see Norberg (1986).

Example 7.6 Define a maxitive capacity by $T(K) = \sup\{f(x): x \in K\}$, where $f : \mathbb{R}^d \to [0, 1]$ is an upper semicontinuous function. Then T describes the distribution of the random set $X = \{x \in \mathbb{R}^d : f(x) \geq \eta\}$, where η is a random variable uniformly distributed on $[0, 1]$.

The concept of a capacity goes back to mathematical physics. Consider the function $k_\alpha(x, y) = C_{d,\alpha} \|x - y\|^{\alpha-d}$, $\alpha < d$, which is said to be the Riesz kernel ($C_{d,\alpha}$ is a normalising constant). Then

$$U_\alpha^\mu(x) = \int_{S_\mu} k_\alpha(x, y) d\mu(y)$$

is said to be the potential of measure μ with support S_μ, see Landkof (1972). The (Riesz) capacity of $K \in \mathcal{K}$ is defined by

$$C_\alpha(K) = \sup\{\mu(K): U_\alpha^\mu(x) \leq 1, \ x \in S_\mu \subseteq K\}. \tag{7.6}$$

In physical terms, $C_\alpha(K)$ is the maximum charge which it is possible to put on K such that the resulting potential is no greater than 1. If $d = 3$, then $C_2(K)$ is the famous *Newton capacity* (well-known from electrostatics). This capacity corresponds to random sets obtained as trajectories of the Brownian motion in $\mathbb{R}^3$, see Matheron (1975).

Problem 7.3 Find out a kernel which yields the capacity functional (7.4).

7.2.4 Choquet theorem

The capacity functional of a random closed set satisfies conditions **(T1)**–**(T4)**. The following important result states that these properties single out those capacities which arise from distributions of random closed sets. Moreover, the capacity functional determines uniquely the distribution of a random closed set.

Theorem 7.1 *Let* $T : \mathcal{K} \to [0, 1]$. *There exists a unique random closed set* X *in* $\mathbb{R}^d$ *with capacity functional* T *(or unique probability on* σ_f*)*

such that $P\{X \cap K \neq \emptyset\} = T(K)$ *if and only if* T *satisfies conditions* **(T1)–(T4)**.

The first proof given by Matheron (1975) was based on the routine application of the measure-theoretic arguments related to extension of measures from algebras to σ-algebras. In fact, the idea goes back to the fundamental paper by Choquet (1953/54) and his theorem on characterisation of positively defined functionals on cones. Other proofs make use of the theory of lattices, see Norberg (1989), and harmonic analysis on semigroups, see Berg *et al.* (1984). Both approaches are possible, since $\mathcal{F}$ is both a lattice and an Abelian semigroup with respect to the union operation. Note that the proof based on lattice theory works also for random sets in non-Hausdorff spaces.

Exercise (see Matheron, 1975 and Molchanov, 1984) Define *joint* capacity functionals of several random closed sets and characterise *independent* random closed sets X and Y using capacity functionals.

Generally speaking, the family of all compact sets is still too large, which makes it difficult to define $T(K)$ for *all* $K \in \mathcal{K}$. It is easy to show, see Matheron (1975), that $T(K)$ for K from the family $\mathcal{B}$ of all finite unions of closed balls in $\mathbb{R}^d$ with rational centres and radii still determines the distribution of the corresponding random closed set. It is however important to obtain further effective methods suitable to construct capacity functionals.

7.2.5 Properties of random sets

Stationary random sets

A random closed set X is said to be *stationary*, if X has the same distribution as $X + a$ (notation $X \overset{d}{\sim} X + a$) for all $a \in \mathbb{R}^d$. By Theorem 7.1, X is stationary if and only if $T_X(K)$ is translation-invariant. It is possible to discern random sets whose distributions are invariant with respect to a certain group acting on $\mathbb{R}^d$. The group SO_d of all rotations is especially important. A random set X is said to be *isotropic*, if $X \overset{d}{\sim} gX$ for all $g \in SO_d$.

If X is stationary, then $P\{a \in X\}$, $a \in \mathbb{R}^d$, does not depend on a and is equal to $P\{0 \in X\}$. By Robbins' theorem (7.2), for any $W \in \mathcal{K}$, and σ-finite measure μ,

$$E\mu(X \cap W) = \int_{\mathbb{R}^d} P\{a \in X \cap W\}\mu(da) = P\{0 \in X\}\mu(W).$$

The constant $p = \mathsf{E}\mu(X \cap W)/\mu(W) = \mathsf{P}\{0 \in X\}$ is called the *volume fraction* of X (or area fraction if X is a subset of the plane). It characterises the part of the volume covered by X.

Problem 7.4 Let X be a non-stationary random subset on the line $\mathbb{R}^1$. Assume that for all $K \in \mathcal{K}$, $T(K+t)$ admits a limit $T_\infty(K)$ as $t \to \infty$, which is non-trivial for at least one $K \in \mathcal{K}$. Then $T_\infty(K)$ is a capacity functional of a stationary random set. Give a nontrivial example of this situation. What will be an analogue of this 'stationarisation' procedure in $\mathbb{R}^d$, $d > 1$? Note that in the same way other transformations (for instance, $K \mapsto cK$) can be used.

Convex random sets

A random set X is said to be *convex*, if $X \in \mathcal{C}$ a.s., where $\mathcal{C}$ is the family of convex sets in $\mathbb{R}^d$. If X takes values from the class $\mathcal{C}_0$ of convex compact sets, then X is said to be a random convex compact set (or *random body*). The following theorem states that compact or convex random sets cannot be stationary.

Theorem 7.2 (see Matheron, 1975)
1. *A random compact set X cannot be stationary unless $X = \emptyset$ a.s.*
2. *A random convex closed set which is stationary is either $\emptyset$ or $\mathbb{R}^d$ almost surely.*

Random convex sets can be characterised using properties of their capacity functionals.

Theorem 7.3 (see Matheron, 1975) *A random closed set X is convex if its capacity functional $T(K)$ is C-additive, i.e.*

$$T(K \cup K') + T(K \cap K') = T(K) + T(K')$$

for all $K, K' \in \mathcal{C}_0$ such that $K \cup K' \in \mathcal{C}_0$.

Theorem 7.4 (see Vitale, 1983 and Molchanov, 1984) *The distribution of a random body X is determined by the values of the (containment) functional $\mathsf{P}\{X \subseteq K\}$ for $K \in \mathcal{C}_0$.*

Problem 7.5 Does the capacity functional $T_X(K)$ for $K \in \mathcal{C}_0$ determine uniquely the distribution of a random body X?

Exercise (see Molchanov, 1993b) Is it possible to characterise the distribution of a random convex closed (non-compact) set X by $\mathsf{P}\{X \subset F\}$ for $F \in \mathcal{C}$?

While intersections of convex sets are also convex, this is no longer true for unions of convex sets. A set $F \in \mathcal{K}$ is said to belong to the *convex ring* if $F = \cup_{i=1}^{n} K_i$ for convex compact sets K_i, $1 \le i \le n$. The *extended convex ring* consists of sets $F \in \mathcal{F}$ such that $F \cap K \in \mathcal{R}$ for all $K \in \mathcal{C}_0$.

Problem 7.6 Find a condition on the capacity functional which guarantees that the corresponding random closed set almost surely takes values from the convex ring (extended convex ring). This would be a useful generalisation of Theorem 7.3.

7.2.6 Random sets and point processes

The concept of a random closed set generalises the notion of a point process. Although every set is composed from its constituent points, the notion 'point process' means that the corresponding set of points is locally finite, i.e. each bounded set contains at most a finite number of points from this set. Further material on point processes can be found in Cox and Isham (1980), Daley and Vere-Jones (1988), and Stoyan *et al.* (1995).

A general point process can be viewed as a random element in the space $\mathcal{N}$ of counting measures endowed with σ-algebra generated by $\{N \in \mathcal{N} : N(K) = n\}$, $n \ge 0$, $K \in \mathcal{K}$.

Exercise (see Ripley, 1981) N is a point process without multiple points (a simple point process) if and only if $X = \{x \in \mathbb{R}^d : N(\{x\}) = 1\}$ is a locally finite random closed set in $\mathbb{R}^d$. Use Theorem 7.1 to deduce that the distribution of N is uniquely determined by the probabilities $P\{N(K) = 0\}$, $K \in \mathcal{K}$.

7.3 Convergence of random set distributions

7.3.1 Convergence in $\mathcal{F}$ and $\mathcal{K}$

The so-called *vague topology* on $\mathcal{F}$ can be defined directly by its base or by description of convergent sequences, see, e.g., Michael (1951) and Matheron (1975). A sequence $\{F_n, n \ge 1\}$ of closed sets converges in $\mathcal{F}$ to F if and only if

(F1) for all $x \in F$ and all sufficiently large n there exists $x_n \in F_n$ such that $x_n \to x$ as $n \to \infty$;

(F2) if $x_{n_k} \in F_{n_k}$, $k \ge 1$, and $x_{n_k} \to x$ as $k \to \infty$, then $x \in F$.

This definition endows $\mathcal{F}$ with the structure of a compact separable Hausdorff space.

Exercise (see Matheron, 1975) Prove that $(F, F') \mapsto F \cup F'$ is a continuous map from $\mathcal{F} \times \mathcal{F}$ to $\mathcal{F}$. Give a counterexample which shows that the map $(F, F') \mapsto F \cap F'$ is *not* continuous. However, the latter function is semicontinuous and therefore measurable. From this one can deduce that $X_1 \cup X_2$ and $X_1 \cap X_2$ are random closed sets if X_1 and X_2 are random closed sets.

The induced topology on $\mathcal{K} \subset \mathcal{F}$ is called the *narrow topology*. It makes $\mathcal{K}$ a locally-compact space and can be metrised by the *Hausdorff metric*

$$\rho_{\mathrm{H}}(K, K_1) = \inf\{\varepsilon > 0 : K \subset K_1^\varepsilon, K_1 \subset K^\varepsilon\}, \quad K, K_1 \in \mathcal{K}, \quad (7.7)$$

see Matheron (1975), Rockafellar (1970). Here $K^\varepsilon = \{x : \rho(x, K) \leq \varepsilon\}$ is the set of points at the distance not greater than ε to K.

7.3.2 Weak convergence of random sets

Weak convergence of random sets is a particular case of weak convergence of probability measures, see Billingsley (1968). Namely the random closed sets $X_n, n \geq 1$, *converge weakly* if the corresponding probability measures converge weakly in the usual sense, i.e.

$$P\{X_n \in \mathcal{A}\} \to P\{X \in \mathcal{A}\} \quad \text{as} \quad n \to \infty \qquad (7.8)$$

for each measurable class $\mathcal{A} \subset \mathcal{F}$ such that $P(\partial \mathcal{A}) = P\{X \in \partial \mathcal{A}\} = 0$ for the boundary of $\mathcal{A}$ with respect to the vague topology on $\mathcal{F}$. However, it is rather difficult to check (7.8) for all $\mathcal{A}$ from Σ. The first natural reduction is to let $\mathcal{A} = \mathcal{F}_K = \{F \in \mathcal{F} : F \cap K \neq \emptyset\}$ for K running through $\mathcal{K}$. It was proved by Lyashenko (1983b) that $P(\partial \mathcal{F}_K) = 0$ if and only if $P\{\mathcal{F}_K\} = P\{\mathcal{F}_{\mathrm{Int}\, K}\}$, where Int K is the topological interior of K. In terms of the capacity functional T_X of the limiting random set X, this is equivalent to $T_X(K) = T_X(\mathrm{Int}\, K)$. The family of such compact sets K is denoted by $\mathcal{S}_T$. Then X_n converges weakly to X if

$$T_{X_n}(K) \to T_X(K) \quad \text{as} \quad n \to \infty$$

for each $K \in \mathcal{S}_T$.

A further reduction of the family of testing compact sets is due to Norberg (1984) and Salinetti and Wets (1986). The class $\mathcal{S}_T$ can be replaced by the class $\mathcal{K}_{ub} \cap \mathcal{S}_T$, where $\mathcal{K}_{ub}$ is the class of finite unions of balls having positive radii. In general, the class $\mathcal{M} \subseteq \mathcal{K}$ is said to *determine the weak convergence* if the pointwise convergence of capacity functionals on $\mathcal{M} \cap \mathcal{S}_T$ yields the weak convergence of distributions of random closed sets.

7.3.3 Probability metrics

It is well-known that the convergence in distribution of random variables can be metrised by the Lévy metric. Further probability metrics and applications are discussed by Zolotarev (1986) and Rachev (1991). *Probability metrics method* is a powerful method which allows us to prove limit theorems for the most convenient metric and then reformulate results for other metrics using inequalities between them. A *probability metric* $\mathsf{m}(\xi, \eta)$ is a numerical function on the space of joint distributions of pairs of random elements such that $\mathsf{m}(\xi, \eta) = 0$ implies that the marginal distributions of ξ and η coincide, $\mathsf{m}(\xi, \eta) = \mathsf{m}(\eta, \xi)$, and $\mathsf{m}(\xi, \eta) \leq \mathsf{m}(\xi, \zeta) + \mathsf{m}(\zeta, \eta)$.

A useful generalisation of classical probability metrics for random variables can be obtained by replacing distribution functions in their definitions with capacity functionals of random sets, see Molchanov (1993b). The *uniform distance* between the random sets X and Y is defined as

$$\mathsf{r}(X, Y; \mathcal{M}) = \sup \{|T_X(K) - T_Y(K)| : K \in \mathcal{M}\},$$

where $\mathcal{M}$ is a subclass of $\mathcal{K}$. The *Lévy metric* is defined as

$$\mathbf{L}(X, Y; \mathcal{M}) = \inf \{r > 0 :$$
$$T_X(K) \leq T_Y(K^r) + r, \ T_Y(K) \leq T_X(K^r) + r, \ K \in \mathcal{M}\}.$$

Theorem 7.5 (see Molchanov, 1993b) *Let the class* $\mathcal{M} \subseteq \mathcal{K}$ *determine the weak convergence of random sets, and let* Int K *for each* $K \in \mathcal{M}$ *be equal to the limit of an increasing sequence* $\{K_n, n \geq 1\} \subset \mathcal{M}$. *Then a sequence* $X_n, n \geq 1,$ *of random sets converges weakly to* X *if and only if, for each* $K_0 \in \mathcal{K}$,

$$\mathbf{L}(X_n, X; \mathcal{M}(K_0)) \to 0 \quad as \quad n \to \infty, \tag{7.9}$$

where $\mathcal{M}(K_0) = \{K \in \mathcal{M} : K \subseteq K_0\}$. *If* X *is a.s. compact, then (7.9) can be replaced by* $\mathbf{L}(X_n, X; \mathcal{K}) \to 0$ *as* $n \to \infty$.

Example 7.7 Let $X = \{\xi\}$ and $Y = \{\eta\}$ be random singletons. Then $\mathbf{L}(X, Y)$ is the Lévy-Prokhorov distance between ξ and η, see Zolotarev

(1986). The uniform distance between X and Y is equal to

$$r(X, Y; \mathcal{M}) = \sup \{|P\{\xi \in K\} - P\{\eta \in K\}| : K \in \mathcal{M}\}$$

Then $r(X, Y; \mathcal{K})$ coincides with the total variation distance between the distributions of ξ and η.

Example 7.8 Let $X = (-\infty, \xi]$ and $Y = (-\infty, \eta]$ be random subsets of $\mathbb{R}^1$, and let $\{\inf K : K \in \mathcal{M}\} = \mathbb{R}^1$. Then $r(X, Y; \mathcal{M})$ coincides with the uniform distance between distribution functions of ξ and η, and $L(X, Y; \mathcal{M})$ equals the Lévy distance between the distributions of ξ and η.

Example 7.9 Let X and Y be the Poisson point processes in $\mathbb{R}^d$ with intensity measures Λ_X and Λ_Y respectively. Then $r(X, Y; \mathcal{K})$ is not greater than the total variation distance between Λ_X and Λ_Y. If X and Y are stationary and have intensities λ_X and λ_Y, then $r(X, Y) = h(\lambda_X/\lambda_Y)$, where $h(x) = |x^c - x^{cx}|$ with $c = 1/(1 - x)$.

7.3.4 Convergence almost surely

A sequence of random sets $X_n, n \geq 1$, converges to X almost surely if $X_n(\omega) \to X(\omega)$ in $\mathcal{F}$ (resp. in $\mathcal{K}$) for almost all $\omega \in \Omega$. Here we consider one example related to this convergence, which gave rise to much research and interesting techniques. Let K be a convex compact set in the plane. Choose n independent random points uniformly distributed in K and denote by P_n their convex hull. Thus, P_n is a random convex polygon, such that $P_n \to \text{conv}(K)$ almost surely as $n \to \infty$. Since the classical paper by Rényi and Sulanke (1963), the rate of this convergence and the corresponding limit theorems have been in the focus of attention of many probabilists, see Schneider (1988), where also numerous references can be found. Recent results are obtained by Cabo and Groeneboom (1994), Dümbgen and Walter (1996), Hsing (1994) and Hueter (1994). These results have close relationships to the problem of estimation of domains, see Ripley and Rasson (1977), and countour estimator for densities and images in Korostelev and Tsybakov (1993).

Exercise (see Salinetti *et al.*, 1986) Define convergence of random closed sets *in probability* and prove that each sequence which converges in probability contains a subsequence which converges almost surely.

7.3.5 Applications to random functions

Random functions give rise to many examples of random sets that appear as graphs, hypo- or epi-graphs, and level sets of random functions. On the other hand, random set theory allows us to handle random functions in a new specific manner.

Convergence of graphs

Often, a sequence of functions does not converge in any of topologies known from the classical theory of functional limit theorems, see Billingsley (1968). At the same time the pointwise convergence is too weak to make conclusions about the convergence of many interesting functionals.

Example 7.10 The sequence $f_n(x) = nx1_{0 \leq x \leq 1/n} + 1_{x > 1/n}$, $0 \leq x \leq 1$, $n \geq 1$, converges point-wisely to $f(x) = 1_{x>0}$, but does not converge in either C- or D-topology (the usual topologies in the study of stochastic processes, Billingsley, 1968).

Example 7.11 The sequence $f_n(x) = \sin nx$, $0 \leq x \leq 1$, fills out the rectangle $[0, 1] \times [0, 1]$, but converges neither in C- nor D-topology. Also $f_n(x) = nx$, $0 \leq x \leq 1$, 'converges' to the vertical line.

The statements above can be made precise if we view graphs of the functions as elements of the topological space of closed subsets of $[0, 1] \times \mathbb{R}$. Then a sequence ξ_n, $n \geq 1$, of random functions yields a sequence of random closed sets

$$X_n = \overline{\text{Graph}\,\xi_n} = \overline{\{(t, \xi_n(t)) : 0 \leq t \leq 1\}}.$$

If $\text{Graph}\,\xi_n$ is not closed, then X_n is equal to its topological closure. The weak convergence of X_n as $n \to \infty$ for different random functions ξ_n has been studied by Lyashenko (1983b,1987). We consider here the simplest result of this kind for step-like functions. Let $\alpha_0, \alpha_1, \alpha_2, \ldots$ be a sequence of i.i.d. random variables with distribution $S(A) = \mathsf{P}\{\alpha_i \in A\}$. For $t \geq 0$ define step-function $\xi(t) = \alpha_{[t]}$ and normalise it to get $\xi_n(t) = b_n \xi(nt)$, where b_n, $n \geq 1$, is a strictly increasing sequence of positive normalising constants, $b_n \to \infty$. Finally, define $X_n = \overline{\text{Graph}\,\xi_n}$.

Theorem 7.6 (see Lyashenko, 1983b) *X_n converges weakly to a nondegenerate random set X if and only if the following conditions hold*

(i) *there exists a dense set L in $\mathbb{R}^2$ such that for all $(x, y) \in L$*

$$\lim_{n \to \infty} nS(b_n^{-1}[x, y]) = v([x, y]) \in [0, +\infty],$$

where v is a measure on Borel subsets of $\mathbb{R}^1$.

(ii) *there exists $(x, y) \in L$ such that $0 < v([x, y]) < +\infty$.*

The limiting random closed set X has the capacity functional

$$T(K) = 1 - \exp\{-(\mu_1 \otimes v)(K)\}.$$

Thus, X is a Poisson process in $[0, \infty) \times \mathbb{R}$ with intensity measure $\mu_1 \otimes v$. Further generalisations are possible for random functions obtained as linear interpolators of the step function generated by the α's. Then the weak limit of $\xi_n = b_n \xi(nt)$ is a random set which consists of vertical half-lines radiating from the points of a Poisson process, see Lyashenko (1987).

Epi-convergence and stochastic optimisation

A sequence of lower semicontinuous functions f_n, $n \geq 1$, is said to epi-converge to f if epi f_n converges to epi f in $\mathcal{F}$ as $n \to \infty$. The stochastic variant of the epi-convergence is defined for a sequence of a.s. lower semicontinuous random functions ξ_n and their epi-graphs $X_n = \text{epi } \xi_n$. We then write $\xi_n \xrightarrow{\text{epi}} \xi$. The epi-convergence ensures the weak convergence of minima of functions, i.e. for each compact set K, $\inf_{x \in K} \xi_n(x)$ converges in distribution to $\inf_{x \in K} \xi(x)$. This convergence of minima explains the use of epi-convergence in stochastic optimisation (Salinetti, 1987). Epi-convergence of random functions is useful in the studies of extremal processes, see Norberg (1987) and Molchanov, (1993b, Section 8.3).

Exercise (see Salinetti and Wets, 1986 and Attouch and Wets, 1990) Write down a formula for the finite-dimensional distributions of $\xi(x)$ through the capacity functional of $X = \text{epi } \xi$. (Note that $\xi_n \xrightarrow{\text{epi}} \xi$ cannot be deduced and does not imply the convergence of the corresponding finite-dimensional distributions.)

Clearly, the optimisation problem $J(x) \to \min$ can be approximated by the solution of minimisation problems $J_n(x) \to \min$ if $J_n \xrightarrow{\text{epi}} J$. Many statistical estimators appear as solutions of some minimisation (or maximisation) problems. Since epi-convergence is the weakest functional convergence which ensures convergence of minimum points, it can be applied to prove strong consistency of estimators.

7.3.6 Set-valued martingales

Set-valued random functions appear naturally in stochastic control problems. Consider the controlled stochastic process $\xi(t, \omega, u)$, $u \in U(t)$, where $U(t)$ is the set of admissible controls at time t. All possible values of $\xi(t, \omega, u)$ can be written as a set-valued random function

$$\Xi(t, \omega) = \{\xi(t, \omega, u) : u \in U(t)\}.$$

Then minimisation of $\sup_t \|\xi(t, \omega, u)\|$ for the controlled process is equivalent to minimisation of $\sup_t \|\Xi(t, \omega)\|$, where $\|\Xi\| = \sup\{\|x\| : x \in \Xi\}$, so that the controlled optimisation is reduced to optimisation without control but for a set-valued process, see Aubin and Frankowska (1990) and Artstein (1984).

The theory of set-valued random functions is not yet well-developed. There are very few classes of functions which have been investigated. Some concepts usual in the classical case (for instance, those involving subtraction and increments), are difficult to formulate for set-valued functions.

Problem 7.7 Define a set-valued process with independent increments.

We consider below *set-valued martingales* in $\mathbb{R}^d$. If X is a random set measurable with respect to σ-algebra Σ, then *conditional expectation* $\mathsf{E}(X|\Sigma')$ of X with respect to $\Sigma' \subset \Sigma$ is defined as the set of all conditional expectations $\mathsf{E}(\xi|\Sigma')$, where $\xi \in X$ a.s. and ξ is a Σ-measurable integrable random vector. It is possible to use other concepts of the conditional expectation, for instance, based on the Doss expectation, see Herer (1987) and Section 7.5.3 below.

Definition 7.1 (*see Hiai and Umegaki, 1977*) *Let* $X_1, X_2, \ldots$ *be a sequence of random compact sets adapted to the filtration* $\Sigma_n, n \geq 1$. *This sequence is called a* martingale *if* $\mathsf{E}(X_{n+1}|\Sigma_n) = X_n$ *a.s., $n \geq 1$; a* supermartingale *if* $\mathsf{E}(X_{n+1}|\Sigma_n) \subseteq X_n$ *a.s., $n \geq 1$; and a* submartingale *if* $\mathsf{E}(X_{n+1}|\Sigma_n) \supseteq X_n$ *a.s., $n \geq 1$.*

Note that there is no direct relationship between set-valued super- and submartingales.

Theorem 7.7 (**see Van Cutsem, 1972**) *If* (X_n, Σ_n) *is a set-valued martingale such that* $\{\|X_n\| : n \geq 1\}$ *is uniformly integrable, then there exists an integrable random closed set X such that* $\mathsf{E}(X|\Sigma_n) = X_n$ *a.s. $n \geq$ 1. If* $\|X_n\| \leq \zeta$, $n \geq 1$, *with* $\mathsf{E}\zeta < \infty$, *then X_n converges in $\mathcal{F}$ almost surely to a random compact set X_∞.*

The optional sampling theorem for set-valued martingales in $\mathbb{R}^d$ has been proved in Aló *et al.* (1979). Various Banach spaces generalisations were considered by Papageorgiou (1985,1987).

7.4 Limit theorems

7.4.1 Minkowski addition

Aumann expectation

The Aumann integral was first defined by Aumann (1965) in the context of the integration of set-valued functions, while its random set meaning and the corresponding expectation was discovered by Artstein and Vitale (1975). In the following X is an almost surely non-empty random compact set in $\mathbb{R}^d$, which is not necessarily convex. A random point ξ is said to be a *selection* of X if $\xi \in X$ almost surely. It is well known from set-valued analysis (Wagner, 1977) that there exists at least one selection of X. If the norm of X

$$\|X\| = \sup\{\|x\| : x \in X\}$$

has a finite expectation (X is said to be *integrable*), then $\mathsf{E}\xi$ exists, so that X possesses at least one integrable selection. The *Aumann expectation* of X is defined as the set of expectations of all possible (integrable) selections, i.e.

$$\mathsf{E}X = \{\mathsf{E}\xi : \xi \text{ is a selection of } X\}.$$

Example 7.12 The family of all selections is often very large. For instance, let X be equal to $\{0, 1\}$ almost surely. Then all selections of X can be obtained as $\xi(\omega) = \mathbf{1}_{\Omega_2}(\omega)$ for all measurable partitions $\Omega = \Omega_1 \cup \Omega_2$ of the underlying probability space. Note that $\mathsf{E}\xi = \mathbf{P}(\Omega_2)$, so that the range of possible expectations of $\xi(\omega)$ depends on the structure of atoms of the underlying probability space.

Example 7.12 explains why two random sets with the same distribution may have different Aumann expectations. Let $\Omega' = \{\omega\}$ be a single-point probability space and $\Omega'' = [0, 1]$. Define two random sets on these spaces: $X_1(\omega) = \{0, 1\}$ for $\omega \in \Omega'$, and $X_2(\omega) = K$ for all $\omega \in \Omega''$. Then X_1 and X_2 have the same distribution (in fact, both sets are non-random), but $\mathsf{E}X_1 = \{0, 1\}$, while $\mathsf{E}X_2 = [0, 1]$. The main difference between Ω' and Ω'' is that Ω'' is not atomic. Roughly speaking, Ω' provides the so-called 'reduced' (or minimum) representation of X.

The following theorem is closely related to Lyapunov's theorem for vector measures and the so-called bang-bang principle in optimisation, see Artstein (1980).

Theorem 7.8 (see Aumann, 1965 and Richter, 1963) *If the basic probability space* $(\Omega, \Sigma, \mathsf{P})$ *contains no atoms and* $\mathsf{E}\|X\| < \infty$, *then* $\mathsf{E}X$ *is convex and, moreover,* $\mathsf{E}X = \mathsf{E}\mathrm{conv}(X)$.

Remember that a random convex compact body X corresponds uniquely to its *support function*

$$h(X, u) = \sup\{\langle x, u \rangle : x \in X\}, u \in \mathbf{S}^{d-1},$$

where $\langle x, u \rangle$ is the scalar product and $\mathbf{S}^{d-1}$ is the unit sphere in $\mathbb{R}^d$. If $\mathsf{E}\|X\| < \infty$, then $h(X, u)$ has finite expectation for all u. The support function is characterised by its sublinear property, which is preserved after taking expectation. Therefore, $\mathsf{E}h(X, u)$, $u \in \mathbf{S}^{d-1}$, is again the support function of a convex set, which is exactly the Aumann expectation of X, if $(\Omega, \Sigma, \mathsf{P})$ is non-atomic.

Theorem 7.9 *If* $\mathsf{E}\|X\| < \infty$, *then* $\mathsf{E}h(X, u) = h(\mathsf{E}X, u)$, $u \in \mathbf{S}^{d-1}$.

If X is a random body in $\mathbb{R}^2$, then its perimeter $U(X)$ is equal to the integral of its support function, so that Theorem 7.9 yields $\mathsf{E}U(X) = U(\mathsf{E}X)$.

The Aumann expectation has a number of applications in probability theory. For instance, remember a characterisation result for order statistics by Hoeffding (1953), which states that if $\eta_1, \eta_2, \ldots$ are i.i.d. random variables such that $\mathsf{E}|\eta_1| < \infty$, then the distribution of η's is characterised by the sequence $\mathsf{E}\max(\eta_1, \ldots, \eta_n)$, $n \geq 1$. Vitale (1987) proved that the nested sequence of Aumann expectations $\mathsf{E}Y_n$ for $Y_n = \mathrm{conv}(\{\eta_1, \ldots, \eta_n\})$ determines the distribution function for a sequence of i.i.d. random *vectors* $\eta_1, \eta_2, \ldots$ (in general taking values in a Banach space).

Problem 7.8 Find a 'good' definition of the variance of a random compact set. Is it sensible to define it as the set of variances for all square integrable selections?

Strong law of large numbers

Consider a sequence $X_1, X_2, \ldots$ of i.i.d. random closed sets. Note that if an infinite sequence of non-constant i.i.d. random sets (or general random elements) is defined on a probability space, then this probability space is non-atomic. The general approach to strong laws of large numbers and limit theorems for random compact sets consists of two steps, see Giné *et al.* (1983):

Step 1: Reduce consideration to the case of random convex compact sets.

Step 2: Derive results for random sets by invoking the corresponding results for probabilities in Banach spaces and applying them to the sequence $h(X_1, \cdot), h(X_2, \cdot), \ldots$ of the support functions of $X_1, X_2, \ldots$. These support functions are considered to be random elements in the space of continuous functions on the unit sphere in $\mathbb{R}^d$.

The first step is justified by the following result which says that the Minkowski addition is a 'convexifying' operation.

Theorem 7.10 (Shapley–Folkman–Starr, see Artstein, 1980) *Let* $K_1, \ldots, K_n \in \mathcal{K}$. *Then*

$$\rho_H(K_1 \oplus \cdots \oplus K_n, \operatorname{conv}(K_1 \oplus \cdots \oplus K_n)) \leq \sqrt{d} \max_{1 \leq i \leq n} \|K_i\|,$$

where d is the dimension of the space.

Step 2 is possible, since the support function of the partial sum $S_n = X_1 \oplus \cdots \oplus X_n$ is equal to the sum of the corresponding support functions

$$h(S_n, u) = h(X_1, u) + \cdots + h(X_n, u).$$

The convergence of compact sets is understood with respect to the Hausdorff metric (7.7). The Hausdorff distance between two compact convex sets is equal to the uniform distance between their support functions

$$\rho_H(K, L) = \sup_{u \in \mathbb{S}^{d-1}} |h(K, u) - h(L, u)|, \quad K, L \in \mathcal{C}_0,$$

and, therefore, the uniform convergence of support functions corresponds to the narrow convergence of the corresponding convex sets.

Theorem 7.11 (see Artstein and Vitale, 1975) *Let $X_1, X_2, \ldots$ be i.i.d. integrable random compact sets in $\mathbb{R}^d$. Then*

$$\rho_H(S_n/n, \mathbf{E}X_1) \to 0 \quad a.s. \text{ as } n \to \infty. \tag{7.10}$$

Note that an analogue of Kolmorogov's three series theorem for random sets has been proven by Lyashenko (1982). Weighted sums of random compact sets were considered by Taylor and Inoue (1985). An ergodic theorem for subadditive families of convex compact random sets has been proved by Schürger (1983). The latter requires a special technique, since a pointwise ergodic theorem is not available in the space of continuous functions of the unit sphere.

Brunn–Minkowski inequality and other applications

To begin with, we use the strong law of large numbers for random sets to prove the *Brunn–Minkowski inequality* for random sets. This inequality in its classical form states that

$$\{\mu_d(\lambda K \oplus (1-\lambda)L)\}^{1/d} \geq \lambda \{\mu_d(K)\}^{1/d} + (1-\lambda)\{\mu_d(L)\}^{1/d},$$
$$(7.11)$$

where $0 \leq \lambda \leq 1$, and K, L are compact sets (in fact, their measurability is enough). This inequality has been used by Anderson (1955) in studying translations of unimodal multivariate density functions.

Theorem 7.12 (see Vitale, 1990) *Let X be an integrable random compact set in $\mathbb{R}^d$. Then*

$$\{\mu_d(\mathsf{E}X)\}^{1/d} \geq \mathsf{E}\{\mu_d(X)\}^{1/d} \qquad (7.12)$$

Following Vitale (1990) we apply Theorem 7.12 to prove the *isoperimetric inequality* in $\mathbb{R}^2$. This inequality states that among all figures with given perimeter, the disk has the maximal area. Clearly, it suffices to prove this for convex sets. Let $K \in \mathcal{C}_0$, and let X be the uniform random rotation (isotropisation) of K. Its Aumann expectation is a disk B whose perimeter is equal to the expected perimeter of X, which coincides with the perimeter of K. Finally, by (7.12),

$$\sqrt{\mu_2(B)} \geq \mathsf{E}\sqrt{\mu_2(X)} = \sqrt{\mu_2(K)},$$

so that the area of K is less than the area of the disk B.

Applications of the strong law of large numbers for unbounded random closed sets to the allocation problem (in mathematical economics) were considered by Artstein and Hart (1981).

Renewal theorem

The *elementary renewal theorem* states that

$$t^{-1}H(t) \to 1/\mathsf{E}\xi_1 \quad \text{as} \quad t \to \infty, \qquad (7.13)$$

where $H(t) = \sum \mathsf{P}\{S_n \leq t\}$ is the renewal function, and $S_n = \xi_1 + \cdots + \xi_n, n \geq 1$, are partial sums of i.i.d. nonnegative random variables, $S_0 = 0$.

Let $X_1, X_2, \ldots$ be i.i.d. random bodies, and let $S_n = X_1 \oplus \cdots \oplus X_n$ be their partial sums ($S_0 = \{0\}$ is the origin). For a closed set K define the *containment renewal function*

$$H(K) = \sum_{n=0}^{\infty} P\{S_n \subset K\}.$$

Let us find the limit of $H(tK)/t$ as $t \to \infty$. Clearly, the most difficult point is to find out what should replace $1/E\xi_1$ in the right-hand side of (7.13). Now we formulate an elementary renewal theorem for random bodies.

Theorem 7.13 (see Molchanov *et al.*, 1995) *Suppose that* $E\|X_1\| < \infty$ *and* $E\rho(0, X_1)^2 < \infty$, *where* $\rho(0, X_1)$ *is the minimum distance between points of* X_1 *and the origin. Consider a convex compact set* K. *If* $0 \in \text{Int } K$, *then*

$$\lim_{t \to \infty} \frac{H(tK)}{t} = \inf_{u \in S_X^+} \frac{h(K, u)}{h(EX_1, u)},$$

where $S_X^+ = \{u \in S^{d-1} : h(EX_1, u) > 0\}$. *If* $0 \notin K$, *then*

$$\lim_{t \to \infty} \frac{H(tK)}{t} = \alpha_K - \min(\alpha_K, \beta_K),$$

where

$$\alpha_K = \inf\left\{\frac{h(K, u)}{h(EX_1, u)} : u \in S^{d-1}; h(K, u) > 0, h(EX_1, u) > 0\right\},$$

$$\beta_K = \inf\left\{\frac{h(K, u)}{h(EX_1, u)} : u \in S^{d-1}; h(K, u) < 0, h(EX_1, u) < 0\right\}.$$

Here $\inf \emptyset = \infty$ *and* $\infty - \infty = 0$.

Example 7.13 Let $X_1 = \{\xi\}$ be a random singleton. If $E\|\xi\|^2 < \infty$ and $0 \in K$, then Theorem 7.13 yields

$$\lim_{t \to \infty} H(tK)/t = \sup\{r : rE\xi \in K\} = 1/g(K, E\xi),$$

where $g(K, x) = \inf\{r \geq 0 : x \in rK\}$ is the *gauge function* of K, see Schneider (1993, p. 43).

Exercise (see Molchanov et al., 1996) Prove a renewal theorem for the *inclusion renewal function* $I(K) = \sum P\{K \subset S_n\}$.

Problem 7.9 (a) It is relatively easy to derive a renewal theorem for the *hitting renewal function* $U(K) = \sum P\{S_n \cap K \neq \emptyset\}$ for X being a segment on the line and a simple K (singleton, segment etc.). Is it possible to derive a renewal theorem in this case on the same level of generality as Theorem 7.13?

(b) Derive a random set analogue of Blackwell's theorem (it states that $H(t+h) - H(t) \to h/E\xi_1$ if ξ_1 is not concentrated on some arithmetic lattice.)

(c) Find out a 'good' analogue of the renewal equation for random sets. Remember that for random variables this equation is

$$H(t) = F(t) + \int_0^t H(t-u)dF(u).$$

Central limit theorem

Similarly to the strong law of large numbers, a central limit theorem for random compact sets in $\mathbb{R}^d$ follows from the corresponding results in Banach spaces. Let X be a *square integrable* random compact set (i.e. $E\|X\|^2 < \infty$). Define its covariance function Γ_X as

$$\Gamma_X(u,v) = E\big[h(X,u)h(X,v)\big] - Eh(X,u)Eh(X,v), \quad u,v \in \mathbf{S}^{d-1}.$$

By definition $\Gamma_X = \Gamma_{\text{conv}(X)}$. The first central limit theorem for special random sets with a finite number of values has been proved by Cressie (1979) using explicit calculations of probability distributions of their Minkowski sums. We will give below the most general central limit theorem for square integrable random compact sets in $\mathbb{R}^d$.

Theorem 7.14 (see Weil, 1982) *Let $X_1, X_2, \ldots$ be i.i.d. square integrable random sets. Then*

$$\sqrt{n}\rho_H(n^{-1}(X_1 \oplus \cdots \oplus X_n), EX_1) \to \sup_{u \in \mathbf{S}^{d-1}} \|\zeta(u)\|, \qquad (7.14)$$

where $\{\zeta(u), u \in \mathbf{S}^{d-1}\}$ is a Gaussian centred random element in the space of continuous functions on $\mathbf{S}^{d-1}$ with the covariance $E\zeta(u)\zeta(v) = \Gamma_{X_1}(u,v)$.

The *law of the iterated logarithm* for a square integrable random compact set $X \subset \mathbb{R}^d$ states, see Giné et al. (1983):

$$\limsup_{n\to\infty} \frac{\sqrt{n}}{\sqrt{2\log\log n}}\rho_H(S_n/n, E\text{conv}(X)) \le \sqrt{E\|\text{conv}(X_1)\|^2}.$$

Gaussian random sets

Gaussian random functions on $\mathbf{S}^{d-1}$ appear naturally in the limit theorem for Minkowski sums of random sets, since the normalised Hausdorff distance between $n^{-1}(X_1 \oplus \cdots \oplus X_n)$ and $\mathsf{E}X_1$ is asymptotically distributed as the maximum of a Gaussian random function ζ on $\mathbf{S}^{d-1}$. It is natural to consider random sets whose support functions are Gaussian.

Problem 7.10 Find out whether (and when) $\zeta(u)$, $u \in \mathbf{S}^{d-1}$ (as in Theorem 7.14) is the support function of a random body in $\mathbb{R}^d$.

A random set X is *Gaussian*, if $\phi(X)$ is a Gaussian random variable for all functionals $\phi : \mathcal{K} \to \mathbb{R}$ that satisfy

1. $\phi(\alpha K \oplus \beta L) = \alpha\phi(K) + \beta\phi(L)$ for all $\alpha, \beta \geq 0$ and $K, L \in \mathcal{K}$.

2. ϕ is Lipschitz continuous with respect to the Hausdorff metric.

Proposition 7.1 (see Vitale, 1984) *A random convex set X is Gaussian if and only if $h(X, u)$ is a Gaussian random function on $\mathbf{S}^{d-1}$.*

Proposition 7.1 yields the following important result giving a characterisation of Gaussian random compact sets and states that Gaussian sets have degenerated shapes.

Theorem 7.15 (see Lyashenko, 1983a and Vitale, 1984) *A random compact set X is Gaussian if and only if there are a fixed $M \in \mathcal{C}_0$ (convex compact set) and a Gaussian random vector ξ such that $\mathrm{conv}(X) = M + \xi$ a.s.*

A nice and short proof was given by Vitale (1984). The main idea is to show that a Gaussian random set can be translated to have a positive support function, and then argue that a Gaussian vector with a.s. positive coordinates is deterministic. This translation of X is given by its Steiner point

$$s(X) = \frac{1}{b_d} \int_{\mathbf{S}^{d-1}} h(X, u)u \, du, \qquad (7.15)$$

see e.g. Schneider, (1993, p. 42), where b_d is the volume of the unit ball in $\mathbb{R}^d$ and the integral is taken with respect to the $(d-1)$-dimensional Hausdorff measure (or surface area).

A similar result is valid for integrable convex compact random sets in a separable Banach space E, see Puri and Ralescu (1987). Random compact p-stable sets have been studied by Giné and Hahn (1985).

Problem 7.11 Find a 'good' definition of Gaussian random sets, which yields sets with non-degenerated shape distributions.

7.4.2 Unions

Union-stable random closed sets

In its very classical form, extreme values theory deals with maxima or minima of i.i.d. random variables, see Galambos (1978). By now, extreme values theory has been extended to dependent random variables, coordinate-wise extremes of random vectors, and maxima of random functions.

It is easy to see that *each* random variable ξ is max-infinitely divisible, i.e. $\xi \overset{d}{\sim} \max(\xi_{1n}, \dots, \xi_{nn})$ for i.i.d. random variables $\xi_{1n}, \dots, \xi_{nn}$. The same concept can be defined for random vectors (although not all vectors are max-infinitely divisible) and random elements in lattices. The infinite divisibility property is determined by the basic lattice operation, and can be established in a rather general framework, see Berg *et al.* (1984) and Norberg (1989).

It is easy to see that maxima (or minima) of random variables can be easily 'translated' into the language of random closed sets. For instance, if $X_i = (-\infty, \xi_i]$ is a random half-line, then $X_1 \cup \cdots \cup X_n$ is the half-line bounded by $\max(\xi_1, \dots, \xi_n)$. The first to study infinite divisibility and stability of random closed sets with respect to their unions was Matheron (1975). Unions of random sets have been further investigated by Molchanov (1993b), where also most of the material below can be found.

Definition 7.2 *(see Matheron, 1975) A random closed set X is said to be* infinitely divisible for unions *(or union-infinitely-divisible), if for any positive integer n,*

$$X \overset{d}{\sim} X_{n1} \cup \cdots \cup X_{nn},$$

where X_{ni}, $1 \le i \le n$, are i.i.d. random closed sets.

Point x is said to be a *fixed point* of X if $P\{x \in X\} = T(\{x\}) = 1$. The set of *all* fixed points of X is denoted by F_X. The random closed set X is said to be *non-trivial* if $P\{X = F_X\} < 1$, i.e. X does not coincide almost surely with the set of its fixed points.

Exercise (see Molchanov, 1993a) Let X be a random closed subset of a locally-compact space satisfying $X \overset{d}{\sim} X_1 \cap \cdots \cap X_n$, $n \ge 1$, for $X_1, \dots, X_n$ being i.i.d. copies of X. Prove that $X = F_X$ almost surely.

Clearly, $T(K) = 1$ as soon as K hits F_X. It is easy to prove that F_X is a closed set. Having replaced $\mathbb{R}^d$ by the space $\mathbb{R}^d \setminus F_X$, we can consider only union-infinitely-divisible random sets without fixed points, as it was done in Matheron (1975). The following theorem provides a slight modification of his result, which can be obtained by the instrumentality of the harmonic analysis on semigroups, see Berg *et al.* (1984), or the theory of lattices, see Norberg (1989).

Theorem 7.16 *The random closed set X is infinitely divisible for unions if and only if its capacity functional is represented as*

$$T(K) = 1 - \exp\{-\Psi(K)\}, \qquad (7.16)$$

where $\Psi(K)$ is a completely alternating upper semicontinuous capacity such that $\Psi(\emptyset) = 0$ and $\Psi(K)$ is finite for all K with $K \cap F_X = \emptyset$.

A random closed set X is said to be *union-stable* if, for each $n \geq 1$,

$$a_n X \overset{d}{\sim} X_1 \cup \cdots \cup X_n, \qquad (7.17)$$

where $a_n > 0$ and $X_1, \ldots, X_n$ are independent copies of X. Union-stable sets without fixed points were characterised by Matheron (1975). Although his characterisation relies essentially on the lack of fixed points, a similar characterisation theorem is valid for general union-stable random sets. It should be noted that the characterisation of union-stable random sets is more difficult than the characterisation of max-stable random variables because of the possible self-similarity of random sets. Namely, if $\xi \overset{d}{\sim} c\xi$ for each $c > 0$, then the random variable ξ is equal to 1 a.s. On the other hand, many non-trivial random sets satisfy $X \overset{d}{\sim} cX$ for all $c > 0$ (then X is said to be *self-similar*). For instance, $X = \{t \geq 0 : w_t = 0\}$ is a self-similar random set, if w_t is the Wiener process.

Theorem 7.17 (see Molchanov, 1993b) *A non-trivial random closed set X is union-stable if and only if its capacity functional T is of the form (7.16), where $\Psi(K)$ is a completely alternating Choquet capacity, $\Psi(\emptyset) = 0$,*

$$\Psi(sK) = s^\alpha \Psi(K), \quad \Psi(K) < \infty, \qquad (7.18)$$

for a certain $\alpha \neq 0$, and $s F_X = F_X$, whatever positive s and $K \cap F_X = \emptyset$ may be.

Corollary 7.18 *A union-stable random closed set X has no fixed points if and only if $\alpha > 0$ in (7.18). If $\alpha < 0$, then F_X is non-empty and $0 \in F_X$.*

Example 7.14 The capacity functional of the Poisson point process Π_Λ with intensity measure Λ is given by

$$T(K) = P\{\Pi_\Lambda \cap K \neq \emptyset\} = 1 - \exp\{-\Lambda(K)\}.$$

Evidently, each Borel measure Λ satisfies the conditions **(T1)–(T4)**, so that by Theorem 7.16, each Poisson point process is infinitely divisible for unions with $\Psi(K) = \Lambda(K)$. Moreover, all infinitely divisible distributions on lattices admit representations via some related Poisson point processes, see Norberg (1989) and Resnick (1987).

Assume that Λ has the density λ with respect to the Lebesgue measure. Then Π_Λ is union-stable if and only if λ is *homogeneous*, i.e. $\lambda(su) = s^{\alpha-d}\lambda(u)$ for a real α, whatever u from $\mathbb{R}^d$ and positive s may be. If $\alpha < d$, then the origin is a fixed point of X. If $\alpha = d$, then Π_Λ is the stationary Poisson point process. This point process is called the scale invariant Poisson process.

Example 7.15 Let $f : \mathbb{R}^d \to [0, \infty]$ be an upper semicontinuous function. Then $\Psi(K) = \sup_{x \in K} f(x)$ is a maxitive capacity, see Example 7.6. The capacity functional (7.16) corresponds to the random closed set $X = \{x : f(x) \geq \xi\}$, where ξ is a random variable exponentially distributed with parameter 1. The random set X is union-stable if and only if the function f is homogeneous, i.e. $f(sx) = s^\alpha f(x)$ for each $s > 0$ and $x \in \mathbb{R}^d$.

Example 7.16 The capacity $\Psi(K) = C(K)$ defined by (7.6) corresponds to a union-stable set if $k(sx, sy) = s^{-\alpha}k(x, y)$ for all $s > 0$ and $x, y \in \mathbb{R}^d$, i.e. for k being the Riesz kernel $C_{d,\gamma}\|x - y\|^{\gamma-d}$ for $\gamma = d - \alpha$. The corresponding random set can be obtained as the union of trajectories of stable processes in $\mathbb{R}^d$ starting at the points of a Poisson process.

Convex-stable sets

Note that union-stable random sets in $\mathbb{R}^d$ with $d > 1$ are usually non-convex. We can modify the stability concept as follows.

Definition 7.3 *Convex random closed set* X *is said to be* convex-stable *if, for every* $n \geq 2$ *and independent copies* $X_1, \ldots, X_n$ *of* X,

$$a_n X \overset{d}{\sim} \text{conv}(X_1 \cup \cdots \cup X_n) \oplus K_n$$

for some $a_n > 0$, $K_n \in \mathcal{K}$. *If* $K_n = \{b_n\}$, $n \geq 1$, *i.e.* K_n *is a singleton for all* $n \geq 1$, *then* X *is said to be* strictly convex-stable.

This definition is due to Giné *et al.* (1990), where X is additionally assumed to be compact with support function having a non-degenerate distribution. Reformulating (7.17) in terms of support functions, we obtain

$$a_n h(X, u) \overset{d}{\sim} \max \{h(X_1, u), \dots, h(X_n, u)\} + h(K_n, u), \quad u \in \mathbf{S}^{d-1}.$$

If X is *compact* and $h(X, u)$ has a *non-degenerate* distribution for each $u \in \mathbf{S}^{d-1}$, then $h(X, \cdot)$ is a random max-stable sample continuous process. Convex-stability of general random closed sets can be characterised through their containment functionals, see Molchanov (1993a,b).

A random set X is said to be *generalised union-stable* if, for each collection $X_1, \dots, X_n$ of independent copies of X, $n \geq 2$,

$$a_n X \overset{d}{\sim} (X_1 \cup \dots \cup X_n) \oplus K_n$$

for certain $a_n > 0$, $K_n \in \mathcal{K}$.

Problem 7.12 Characterise generalised union-stable sets for arbitrary compacts K_n and, in particular, for circular K_n. It is difficult, since the characterisation problem cannot be reduced to examination of max-stable support functions. The situation is worse in case of unbounded X, since then X may coincide in distribution with $X + u$ for some $u \neq 0$.

Weak convergence of unions

Let us consider limit theorems for *normalised unions of random sets*, where union-stable sets appear as weak limits. Let $X_1, \dots, X_n$ be independent identically distributed random closed sets with the capacity functional T, and let $Z_n = X_1 \cup \dots \cup X_n$ be their union. We study the weak convergence of $a_n^{-1} Z_n$ where a_n, $n \geq 1$, is a suitable sequence of positive reals. It is evident that the limiting random closed set Z (if it exists) is union-stable. Hence its capacity functional $\tilde{T}$ is characterised by Theorem 7.17. It is possible to show that the corresponding parameter α is positive in case $a_n \to 0$ and is negative if $a_n \to \infty$.

Naturally, while handling unions of random sets we use methods similar to the theory of extreme values. However, the direct generalisation fails due to specific features of capacities. For example, the function $1 - T(xK)$, $x > 0$, plays in our consideration the same role as the distribution function in the theory of extreme values, but this function is no longer monotone and even may not tend to 1 as $x \to \infty$. We consider only the case $a_n \to \infty$ as $n \to \infty$. The limiting random set

Z has a fixed point at the origin, so that Corollary 7.18 yields $\alpha < 0$. For each compact set K put

$$a_n(K) = \sup\{x :\ T(xK) \geq 1/n\}\,,$$

where $a_n(K) = 0$ in case $T(xK) < 1/n$ for all $x > 0$. Set $\tau_K(x) = T(xK)$, $x \geq 0$, and define a class of compact sets by $\mathcal{T} = \{K : \liminf_{x \to \infty} \tau_K(x) = 0\}$. Remember that a measurable function $f : [0, \infty) \to [0, \infty)$ is *regularly varying*, see Seneta (1976), with the *index* α if, for each $x > 0$,

$$\lim_{t \to \infty} f(tx)/f(t) = x^\alpha$$

Theorem 7.19 (see Molchanov, 1993b) *Assume that for each K from $\mathcal{T}$ there exists the limit of $a_n(K)/a_n$ (which is not necessary finite), and let $\tau_K(x)$ be a regularly varying function with negative index α. Then $a_n^{-1} Z_n$ converges weakly to the union-stable set Z with the capacity functional $\tilde{T}(K) = 1 - \exp\{-\Psi(K)\}$, where*

$$\Psi(K) = \begin{cases} \lim(a_n(K)/a_n)^{-\alpha} & ,\quad K \in \mathcal{T}, \\ \infty & ,\quad otherwise. \end{cases}$$

If the function $-\log T(xK)$ is regularly varying, then, under a suitable normalisation, the union of random set has a deterministic limit, see Molchanov (1993b). The corresponding technique involves regularly varying and maxitive capacities, and generalises results on almost sure stability of normalised random samples, see Davis *et al.* (1988).

Some applications of limit theorems for unions to maxima of random functions and intersections of random half-planes have been considered in Molchanov (1993b).

7.5 Statistical inference for random sets

The non-linear structure of the record space $\mathcal{F}$ causes many complications in statistics of random closed sets. The main reasons can be explained as follows.

- There is no 'good' subtraction on the space $\mathcal{F}$. The space $\mathcal{F}$ endowed with the union operation or Minkowski addition has only a structure of a semigroup. This causes problems when defining expectations and especially for higher moments including the variance, see Stoyan and Stoyan (1994). Usually, it is impossible to centre distributions of random sets in order to get random sets with zero expectation.
- Very often random set distributions are determined by set-valued parameters. For example, if $X = M\xi$ with ξ being a random variable,

then deterministic set M becomes a set-valued parameter. Thus, the usual parameter θ in statistical inference is a set and the family Θ of its possible values is a collection of sets.

- The distribution of a random set is determined by the capacity functional $T_X(K)$, which is *not* a measure on $\mathcal{K}$. There is no 'good replacement' for the Lebesgue measure on $\mathcal{F}$, and it is difficult to define a sensible probability density function for random closed sets. A general problem is the efficiency of estimators, and the absence of a replacement for the maximum likelihood method.

7.5.1 Estimation of the capacity functional

As in the theory of empirical processes, see Shorack and Wellner (1986), it is natural to estimate $T_X(K)$ by the *empirical capacity functional*

$$T_n^*(K) = \frac{1}{n} \sum_{i=1}^{n} \mathbf{1}_{X_i \cap K \neq \emptyset}$$

provided a sample of i.i.d. sets $X_1, \dots, X_n$ is given. Clearly, $T_n^*(K) \to T(K)$ a.s. as $n \to \infty$ for each $K \in \mathcal{K}$. If

$$\sup_{K \in \mathcal{M}, \, K \subseteq K_0} |T_n^*(K) - T(K)| \to 0 \quad \text{a.s. as} \quad n \to \infty, \qquad (7.19)$$

for some $\mathcal{M} \subseteq \mathcal{K}$ and each $K_0 \in \mathcal{K}$, then X is said to satisfy the *Glivenko-Cantelli theorem* over the class $\mathcal{M}$. The classical Glivenko-Cantelli theorem can be obtained for the case when $X = [\xi, \infty)$ and $\mathcal{M}$ is a class of singletons. However, there are some peculiar examples of random sets which do not satisfy (7.19) even for a simple class $\mathcal{M}$.

Example 7.17 Let $X = \xi + M \subset \mathbb{R}$, where ξ is a Gaussian random variable, and M is a nowhere dense subset of $[0, 1]$ having positive Lebesgue measure. The complete metric space $[0, 1]$ cannot be covered by a countable union of nowhere dense sets, so that, for each $n \geq 1$, there exists a point $x_n \in [0, 1] \setminus \cup_{i=1}^{n} X_i$. Thus, $T_n^*(\{x_n\}) = 0$, while $T_X(\{x_n\}) \geq \varepsilon > 0, n \geq 1$, and (7.19) does not hold for $\mathcal{M} = \{\{x\} : x \in [0, 1]\}$. There are examples of random sets that do not satisfy (7.19) even for $\mathcal{M}$ being a countable family of singletons, see Molchanov (1990a).

In order to prove (7.19) for random sets we have to impose some conditions on realisations of random sets. The random set X is called *regular closed* if $X = \overline{\text{Int}\, X}$ with probability 1, i.e. X almost surely coincides with the closure of its interior.

Theorem 7.20 (see Molchanov, 1987) *Let X be a regular closed random set, and let*

$$T_X(K) = \mathsf{P}\{(\text{Int } X) \cap K \neq \emptyset\} \qquad (7.20)$$

for all $K \in \mathcal{M}$. Then (7.19) holds for each $K_0 \in \mathcal{K}$. If $\mathcal{M} = \mathcal{K}$, then (7.20) is also a necessary condition for (7.19) provided X is a.s. continuous, i.e. $\mathsf{P}\{X \in \partial X\} = 0$ for all $x \in \mathbb{R}^d$

The conditions are especially simple if X is a stationary random closed set. Then (7.19) holds for $\mathcal{M} = \mathcal{K}$ if X is regular closed (and only if, provided X is a.s. continuous), see Molchanov (1989).

The random set setting can be very natural in multivariate statistics. For instance, if $X = \{\xi\}$ is a random singleton, then its set-valued quantile can be defined as $\{x : F(x) \geq p\}$, while the *empirical quantile* is a random set obtained by replacing the distribution function F with its empirical counterpart, see Molchanov (1990b).

7.5.2 Statistics of compact sets

Statistical studies of compact random sets are difficult because of the lack of models (or distributions of random sets) which allow evaluations and provide sets of sufficiently variable shape. Clearly, simple random sets, like random singletons and balls, are ruled out, if we would like to have models with really *random shapes*. At the first approximation, one can characterise shape of a compact set K by numerical parameters, called *shape ratios*, see Stoyan and Stoyan (1994, pp.103–107). For instance, the *area-perimeter ratio* (or compacity) is given by $4\pi A(K)/U(K)^2$, where $A(K)$ is the area of K and $U(K)$ its perimeter, *circularity shape ratio* is the quotient of the diameter of the circle with area $A(K)$ and the diameter of the minimum circumscribed circle of K. Further shape ratios are discussed in Tuzikov *et al.* (1996). All these shape ratios are motion-invariant functionals, so that their values do not depend on translations/rotations and scale transformations of K.

Typically, the starting point is a sample of i.i.d realisations of a random compact set. If positions of sets are known, then we speak about statistics of *sets*, in contrast to statistics of *figures* when locations/orientations of sets are not given. In the engineering literature it is usual to perform statistical analysis of a sample of sets $K_1, \ldots, K_n$ by computing several shape ratios for each set from the sample. This yields a multivariate sample, which can be analysed using methods from multivariate statistics. The 'only' problem is that distributions of shape ratios are not known theoretically for most models of random sets, and even their lower moments are known only in very few special cases.

Typical polygons in tessellations

Below we discuss several models of planar random polygons. A *Poisson polygon* is the 'typical' random polygon X generated by a stationary and isotropic Poisson line tessellation of intensity λ, see Matheron (1975), Stoyan *et al.* (1995) and Stoyan and Stoyan (1994). To obtain this typical polygon, all polygons generated by the family of lines are shifted in such a way that their centres of gravity lay in the origin and are interpreted as realisations of the 'typical' random polygon X.

Consider the Dirichlet (or Voronoi) mosaic generated by a stationary Poisson point process of intensity λ. For each point x_i we construct the open set consisting of all points of the plane whose distance to x_i is less than the distances to other points. If shifted by x_i, the closures of these sets give realisations of the so-called *Poisson–Dirichlet polygon*, see Stoyan and Stoyan (1994, pp. 125–128).

The distributions of both polygons depend only on one parameter, which characterises the polygons' mean size. For example, the Poisson–Dirichlet polygon X has the same distribution as $\lambda^{-1/2} X_1$, where X_1 is the Poisson–Dirichlet polygon obtained by a Poisson point process with unit intensity. Several important numerical parameters of the Poisson polygon and the Poisson–Dirichlet polygon are known either theoretically or obtained by numerical integration or simulations, see Stoyan and Stoyan (1994). It is however not clear how to compute the capacity functionals for both random polygons.

One can use random polygons to obtain further models of random sets. For instance, a *rounded polygon* is defined by $Y = X \oplus B_\xi(0)$, where ξ is a positive random variable, and $B_\xi(0)$ is the disk of radius ξ centred at the origin.

Exercise(see Stoyan and Stoyan, 1994) Compute the Aumann expectation of the Poisson polygon and Poisson–Dirichlet polygon.

Finite convex hulls

Another model of a random isotropic polygon is the convex hull of N independent points uniformly distributed within the disk $B_r(0)$ of radius r centred at the origin, see Stoyan and Stoyan (1994, p. 135). For given N, the containment functional of X is given by

$$P\{X \subset K\} = (\mu_d(K)/\mu_d(B_r(0)))^N ,$$

where K is a convex subset of $B_r(0)$. However, further exact distributional characteristics are not so easy to find; mostly only asymptotic properties for large N can be investigated, see the references in Section 7.3.4. The values of r and N are the two model parameters.

Convex-stable sets

Convex-stable sets (more precisely, strictly convex-stable sets) appear as weak limits of scaled convex hulls $a_n^{-1}\mathrm{conv}(Z_1 \cup \cdots \cup Z_n)$ of i.i.d. random compact sets $Z_1, Z_2, \ldots$ having a regularly varying distribution in a certain sense. We will deal only with the simplest model where X is the weak limit of convex hulls of random singletons $Z_i = \{\xi_i\}$. Note that the corresponding distribution has an unbounded support. This case has been studied by Aldous *et al.* (1991), Brozius (1989) and Brozius and de Haan (1987), while statistical applications have been considered by Molchanov and Stoyan (1996). In order to obtain non-degenerate limit distributions, the probability density f of the ξ_i's must be *regular varying* in $\mathbb{R}^d$, i.e. for any vector $e \neq 0$,

$$f(tu_t)/f(te) \to \phi(u) \neq 0, \infty \qquad (7.21)$$

for $u_t \to u \neq 0$ as $t \to \infty$. In the isotropic case one can use the density

$$f(u) = \frac{c}{c_1 + \|u\|^{\alpha+d}}, \quad u \in \mathbb{R}^d,$$

for some $\alpha > 0$. The constant c is a scaling parameter, while the normalising parameter c_1 is chosen in such a way that f is a probability density function. Then (7.21) is valid, $a_n \sim n^{1/\alpha}$ as $n \to \infty$, and, for $\|e\|^{\alpha+d} = c$, (7.21) yields $\phi(u) = c\|u\|^{-\alpha-d}$. Thus, for $a_n = n^{1/\alpha}$, the random set $a_n^{-1}\mathrm{conv}(\xi_1, \ldots, \xi_n)$ converges weakly as $n \to \infty$ to the isotropic convex-stable random set X with the containment functional

$$P\{X \subset K\} = \exp\left\{-\int_{\mathbb{R}^d \setminus K} \phi(u)\, du\right\}. \qquad (7.22)$$

The model has two parameters: the *size parameter* c and the *shape parameter* α. The choice of a *positive* α implies that X is almost surely a compact convex polygon, see Davis *et al.* (1987). The limiting random set X is the convex hull of a *scale invariant Poisson point process*, see Daley and Vere-Jones (1988, p. 325) and Example 7.14.

Exercise (see Molchanov, 1993b and Molchanov and Stoyan, 1996)
1. Compute the Aumann expectation of X and the distribution of

$h(X, u)$.

2. Prove that $h(X, u)$ and $h(X, -u)$ are independent.

3. Give an algorithm for simulation convex-stable polygons.

Statistical inference for convex-stable sets (and for other models of compact random sets) is based on the method of moments applied to some functionals, see Molchanov and Stoyan (1996). For instance, in the planar case the equation

$$\mathsf{E}U(X)/\sqrt{\mathsf{E}A(X)} = 2\pi\Gamma(1 - \alpha^{-1})\left(\pi\alpha\Gamma(2 - 2\alpha^{-1})/(\alpha - 1)\right)^{-1/2}$$

yields an estimate of α.

Union-stable random sets may serve as a model of random closed sets, since the capacity functionals of union-stable random sets are expressed by explicit formulae and union-stable sets can be easily simulated as scaled unions of 'simple' random sets.

Other possible models for compact random sets include radius-vector perturbed sets, weak limits of intersections of half-planes, see Molchanov (1993b) and set-valued growth processes, see Stoyan and Stoyan (1994, p. 139). It is typical also to consider models of random fractal sets, see Falconer (1986) and Stoyan and Stoyan (1994). Usually they are defined by some iterative random procedure or through level sets or graphs of random functions.

7.5.3 Expectations of random sets

To explain why an expectation of random compact (closed) set is not straightforward to define, consider random closed set X which is with probability $1/2$ equal to $[0, 1]$ and with probability $1/2$ to $\{0, 1\}$. What can be a 'reasonable' expectation of X?

The main problem in defining expectations is explained by the fact that the space $\mathcal{F}$ is not linear. The usual trick is to 'linearise' $\mathcal{F}$, i.e. embed $\mathcal{F}$ into a linear space. Then one defines expectation in this linear space and, if possible, 'maps' it back to $\mathcal{F}$. For instance, the Aumann expectation stems from the Hörmander embedding theorem, which states that the family of compact closed subsets of a Banach space E can be embedded, by an isometry j, into the space of all bounded functionals on the unit ball in the adjoint space E^*. This isometry $j(K)$ is given by the support function $h(K, \cdot)$. For general sets it is difficult to construct a similar embedding and linearity is usually lost.

Fréchet expectation. In application to $\mathcal{K}$ with the Hausdorff metric, the set $K = K_0 \in \mathcal{K}$ which minimises $\mathsf{E}\rho_H(X, K)^2$ for $K \in \mathcal{K}$ is said to be the *Fréchet expectation* of the random compact set X, and $\mathsf{E}\rho_H(X, K_0)^2$ is called the variance of X, see Fréchet (1948). In general, X may have several expectations. Unfortunately, in most practical cases it is not possible to solve the basic minimisation problem, since the parameter space $\mathcal{K}$ is too rich.

Doss expectation. Let $\rho(x, F) = \inf\{\rho(x, y) : y \in F\}$ be the minimum distance between x and F, and $\rho_H(x, F) = \sup\{\rho(x, y) : y \in F\}$ be the Hausdorff distance between $\{x\}$ and $F \in \mathcal{F}$. The *Doss expectation* of X is defined by

$$\mathsf{E}_D X = \{y : \rho(x, y) \leq \mathsf{E}\rho_H(x, X) \text{ for all } x \in \mathbb{R}^d\},$$

see Herer (1986). This expectation can also be defined for general metric spaces. Clearly, $\mathsf{E}_D X$ is the intersection of all balls $B_{\mathsf{E}\rho_H(x, F)}(x)$ for $x \in \mathbb{R}^d$.

Exercise (see Doss, 1949) Prove that $\mathsf{E}_D X$ is convex and $\mathsf{E}_D \supseteq \mathsf{E}X$, where $\mathsf{E}X$ is the Aumann expectation of X (if X is integrable).

Vorob'ev expectation. Let $\xi_X(x) = \mathbf{1}_X(x)$ be the *indicator function* of X. Then $\mathsf{E}\xi_X(x) = p_X(x) = \mathsf{P}\{x \in X\}$ is the *coverage function*. In general, $p_X(x)$ is not an indicator function. Assume that $\mathsf{E}\mu_d(X) < \infty$. A set-theoretic mean is defined by Vorob'ev (1984) by $L_p = \{x \in \mathbb{R}^d : p_X(x) \geq p\}$ for p determined from the inequality

$$\mu_d(L_q) \leq \mathsf{E}\mu_d(X) \leq \mu_d(L_p), \quad \text{for all} \quad q > p.$$

The set $L_{1/2} = \{x \in \mathbb{R}^d : p_X(x) \geq 1/2\}$ has properties of a *median*, see Stoyan and Stoyan (1994, p. 115).

This approach considers indicator functions as elements of $L^2(\mathbb{R}^d)$. Hence singletons as well as sets of almost surely vanishing Lebesgue measure are considered as uninteresting, since the corresponding indicator function $\mathbf{1}_X(x)$ vanishes almost surely for all x as soon as the corresponding distribution is atomless.

Exercise (see Stoyan and Stoyan, 1994 and Kovyazin, 1986) **1.** Let L_p be the Vorob'ev expectation of X. Prove that, for all Borel sets B with $\mu_d(B) = \mu_d(L_p), \mathsf{E}\mu_d(X \triangle L_p) \leq \mathsf{E}\mu_d(X \triangle B)$, where '$\triangle$' denotes the symmetric difference. **2.** Prove that $\mathsf{E}\mu_d(X \triangle L_{1/2}) \leq \mathsf{E}\mu_d(X \triangle B)$ for each bounded B. This property is similar to the classical property of the median which minimises the first absolute central moment. **3.** Formulate a law of large numbers where the Vorob'ev expectation appears as the limit.

Radius-vector mean. Let X be *star-shaped* with respect to the origin 0, i.e. X contains the segment $[0, x]$ for each $x \in X$. Then X is determined by its *radius-vector function* of X defined by

$$r_X(u) = \sup\{t : tu \in X, \ t \geq 0\}, \quad u \in S^{d-1}$$

The expected values $E r_X(u), u \in S^{d-1}$, yield the radius-vector function of a deterministic shrinkable set, which is called the *radius-vector mean* of X, see Stoyan and Stoyan (1994, p.111). Radius-vector functions are very popular in the engineering literature, where it is usual to apply Fourier methods for shape description, see Beddow and Mellow (1980) and Stoyan and Stoyan, (1994).

Distance function average. Let $\xi_X(x) = \rho(x, X)$ be the *distance function* of X, i.e. $\xi_X(x)$ equals the Euclidean distance from x to the nearest point of X. A suitable level set of the mean distance function $\bar{d}(x) = E\rho(x, X)$ serves as the mean of X. First, we threshold this function to get family of sets $X(\varepsilon) = \{x : \bar{d}(x) \geq \varepsilon\}, \quad \varepsilon > 0$. Then the *distance average* $\bar{X}$ is the set $X(\varepsilon)$, where ε is chosen to minimise

$$\|\bar{d}(\cdot) - \rho(\cdot, X(\varepsilon))\|_\infty = \sup_{x \in R^d} |\bar{d}(x) - \rho(x, X(\varepsilon))|,$$

see Baddeley and Molchanov (1995) for details and further generalisations. This approach makes it possible to deal with sets of zero Lebesgue measure, since even in this case the distance functions are non-trivial (in contrast to indicator functions used to define the Vorob'ev expectation).

7.5.4 Averaging of figures

Very often an observer deals with samples of figures rather with samples of sets. This means that positions of the sets are irrelevant for the problem and the aim is to find the average shape of the sets in the sample. Such a situation appears in studies of particles (dust powder, sand grains, abrasives etc.). The problem is difficult since orientation and location of the particles are arbitrary; directly congruent particles are considered as identical. Consequently, the definitions of mean values of random compact sets are not directly applicable for random figures. For instance, the images of particles are isotropic sets, whence the corresponding set-valued expectations are balls or discs.

To come to the space of figures we introduce an equivalence relationship on the space $\mathcal{K}$ of compact sets. Namely, two compact sets are
equivalent if they can be superimposed by a rigid motion (scale transformations are excluded!). Since the space $\mathcal{K}$ is already non-linear,
considering the factor space $\mathcal{K}/\sim$ worsens the situation even more. The
approach below is inspired by studies of shapes and landmark configurations, see Bookstein (1986), Dryden and Mardia (1993), Le and
Kendall (1993) and Ziezold (1994). Landmarks are characteristic points
of planar objects, such as the tips of the nose and the chin, if human
profiles are studied. However, for the study of particles such landmarks
are not natural. Perhaps they could be points of extremal curvature or
other interesting points on the boundary, but for a useful application
of the landmark method the number of landmarks per object has to be
constant, and this may lead to difficulties or unnatural restrictions. For
the practically important problem of determining an empirical mean of
a sample of figures the following general idea seems to be natural:

> Give the figures particular locations and orientations such that they are in
> a certain sense 'close together'; then consider the new sample as a sample
> of sets and, finally, determine a set-theoretic mean.

As we have seen, we usually linearise $\mathcal{K}$, i.e. replace sets by some
functions. Then motions of sets correspond to transformations of functions considered to be elements of a general Hilbert space, see Stoyan
and Molchanov (1997). For example, consider convex compact sets
described by support functions. The group of proper motions of sets
corresponds to the group G acting on $L^2(\mathbf{S}^{d-1})$ as follows

$$gh(K, u) = \tilde{g}h(K, u) + \langle l, u \rangle, \quad u \in \mathbf{S}^{d-1}, \; l \in \mathbb{R}^d, \qquad (7.23)$$

where $\tilde{g}h(K, u)$ is the support function of the set $\tilde{g}K$ obtained as a
rotation of K. It was proven by Stoyan and Molchanov (1997) that the
'optimal' translations of convex sets are given by their Steiner points
(7.15), so that the Steiner points of all figures must be superimposed.

Problem 7.13 Find an effective way to characterise optimal position
for the sample $\mathbf{1}_{K_1}(\cdot), \dots, \mathbf{1}_{K_n}(\cdot)$ of indicator functions considered to
be elements of $L^2(\mathbb{R}^d)$.

7.5.5 Boolean models

Evidently, completely observable i.i.d. samples of random sets not only
provide information on their means, but also on their distributions. The
situation gains new features when i.i.d. random sets are not directly

observable. For example, an observer very often views only a clump of i.i.d. random sets placed in the space. Models of clumps are used to produce sets of complicated shape from 'simple' components. Sometimes it is possible to discern separate grains by manual work and then to reduce the problem to the analysis of an i.i.d. sample. However, for high intensities this is rather difficult.

The heuristic notion of a clump can be described mathematically as the *Boolean model*. This (the most used) model of overlapping particle systems can be defined as follows. First, take i.i.d. random closed sets $\Xi_0, \Xi_1, \Xi_2, \ldots$. Then consider the stationary Poisson point process $\Pi_\lambda = \{x_i, i \geq 1\}$ in $\mathbb{R}^d$ with intensity λ. Finally, place each Ξ_i at the position of the corresponding point x_i and take their union

$$\Xi = \bigcup_{i:x_i \in \Pi_\lambda} (\Xi_i + x_i). \tag{7.24}$$

This set is said to be the Boolean model, see Matheron (1975) and Stoyan *et al.* (1995). The points x_i are called 'germs', and the random set Ξ_0 is called the (typical) 'grain'. The key objective of statistical inference for the Boolean model can be described as follows.

Given an observation of a Boolean model estimate its parameters: the intensity of the germ process and the distribution of the grain.

As in statistics of point processes, all properties of estimators are formulated for an observation window W growing without bounds (although, in practice, the window is fixed), see Daley and Vere-Jones (1988, p. 332).

Statistics of the Boolean model began with estimation of λ and mean values of the Minkowski functionals of the grain (mean area, volume, perimeter, etc.). In general, these values are not enough to retrieve the distribution of the grain Ξ_0, although sometimes it is possible to find an appropriate distribution if some parametric family is given. Clearly, without the distribution of the grain it is impossible to simulate the underlying Boolean model, and, therefore, to use tests based on simulations. For example, if the grain is a ball, then, in general, it is impossible to determine its radius distribution by the corresponding moments up to the dth order. However, if the distribution belongs to some parametric family, say log-normal, then it is determined by these moments. Even in the parametric setup now most studies end with proof of strong consistency. Results concerning asymptotic normality are still rather exceptional, see Heinrich (1993) and Molchanov and Stoyan (1994), and there are no theoretical studies of efficiency.

By the Choquet theorem, the distribution of Ξ_0 is determined by the corresponding capacity functional $T_{\Xi_0}(K)$. The capacity functional of the Boolean model Ξ can be evaluated as

$$T_{\Xi}(K) = P\{\Xi \cap K \neq \emptyset\} = 1 - \exp\{-\lambda E\mu_d(\Xi_0 \oplus \check{K})\}, \qquad (7.25)$$

see Hall (1988), Matheron (1975) and Stoyan et al. (1995). Here $\check{K} = \{-x : x \in K\}$. By the Fubini theorem, we get from (7.25)

$$T_{\Xi}(K) = 1 - \exp\left\{-\lambda \int_{\mathbb{R}^d} T_{\Xi_0}(K + x)\, dx\right\} \qquad (7.26)$$

The functional in the left-hand side of (7.26) is determined by the whole set Ξ and can be estimated from observations of Ξ. However, it is unlikely that the integral equation (7.26) can be solved directly.

Below we give a short review of some known estimation methods. Mostly, we will present only relations between *observable* characteristics and parameters of the Boolean model, bearing in mind that replacing these observable characteristics by their empirical counterparts provides estimators for the corresponding parameters. Further methods are discussed in Molchanov (1997).

In the simplest case, the accessible information about Ξ is only the volume fraction p covered by Ξ. Because of stationarity and (7.25)

$$p = T(\{0\}) = 1 - \exp\{-\lambda E\mu_d(\Xi_0)\}, \qquad (7.27)$$

so that an estimator of $\lambda E\mu_d(\Xi_0) = \lambda \bar{V}(\Xi_0)$ can be obtained. Estimators of p can be defined as, e.g., $\hat{p} = \mu_d(\Xi \cap W)/\mu_d(W)$ or $\hat{p} = \text{card}(\Xi \cap W \cap Z)/\text{card}(W \cap Z)$, where W is a window and Z is a lattice in $\mathbb{R}^d$, see Mase (1982).

Two-point covering probabilities determine the *covariance* of Ξ

$$C(v) = P\{\{0, v\} \subset \Xi\} = 2p - T(\{0, v\}).$$

Then the function

$$q(v) = 1 + \frac{C(v) - p^2}{(1 - p)^2} = \exp\{\lambda E\mu_d(\Xi_0 \cap (\Xi_0 - v))\} \qquad (7.28)$$

yields the set-covariance function of the grain $\gamma_{\Xi_0}(v) = E\mu_d(\Xi_0 \cap (\Xi_0 - v))$, see Stoyan and Stoyan (1994). This fact has been used in Molchanov (1994) to estimate the shape of a deterministic convex grain. It should be noted that the covariance is quite flexible in applications, since it only depends on the capacity functional on two-point compacts. For example, almost all covariance-based estimators are easy to reformulate for censored observations, see Molchanov (1992, 1997). Applications of the covariance function to statistical estimation of the

Boolean model parameters were discussed also by Hall (1988), Serra (1982) and Stoyan *et al.* (1995).

Historically, the first statistical method for the Boolean model was the *minimum contrast method* for contact distribution functions or the covariance, see Serra (1982), where a number of references to the works of the Fontainebleau school in the seventies are given, and also Diggle (1981). Its essence is to determine the values of T_Ξ for some sub-families of compact sets (balls, segments or two-point sets). Then the right-hand side of (7.25) can be expressed by means of known integral geometric formulae. In particular, when $K = B_r(0)$ is a ball of radius r, then the Steiner formula gives the expansion of $\mathsf{E}\mu_d(\Xi_0 \oplus B_r(0))$ as a polynomial of dth order whose coefficients are expressed through the Minkowski functionals of Ξ_0. The next step is to replace the left-hand side of (7.25) by its empirical counterpart and approximate it by a polynomial, see Cressie (1991), Hall (1988) and Heinrich (1993). Finally, the polynomial's coefficients yield estimators of the unknown parameters.

Now consider another estimation method which could be named the *method of intensities*. First, note that (7.27) is an equality relating spatial averages to parameters of the Boolean model. The volume fraction is the simplest spatial average. It is possible to consider other spatial averages, for example, mean surface area per unit volume, the specific Euler-Poincaré characteristics or, more general, densities of extended Minkowski measures, see Weil and Wieacker (1984). According to the method of intensities, estimators are chosen as solutions of the equations relating these spatial averages (their intensities) to estimated parameters.

A particular implementation of the method of intensities depends on the way of extending of the Minkowski functionals onto the extended convex ring (the family of locally finite unions of convex compact sets). The *additive extension*, see Schneider (1993), was used by Kellerer (1985) and Weil (1988). Another technique, the so-called *positive extension*, see Matheron (1975) and Schneider (1993), has been applied to statistics of the Boolean model by Cressie (1991), Serra (1982) and Stoyan *et al.* (1987). It goes back to Haas, Matheron and Serra (1967) and De Hoff (1967).

In the planar case the latter approach is based on (7.27) and the following relationships

$$L_A = (1 - p)\bar{U}(\Xi_0)\lambda, \tag{7.29}$$
$$N_A^+ = (1 - p)\lambda \tag{7.30}$$

are used to express the intensity λ, the mean perimeter of the grain $EU(\Xi_0)$ and the mean area $EA(\Xi_0)$ through the following *observable* values: the area fraction p, the specific boundary length L_A and the intensity N_A^+ of the point process of (say lower) positive tangent points. The specific boundary length is the expected length of the boundary of Ξ within a unit area. If we mark each lower-left tangent point of all grains, then some of these points will be covered by other grains, while others will be exposed. Then N_A^+ is the expected number of these exposed points in a unit area. This method yields biased but strong consistent and asymptotically normal estimators, see Molchanov and Stoyan (1994). For instance, the asymptotic variance of the intensity estimator obtained from (7.30) is equal to $\lambda/(1 - p)$. Higher-order characteristics of the point process of tangent points are ingredients to construct estimators of the distribution of the grain, see Molchanov (1995).

Problem 7.14 How it is possible to compute 'information' contained in a realisation of the Boolean model? Find the lowest possible bound for the variance of an unbiased intensity estimator. Note that it is very easy to find an example of 'super-efficient' estimators for the grain's size.

7.6 Concluding remarks

As this survey shows, random sets theory uses techniques from various areas of mathematics: topology, capacity theory, convex and integral geometry and optimisation. Within probability theory and mathematical statistics, random sets have a number of relationships to point processes, stationary random fields, local properties of stochastic processes, limit theorems in functional spaces, the theory of extreme values, empirical measures and spatial statistics.

On the other hand, random sets have found a number of applications in the several areas including those, mentioned in the 'definitions' in the beginning of this survey. For example, random sets are widely used in biometry to model images that appears in the studies of tumour growth by Cressie and Hulting (1992), spatial patterns of heather in the countryside, see Diggle (1981), nerve fibre degeneration in Ayala and Simo (1995), the microstructure of dough, see Bindrich and Stoyan (1991), milk processing, Hansen (1995), and forest fires, Vorob'ev (1984). A strong law of large numbers for random sets can be applied to the allocation problem in mathematical economics, see Artstein and Hart (1981) and also Hildenbrand (1974). Applications in geology and material science started with models for metal crystallisation by Kolmogorov (1937) and include models for photographic emulsions (Lyashenko, 1986), defects in polymers (Bulinskaya *et al.*, 1986) and polymer crystallisation (Capasso *et al.*, 1996), structure of materials (Ohser and Tscherny, 1988 and Saxl *et al.*, 1994), alloy structures (Quenec'h *et al.*, 1994), geological deposits (Chessa, 1995), and the microstructure of paper (Molchanov *et al.*, 1993). Furthermore, random sets are widely used to model noise (Dougherty and Handley, 1995), filters (Goutsias, 1993) and prior distributions (Van Lieshout, 1995) in image analysis. Finally, random sets appear in statistical studies of set-valued estimators, see Meister and Moeschlin (1988).

References

Adler, R.J. (1981) *The Geometry of Random Fields*. Wiley, New York.

Aldous, D., Fristedt, B., Griffin, Ph.S. and Pruitt, W.E. (1991) The number of extreme points in the convex hull of a random sample. *J. Appl. Probab.*, **28**, 287–304.

Aló, R.A., Korvin, A. de and Roberts, C. (1979) The optional sampling theorem for convex set valued martingales. *J. Reine Angew. Math.*, **310**, 1–6.

Anderson, T.W. (1955) The integral of a symmetric unimodal function over a symmetric convex set and some probability inequalities. *Proc. Amer. Math. Soc.*, **6**, 170–176.

Artstein, Z. (1980) Discrete and continuous bang-bang and facial spaces or: look for the extreme points. *SIAM Rev.*, **22**, 172–185.

Artstein, Z. (1984) Limit laws for multifunctions applied to an optimization problem. In: *Multifunctions and Integrands* (Edited by G. Salinetti), vol. 1091 of *Lect. Notes Math.*, 66–79, Springer, Berlin.

Artstein, Z. and Hart, S. (1981) Law of large numbers for random sets and allocation processes. *Math. Oper. Res.*, **6**, 485–492.

Artstein, Z. and Vitale, R.A. (1975) A strong law of large numbers for random compact sets. *Ann. Probab.*, **3**, 879–882.

Attouch, H. and Wets, R.J.-B. (1990) Epigraphical processes: law of large numbers for random LSC functions. *Sém. Anal. Convexe*, **20**, 29 pp.

Aubin, J.-P. and Frankowska, H. (1990) *Set-Valued Analysis*, vol. 2 of *System and Control, Foundation and Applications*. Birkhäuser, Boston.

Aumann, R.J. (1965) Integrals of set-valued functions. *J. Math. Anal. Appl.*, **12**, 1–12.

Ayala, G. and Simo, A. (1995) Bivariate random closed sets and nerve fibre degeneration. *Adv. in Appl. Probab.*, **27**, 293–305.

Baddeley, A.J. and Molchanov, I.S. (October 1995) Averaging of random sets based on their distance functions. Tech. Rep. BS-R9528, Centrum voor

Wiskunde en Informatica, Amsterdam. To appear in *J. Math. Imaging and Vision*.

Baddeley, A.J. and Molchanov, I.S. (1997) On the expected measure of a random set. In: *Advances in Theory and Applications of Random Sets* (Edited by D. Jeulin et al.), 3–20, Singapore, Ecole Nationale Supérieure des Mines de Paris. Fontainebleau, World Scientific Publishing Co.

Beddow, J.K. and Mellow, T. (1980) *Testing and Characterization of Powder and Fine Particles*. Heyden & Sons, London.

Berg, C., Christensen, J.P.R. and Ressel, P. (1984) *Harmonic Analysis on Semigroups*. Springer, Berlin.

Billingsley, P. (1968) *Convergence of Probability Measures*. Wiley, New York.

Bindrich, U. and Stoyan, D. (1991) Stereology for pores in white bread: Statistical analyses for the Boolean model by serial sections. *J. Microscopy*, **162**, 231–239.

Bookstein, F.L. (1986) Size and shape spaces for landmark data in two dimensions (with discussion). *Statist. Sci.*, **1**, 181–242.

Brozius, H. (1989) Convergence in mean of some characteristics of the convex hull. *Adv. in Appl. Probab.*, **21**, 526–542.

Brozius, H. and Haan, L. de (1987) On limiting laws for the convex hull of a sample. *J. Appl. Probab.*, **24**, 852–862.

Bulinskaya, E.V., Molchanov, S.A. and Feraig, N. (1986) Parameters estimates for complex random sets. I. *Moscow University Math. Bulletin*, **41**, 16–21.

Cabo, A. and Groeneboom, P. (1994) Limit theorems for functionals of convex hulls. *Probab. Theor. Relat. Fields*, **100**, 31–55.

Capasso, V., Micheletti, A., De Giosa, M. and Mininni, R. (1996) Stochastic modelling and statistics of polymer crystallization processes. *Surv. Math. Ind.*, **6**, 109–132.

Chessa, A.G. (1995) *Conditional Simulation of Spatial Stochastic Models for Reservoir Heterogeneity*. Ph.D. thesis, Delft University of Technology.

Choquet, G. (1953/54) Theory of capacities. *Ann. Inst. Fourier*, **5**, 131–295.

Cox, D.R. and Isham, V. (1980) *Point Processes*. Chapman and Hall, London.

Cramér, H. and Leadbetter, M.R. (1967) *Stationary and Related Stochastic Processes*. Wiley, New York.

Cressie, N.A.C. (1979) A central limit theorem for random sets. *Z. Wahrsch. verw. Gebiete*, **49**, 37–47.

Cressie, N.A.C. (1991) *Statistics for Spatial Data*. Wiley, New York.

Cressie, N.A.C. and Hulting, F.L. (1992) A spatial statistical analysis of tumor growth. *J. Amer. Statist. Assoc.*, **87**, 272–283.

Cutsem, B. van (1972) Martingales de convexes fermés aléatoires en dimension finie. *Ann. Inst. H.Poincaré, Sect. B, Prob. et Stat.*, **8**, 365–385.

Daley, D.J. and Vere-Jones, D. (1988) *An Introduction to the Theory of Point Processes*. Springer, New York.

Davis, R.A., Mulrow, E. and Resnick, S.I. (1987) The convex hull of a random sample in $\mathbf{R}^2$. *Comm. Statist. Stochastic Models*, **3**, 1–27.

Davis, R.A., Mulrow, E. and Resnick, S.I. (1988) Almost sure limit sets of random samples in $\mathbf{R}^d$. *Adv. in Appl. Probab.*, **20**, 573–599.

DeHoff, R.T. (1967) The quantitative estimation of mean surface curvature. *Transactions of the American Institute of Mining, Metallurgical and Petroleum Engineering*, **239**, 617.

Diggle, P.J. (1981) Binary mosaics and the spatial pattern of heather. *Biometrics*, **37**, 531–539.

Doss, S. (1949) Sur la moyenne d'un élément aléatoire dans un espace distancié. *Bull. Sci. Math.*, **73**, 48–72.

Dougherty, E.R. and Handley, J.C. (1995) Recursive maximum-likelihood estimation in one-dimensional discrete Boolean random set model. *Signal Processing*, **43**, 1–15.

Dryden, I.L. and Mardia, K.V. (1993) Multivariate shape analysis. *Sankhya A*, **55**, 460–480.

Dümbgen, L. and Walter, G. (1996) Rates of convergence for random approximations of convex sets. *Adv. in Appl. Probab.*, **28**, 384–393.

Dynkin, E.B. (1961) *Die Grundlagen der Theorie der Markoffschen Prozesse*. Springer, Berlin, Translation from Russian *Osnovaniya teorii Markovskikh protsessov* (1959).

Falconer, K.J. (1986) Random fractals. *Math. Proc. Cambridge Philos. Soc.*, **100**, 559–582.

Falconer, K.J. (1990) *Fractal Geometry*. Wiley, Chichester.

Fréchet, M. (1948) Les éléments aléatoires de nature quelconque dans un espace distancié. *Ann. Inst. H.Poincaré, Sect. B, Prob. et Stat.*, **10**, 235–310.

Fristedt, B. (1996) Intersections and limits of regenerative sets. In: *Random Discrete Structures* (Edited by D. Aldous and R. Pemantle), vol. 76 of *The IMA Volumes in Mathematics and its Applications*, 121–151, Springer, New York.

Galambos, J. (1978) *The Asymptotic Theory of Extreme Order Statistics*. Wiley, New York.

Giné, E. and Hahn, M.G. (1985) Characterization and domains of attraction of p-stable compact sets. *Ann. Probab.*, **13**, 447–468.

Giné, E., Hahn, M.G. and Vatan, P. (1990) Max-infinitely divisible and max-stable sample continuous processes. *Probab. Theor. Relat. Fields*, **87**, 139–165.

Giné, E., Hahn, M.G. and Zinn, J. (1983) Limit theorems for random sets: application of probability in Banach space results. *Lect. Notes Math.*, **990**, 112–135.

Goutsias, J. (1993) Binary random fields, random set theory, and the morphological analysis of shape. In: *Image Algebra and Morphological Image Processing, IV (San Diego, CA, 1993)*, vol. 2030 of *Proc. SPIE*, 54–64, SPIE, Bellingham, WA.

Haas, A., Matheron, G. and Serra, J. (1967) Morphologie mathematique et granulometries en place. *Ann. Mines*, **11,12**, 736–753 and 767–782.

Hall, P. (1988) *Introduction to the Theory of Coverage Processes*. Wiley, New York.

Hansen, M. (1995) *Spatial Statistics and the Variation of the Protein Network during Milk Processing*. Ph.D. thesis, Royal Veterinary and Agricultural University, Copenhagen.

Heinrich, L. (1993) Asymptotic properties of minimum contrast estimators for parameters of Boolean models. *Metrika*, **31**, 349–360.

Herer, W. (1986) Esperance mathematique au sens de Doss d'une variable aleatoire a valeurs dans un espace metrique. *C. R. Acad. Sci., Paris, Ser. I*, **302**, 131–134.

Herer, W. (1987) Martingales with values in closed bounded subsets of a metric space. *C. R. Acad. Sci., Paris, Ser. I*, **305**, 275–278, In French.

Hiai, F. and Umegaki, H. (1977) Integrals, conditional expectations, and martingales of multivalued functions. *J. Multivariate Anal.*, **7**, 149–182.

Hildenbrand, W. (1974) *Core and Equilibria for Large Economy*. Princeton University Press, Princeton.

Hoeffding, W. (1953) On the distribution of the expected values of the order statistics. *Ann. Statist.*, **24**, 93–100.

Hoffman-Jørgensen, J. (1969) Markov sets. *Math. Scand.*, **24**, 145–166.

Hsing, T. (1994) On the asymptotic distribution of the area outside a random convex hull in a disk. *Ann. Appl. Probab.*, **4**, 478–493.

Huber, P.J. (1981) *Robust Statistics*. Wiley, New York.

Hueter, I. (1994) The convex hull of a normal sample. *Adv. in Appl. Probab.*, **26**, 855–875.

Itô, K. and McKean, H.P. (1965) *Diffusion Processes and Their Sample Paths*. Springer, Berlin.

Kendall, D.G. (1985) Exact distributions for shapes of random triangles in convex sets. *Adv. in Appl. Probab.*, **17**, 308–329.

Kingman, J.F.C. (1973) Homecomings of Markov processes. *Adv. in Appl. Probab.*, **5**, 66–102.

Kolmogorov, A.N. (1937) On statistical theory of metal crystallization. *Izvestia Academy of Science, USSR, Ser. Math.*, **3**, 355–360, In Russian.

Korostelev, A.P. and Tsybakov, A.B. (1993) *Minimax Theory of Image Restoration*. Springer, New York.

Kovyazin, S.A. (1986) On the limit behavior of a class of empirical means of a random set. *Theory Probab. Appl.*, **30**, 814–820.

Krylov, N.V. and Yushkevitch, A.A. (1964) Markov random sets. *Theory Probab. Appl.*, **9**, 738–743, In Russian.

Landkof, N.S. (1972) *Foundations of Modern Potential Theory*. Springer, Berlin.

Le, H. and Kendall, D.G. (1993) The riemannian structure of euclidean shape space: a novel environment for statistics. *Ann. Statist.*, **21**, 1225–1271.

Lévy, P. (1992) *Processus stochastiques et mouvement brownien*. Édition Jacques Gabay, Sceaux, Reprint of the second (1965) edition.

Lieshout, M.N.M. van (1995) *Stochastic geometry models in image analysis and spatial statistics*, vol. 108 of *CWI Tract*. Stichting Mathematisch Centrum, Centrum voor Wiskunde en Informatica, Amsterdam.

Lyashenko, N.N. (1982) Limit theorems for sum of independent compact random subsets. *J. Soviet Math.*, **20**, 2187–2196.

Lyashenko, N.N. (1983a) Statistics of random sets. In: *Applied Statistics*, vol. 45 of *Uch. Zapiski Stat. CEMI*, 40–59, CEMI, In Russian.

Lyashenko, N.N. (1983b) Weak convergence of step-functions in the space of closed sets. *Zapiski Nauch. Seminarov LOMI*, **130**, 122–129, In Russian.

Lyashenko, N.N. (1986) Random 'grainy' patterns. I. *Zap. Nauchn. Sem. Leningrad. Otdel. Mat. Inst. Steklov. (LOMI)*, **153**, 73–96.

Lyashenko, N.N. (1987) Graphs of random processes as random sets. *Theory Probab. Appl.*, **31**, 72–80.

Maissonneuve, B. (1974) *Systèmes Régénératifs*, vol. 15 of *Astérisque*.

Mase, S. (1982) Asymptotic properties of stereological estimators of volume fraction for stationary random sets. *J. Appl. Probab.*, **19**, 111–126.

Matheron, G. (1975) *Random Sets and Integral Geometry*. Wiley, New York.

Meister, H. and Moeschlin, O. (1988) Unbiased set-valued estimators with minimal risk. *J. Math. Anal. Appl.*, **130**, 426–438.

Michael, E. (1951) Topologies on spaces of subsets. *Trans. Amer. Math. Soc.*, **71**, 152–182.

Molchanov, I.S. (1984) A generalization of the Choquet theorem for random sets with a given class of realizations. *Theory Probab. Math. Statist.*, **28**, 99–106, Translation from *Teor. Veroyatn. Mat. Stat.* **28**, 86-93, 1983.

Molchanov, I.S. (1987) Uniform laws of large numbers for empirical associated functionals of random closed sets. *Theory Probab. Appl.*, **32**, 556–559.

Molchanov, I.S. (1989) On convergence of empirical accompanying functionals of stationary random sets. *Theory Probab. Math. Statist.*, **38**, 107–109, Translation from *Teor. Veroyatn. Mat. Stat.* 38: 97–99, 1988.

Molchanov, I.S. (1990a) A characterization of the universal classes in the Glivenko–Cantelli theorem for random closed sets. *Theory Probab. Math. Statist.*, **41**, 85–89, Translation from *Teor. Veroyatn. Mat. Stat.* 41: 74–78, 1989.

Molchanov, I.S. (1990b) Empirical estimation of distribution quantiles of random closed sets. *Theory Probab. Appl.*, **35**, 594–600.

Molchanov, I.S. (1992) Handling with spatial censored observations in statistics of Boolean models of random sets. *Biometrical J.*, **34**, 617–631.

Molchanov, I.S. (1993a) Limit theorems for convex hulls of random sets. *Adv. in Appl. Probab.*, **25**, 395–414.

Molchanov, I.S. (1993b) *Limit Theorems for Unions of Random Closed Sets*, vol. 1561 of *Lect. Notes Math.*. Springer, Berlin.

Molchanov, I.S. (1994) On statistical analysis of Boolean models with non-random grains. *Scand. J. Statist.*, **21**, 73–82.

Molchanov, I.S. (1995) Statistics of the Boolean model: From the estimation of means to the estimation of distributions. *Adv. in Appl. Probab.*, **27**, 63–86.

Molchanov, I.S. (1997) *Statistics of the Boolean Model for Practitioners and Mathematicians*. Wiley, Chichester.

Molchanov, I.S., Omey, E. and Kozarovitzky, E. (1995) An elementary renewal theorem for random convex compact sets. *Adv. in Appl. Probab.*, **27**, 931–942.

Molchanov, I.S. and Stoyan, D. (1994) Asymptotic properties of estimators for parameters of the Boolean model. *Adv. in Appl. Probab.*, **26**, 301–323.

Molchanov, I.S. and Stoyan, D. (1996) Statistical models of random polyhedra. *Comm. Statist. Stochastic Models*, **12**, 199–214.

Molchanov, I.S., Stoyan, D. and Fyodorov, K.M. (1993) Directional analysis of planar fibre networks: Application to cardboard microstructure. *J. Microscopy*, **172**, 257–261.

Norberg, T. (1984) Convergence and existence of random set distributions. *Ann. Probab.*, **12**, 726–732.

Norberg, T. (1986) Random capacities and their distributions. *Probab. Theor. Relat. Fields*, **73**, 281–297.

Norberg, T. (1987) Semicontinuous processes in multi-dimensional extreme-value theory. *Stochastic Process. Appl.*, **25**, 27–55.

Norberg, T. (1989) Existence theorems for measures on continuous posets, with applications to random set theory. *Math. Scand.*, **64**, 15–51.

O'Brien, G. L. and Vervaat, W. (1995) Compactness in the theory of large deviations. *Stochastic Process. Appl.*, **57**, 1–10.

Ohser, J. and Tscherny, H. (1988) *Grundlagen der quantitativen Gefügeanalyse*, vol. 264 of *Freiberger Forschungshefte, Reihe B*. Deutscher Verlag für Grundstoffindustrie, Leipzig.

Papageorgiou, N.S. (1985) On the theory of Banach space valued multifunctions I, II. *J. Multivariate Anal.*, **17**, 185–206, 207–227.

Papageorgiou, N.S. (1987) A convergence theorem for set-valued supermartingales with values in a separable Banach space. *Stochastic Anal. Appl.*, **5**, 405–422.

Peres, Y. (1996) Intersection-equivalence of Brownian paths and certain branching processes. *Comm. Math. Phys.*, **177**, 417–434.

Puri, M.L., Ralescu, D.A. and Ralescu, S.S. (1987) Gaussian random sets in Banach space. *Theory Probab. Appl.*, **31**, 598–601.

Quenec'h, J.-L., Chermant, J.-L., Coster, M. and Jeulin, D. (1994) Liquid phase sintered materials modelling by random closed sets. In: *Mathematical Morphology and Its Applications to Image Processing* (Edited by J. Serra and P. Soille), 225–232, Kluwer, Dordrecht.

Rachev, S.T. (1991) *Probability Metrics and the Stability of Stochastic Models*. Wiley, Chichester.

Rényi, A. and Sulanke, R. (1963) Über die konvexe Hülle von n zufällig gewällten Punkten. *Z. Wahrsch. verw. Gebiete*, **2**, 75–84.

Resnick, S.I. (1987) *Extreme Values, Regular Variation and Point Processes*.

Springer, Berlin.

Richter, H. (1963) Verallgemeinerung eines in der Statistik benötigten Satzes der Maßtheorie. *Math. Ann.*, **150**, 85–90 and 440–441.

Ripley, B.D. (1981) *Spatial Statistics*. Wiley, New York.

Ripley, B. and Rasson, J.-P. (1977) Finding the edge of a Poisson forest. *J. Appl. Probab.*, **14**, 483–491.

Robbins, H.E. (1944) On the measure of a random set. I. *Ann. Math. Statist.*, **15**, 70–74.

Rockafellar, R.T. (1970) *Convex Analysis*. Princeton University Press, Princeton, NJ.

Salinetti, G. (1987) Stochastic optimization and stochastic processes: the epigraphical approach. *Math. Res.*, **35**, 344–354.

Salinetti, G., Vervaat, W. and Wets, R.J.-B. (1986) On the convergence in probability of random sets (measurable multifunctions). *Math. Oper. Res.*, **11**, 420–422.

Salinetti, G. and Wets, R.J.-B. (1986) On the convergence in distribution of measurable multifunctions (random sets), normal integrands, stochastic processes and stochastic infima. *Math. Oper. Res.*, **11**, 385–419.

Saxl, I., Pelikan, K., Rataj, J. and Besterci, M. (1994) *Quantification and Modelling of Heterogeneous Systems*. Cambridge Interscience Publishing, Cambridge.

Schneider, R. (1988) Random approximations of convex sets. *J. Microscopy*, **151**, 211–227.

Schneider, R. (1993) *Convex Bodies. The Brunn–Minkowski Theory*. Cambridge University Press, Cambridge.

Schürger, K. (1983) Ergodic theorems for subadditive superstationary families of convex compact random sets. *Z. Wahrsch. verw. Gebiete*, **62**, 125–135.

Seneta, E. (1976) *Regularly Varying Functions*, vol. 508 of *Lect. Notes Math.*. Springer, Berlin.

Serra, J. (1982) *Image Analysis and Mathematical Morphology*. Academic Press, London.

Shorack, G.R. and Wellner, J.A. (1986) *Empirical Processes with Applications to Statistics*. Wiley, New York.

Stoyan, D., Kendall, W.S. and Mecke, J. (1995) *Stochastic Geometry and Its Applications*. Wiley, Chichester, 2nd edn.

Stoyan, D. and Molchanov, I.S. (1997) Set-valued means of random particles. *Journal of Mathematical Imaging and Vision*, **7**, 111–121.

Stoyan, D. and Stoyan, H. (1994) *Fractals, Random Shapes and Point Fields*. Wiley, Chichester.

Taylor, R.L. and Inoue, H. (1985) Convergence of weighted sums of random sets. *Stochastic Anal. Appl.*, **3**, 379–396.

Tuzikov, A.V., Margolin, G.L. and Grenov, A.I. (1996) Convex set symmetry measurement via Minkowski addition. *Journal of Mathematical Imaging and Vision*, To appear.

Vitale, R.A. (1983) Some developments in the theory of random sets. *Bull. Inst. Intern. Statist.*, **50**, 863–871.

Vitale, R.A. (1984) On Gaussian random sets. In: *Stochastic Geometry, Geometric Statistics, Stereology* (Edited by R. Ambartzumian and W. Weil), 222–224, Teubner, Leipzig, Teubner Texte zur Mathematik, B.65.

Vitale, R.A. (1987) Expected convex hulls, order statistics, and Banach space probabilities. *Acta Appl. Math.*, **9**, 97–102.

Vitale, R.A. (1990) The Brunn–Minkowski inequality for random sets. *J. Multivariate Anal.*, **33**, 286–293.

Vorob'ev, O.Yu. (1984) *Srednemernoje Modelirovanie (Mean-Measure Modelling)*. Nauka, Moscow, In Russian.

Wagner, D. (1977) Survey of measurable selection theorem. *SIAM J. Control Optim.*, **15**, 859–903.

Weil, W. (1982) An application of the central limit theorem for Banach-space-valued random variables to the theory of random sets. *Z. Wahrsch. verw. Gebiete*, **60**, 203–208.

Weil, W. (1988) Expectation formulas and isoperimetric properties for non-isotropic Boolean models. *J. Microscopy*, **151**, 235–245.

Weil, W. and Wieacker, J.A. (1984) Densities for stationary random sets and point processes. *Adv. in Appl. Probab.*, **16**, 324–346.

Ziezold, H. (1994) Mean figures and mean shapes applied to biological figures and shape distributions in the plane. *Biometrical J.*, **36**, 491–510.

Zolotarev, V.M. (1986) *Modern Theory of Summation of Independent Random Variables*. Nauka, Moscow, In Russian.

CHAPTER 8

General shape and registration analysis

Ian Dryden

Department of Statistics, University of Leeds,

Leeds, LS2 9JT, U.K.

The paper reviews various topics in the shape analysis of landmark data. In particular, matching configurations using regression is emphasized. Connections with general shape spaces and shape distances are discussed. Kendall's shape space for point configurations and the affine shape space are considered in particular detail. Matching two configurations and the extension to generalized matching are illustrated with applications in electrophoresis and biology. Shape distributions are briefly discussed and inference in tangent spaces is considered. Finally, some robustness and smoothing issues are highlighted.

8.1 Introduction

The geometrical description of an object can be decomposed into registration and shape information. For example, an object's location, rotation and size could be the registration information and the geometrical information that remains is the object's shape. An object's shape is invariant under registration transformations and two objects have the same shape if they can be registered to match exactly.

Depending on the application the registration information may be irrelevant, of some interest or highly important. For example, if we wish to estimate an average shape and the variability in shape in a population of objects, then the registration information is not important and can be ignored. If we wish to locate and identify an object in an image then the particular registration of the object in the image is of interest, as well as its shape. If we wish to find a match between two different types of images then the registration information is of primary interest.

The range of applications of shape analysis is vast. For example, studies of shape can be found in archaeology, biology, chemistry, geography, image analysis, medicine and engineering. In fact, any application where the geometrical comparison of objects is required will require the use of shape analysis.

The pioneers of the subject of shape analysis are Kendall (1977, 1984, 1989) and Bookstein (1978, 1986, 1989, 1991). Some references and reviews include Goodall (1991), Le and Kendall (1993), Kent (1994, 1995), Dryden and Mardia (1993), Small (1988), Stoyan *et al.* (1995), Stoyan and Stoyan (1994) and Mardia (1995). Two recent books on the topic are Small (1996) and Dryden and Mardia (1997).

In Section 8.2 we describe the matching of two configurations using regression, making connections with general shape spaces and shape distances. In particular, we consider shape matching and affine shape matching. A particular application of matching electrophoresis gels is described. In Section 8.3 we consider generalized matching where a random sample of objects is available, and we study shape in two dimensions and affine shape in particular detail. An application studying the shape of mouse vertebrae is described to illustrate the work. In Section 8.4 we consider some shape distributions which can be used for inference. In Section 8.5 we extend the work to consider robustness issues and in Section 8.6 we consider smoothed matching. We conclude with a brief discussion.

8.2 Matching with regression

8.2.1 Introduction

We first consider the problem of matching one configuration of points to another using regression. We consider some common situations, and make connections with calculations of distances in shape spaces.

Let M^k be the Euclidean space of k labelled points in m real dimensions, where particular sub-spaces might be excluded (such as co-incident points or deficient rank configurations). We use the Euclidean metric on $\mathbb{R}^m$, and so if $Y \in M^k$ is a $k \times m$ matrix, the Euclidean norm of the matrix Y is given by $\|Y\| = \sqrt{\text{trace}(Y^T Y)}$. Here Y^T denotes the transpose of the matrix Y.

Let G be a set of transformations (registrations) which act on $\mathbb{R}^m$. Consider T, Y to be $k \times m$ matrices representing the Cartesian coordinates of two configurations of k points in m dimensions i.e. $T, Y \in M^k$. We write the transformation of T by $g \in G$ as $g(T)$. Two configurations have the same **general shape** if there exists a $g_0 \in G$ such that $Y = g_0(T)$. We shall mainly consider the case where G is a group, although it need not be.

The regression equation of Y on T is written as

$$Y = g(T) + E$$

where $E \in \mathbb{R}^{km}$ is the error matrix and the elements of $g \in G$ are the regression parameters, also known as the registration parameters. The estimates of the regression parameters can be obtained by minimizing an objective function $s^2(E)$ and a popular choice is the least squares objective function $s^2(E) = \|E\|^2 = \|Y - g(T)\|^2$. The **match** of T onto Y is given by $\hat{Y} = \hat{g}(T)$ where

$$\hat{g} = \arg\inf_{g \in G} s^2(Y - g(T))$$

and $\hat{g}$ are the fitted regression parameters. The adequacy of the match can be examined by inspecting the **residual matrix** $R = Y - \hat{g}(T)$ and using regression diagnostics. The **residual discrepancy measure** is given by

$$D_G(Y, T) = s(Y - \hat{g}(T))$$

where $s(E) = +\sqrt{s^2(E)} \geq 0$, and in general $D_G(Y, T) \neq D_G(T, Y)$. A symmetric residual discrepancy measure can sometimes be obtained for the set of transformations G:

$$d_G(Y, T) = \inf_{g \in G} s(h(Y) - g(T))$$

where $h \in G$ is chosen to ensure $d_G(Y, T) = d_G(T, Y)$. Note that h depends on Y and might need to depend on T in general.

8.2.2 General shape spaces

If the set of transformations G is a topological group which acts smoothly on $M^k = \mathbb{R}^{km} \setminus \mathcal{A}$ (where $\mathcal{A}$ is a set of special cases, such as coincident points or the set of deficient rank configurations), then the orbit space of M^k under the action of G is called the **general shape space** (Carne, 1990). All configurations which have the same shape (under G) form an equivalence class, and the union of all the equivalence classes is the orbit space.

The set of Euclidean similarity transformations (translation, rotation and isotropic scaling) of a configuration T ($k \times m$ matrix) is given by

$$\{rT\Gamma + 1_k c^T : r \in \mathbb{R}^+, \Gamma \in SO(m), c \in \mathbb{R}^m\}$$

where c is a translation m-vector, 1_k is the k-vector of ones, Γ is an $m \times m$ special orthogonal matrix (with determinant $+1$) and $r > 0$ denotes an isotropic scaling. If $M^k = \mathbb{R}^{km} \setminus C$, where C is the set of coincident points, and G is the set of Euclidean similarity transformations then the general shape space is **Kendall's shape space**, Σ_m^k. A suitable distance in Σ_m^k is the Riemannian metric ρ, derived by Kendall (1984). Let $\check{T}, \check{Y}$ be the **preshapes** of two configurations $T, Y \in M^k$ which are obtained by centering so that $1_k^T \check{T} = 0 = 1_k^T \check{Y}$ and rescaling so that $\|\check{T}\| = 1 = \|\check{Y}\|$. The Riemannian metric ρ in shape space is given by

$$\rho(T, Y) = \arccos\{\text{trace}(\Lambda)\}, \tag{8.1}$$

where the matrix Λ is the diagonal $m \times m$ matrix with positive elements given by square roots of the eigenvalues of $\check{T}^T \check{Y} \check{Y}^T \check{T}$, except the smallest diagonal element which is negative if we have $\det(\check{Y}^T \check{T}) < 0$. Related distances in Kendall's shape space are the **full Procrustes distance**

$$d_1(Y, T) = \inf_{c \in \mathbb{R}^m, \Gamma \in SO(m), r \in \mathbb{R}^+} \|\check{Y} - [1_k, \check{T}][c, r\Gamma^T]^T\|$$

$$= \left(1 - \{\text{trace}(\Lambda)\}^2\right)^{1/2} = \sin\rho, \tag{8.2}$$

and the **partial Procrustes distance**

$$d(Y, T) = \inf_{c \in \mathbb{R}^m, \Gamma \in SO(m)} \|\check{Y} - [1_k, \check{T}][c, \Gamma^T]^T\|$$

$$= \sqrt{2}(1 - \text{trace}(\Lambda))^{1/2} = \sqrt{2}(1 - \cos\rho)^{1/2} \tag{8.3}$$

The term 'full' is used as the minimization is over the full set of similarity transformations and the term 'partial' is used as the minimization is only over translation and rotation (Kent, 1992).

Another general shape space of interest is the **size-and-shape space** (Kendall, 1989) denoted by $S\Sigma_m^k$ where G is the isometry group (translation and rotation only). The squared partial Procrustes distance between the size and shapes of Y, T is given by

$$d_S^2(Y, T) = \|CY\|^2 + \|CT\|^2 - 2\|CY\|\|CT\|\cos\rho, \tag{8.4}$$

where C is the $k \times k$ centering matrix. which also results from a calculation of a symmetric residual discrepancy measure with regression.

The **affine shape space** is another important case (the general affine space of Ambartzumian, 1982; 1990) which has $M^k = \mathbb{R}^{km} \setminus \mathcal{D}$, where $\mathcal{D}$ is the subspace of configurations which are less than full rank m, and G is the affine transformation group. The set of affine transformations of a configuration T $(k \times m)$ is given by

$$\{TA + 1_k c^T : A \in GL(m)\}$$

where A is an $m \times m$ matrix in the general linear group of invertible $m \times m$ matrices $GL(m)$, and c is a translation m-vector. Note that if $k \leq m + 1$ all configurations have the same affine shape, so we require $k > m + 1$ for non-trivial affine shape. Ambartzumian (e.g. 1982, 1990) and co-workers have considered the geometry and probability measures in this and other affine shape spaces in some detail. Affine shape has also been considered by many workers in computer vision, including Bhavnagri (1995) and Sparr (1991). Since G is a group we could study the general shape space known as the **affine shape space** (the orbit space of M^k under the action of the affine transformation group). If configurations of less than full rank are ignored, the affine shape space is the Grassmann manifold of m-planes in $\mathbb{R}^{m(k-1)}$ (e.g. Harris, 1992).

The **preshape space** has G given by translation and scale only (Kendall, 1984) and this general shape space is equivalent to the unit sphere in $(k - 1)m$ dimensions $S^{(k-1)m}$.

Carne (1990) considered more general shape spaces where M^k is the set of k-tuples in a differentiable Riemannian manifold and G is a Lie group. We restrict our attention to points in Euclidean space in this paper. Table 8.1 summarizes the particular matching situations of interest and the name of the associated general shape space.

Name	M^k	Set of transformations G	Shape space
2D shape	$\mathbb{C}^k \setminus \mathcal{C}$	Translation, Scale, Rotation	Kendall's (2D) shape Σ_2^k
Shape	$\mathbb{R}^{km} \setminus \mathcal{C}$	Translation, Scale, Rotation	Kendall's shape Σ_m^k
Affine shape	$\mathbb{R}^{km} \setminus \mathcal{D}$	Linear transformations	Ambartzumian's general affine shape
Preshape	$\mathbb{R}^{km} \setminus \mathcal{C}$	Translation, Scale	Preshape sphere $S^{(k-1)m}$
Size and shape	$\mathbb{R}^{km} \setminus \mathcal{C}$	Translation, Rotation	Size and shape space $S\Sigma_m^k$

Table 8.1 *The particular sets of transformations of interest in this paper, where $\mathcal{C}$ is the set of coincident points and $\mathcal{D}$ is the set of configurations of deficient rank (rank$[1_k, T] < m + 1$).*

8.2.3 Shape matching in two dimensions

Consider the space of $k \geq 3$ points in $\mathbb{R}^2 \equiv \mathbb{C}$. It is convenient to consider complex notation in this case and so $M^k = \mathbb{C}^k \setminus \mathcal{C}$. We consider the set of transformations G as the Euclidean similarity group (translation, rotation and scale).

Consider matching two configurations $T, Y \in \mathbb{C}^k$ (complex k-vectors). In order to find the match of $T \in \mathbb{C}^k$ onto $Y \in \mathbb{C}^k$ we have the regression equation

$$Y = rTe^{i\alpha} + 1_k c + E = XB + E$$

where $c \in \mathbb{C}$ is the location, $r \in \mathbb{R}^+$ is the scale, $\alpha \in [0, 2\pi)$ is the angle of rotation, $X = [1_k, T]$ is the design matrix, $B = [c, re^{i\alpha}]^T \in \mathbb{C}^2$ are the regression parameters (here in linear regression) and E is the complex k-vector of errors. The most straightforward approach to estimating the regression parameters B is a least squares approach, i.e. minimize E^*E where E^* is the complex conjugate of the transpose of E. The solution is given by

$$\hat{B} = (X^*X)^{-1}X^*Y$$

assuming that X has rank 2 (i.e. the points in T are not all coincident). A measure of residual discrepancy between Y and T is

$$D_G(Y, T) = \|Y - X\hat{B}\| = \|(I_k - X(X^*X)^{-1}X^*)Y\|,$$

and $D_G^2(Y, T)$ is known as the Procrustes sum of squares (e.g. Goodall, 1991). A symmetric measure of residual discrepancy can be obtained by initially centering and rescaling Y, T to their preshapes, denoted by $\check{Y}, \check{T}$. The preshape design matrix is denoted by $\check{X} = [1_k, \check{T}]$. A symmetric measure of residual discrepancy is given by

$$d_{GF}(Y, T) = \inf \|\check{Y} - \check{X}B\| = \|(I_k - \check{X}\check{X}^*)\check{Y}\| = \left(1 - \check{Y}^*\check{T}\check{T}^*\check{Y}\right)^{1/2},$$
$$(8.5)$$

where the minimization is over the **full** set of transformations (translation, scale and rotation). Note that $D_G(Y, T) = \|Y\|d_{GF}(Y, T)$. An alternative symmetric measure of residual discrepancy d_{GP} involves matching the preshapes over translation and rotation only (**partial** matching). It can be seen quite simply that

$$d_{GP}^2(Y, T) = \inf_{c, \alpha} \|\check{Y} - \check{X}[c, e^{i\alpha}]^T\|^2 = 2(1 - \|\check{Y}^*\check{T}\|) \qquad (8.6)$$

where $\hat{c} = 0$ and $\hat{\alpha} = -\arg(\check{Y}^*\check{T})$.

As G is a group we can alternatively consider the general shape space in this situation, which is Kendall's shape space Σ_2^k. It can be seen (e.g. from Kendall, 1984) that the symmetric residual discrepancy measures in equations (8.5), (8.6) are equal to the full and partial Procrustes distances in equations (8.2), (8.3) respectively. This result can be seen by using $\rho = \arccos(\|\check{Y}^*\check{T}\|)$ and so

$$d_{GF}^2(Y, T) = \sin^2 \rho = d_1^2(Y, T), \ d_{GP}^2(Y, T) = 2(1-\cos\rho) = d^2(Y, T).$$

So, the full and partial Procrustes distances in Kendall's shape space for two dimensional points can be obtained as the solution to complex linear regression problems, as is quite well known.

8.2.4 Shape matching in m dimensions

We now consider the particular case of matching point configurations in $m \geq 2$ dimensions using the Euclidean similarity transformations. The previous section is the special case $m = 2$. Consider $k \geq m + 1$ points in m dimensions, so we have $M^k = \mathbb{R}^{km} \setminus C$ with the Euclidean metric. The registration group G is taken as the full set of similarity transformations. Matching two configurations Y and T (both $k \times m$ matrices) we have

$$Y = rT\Gamma + 1_k c^T + E$$

where $c \in \mathbb{R}^m$ is the location ($m \times 1$ vector), $r \in \mathbb{R}^+$ is the scale, $\Gamma \in SO(m)$ is the rotation, and E is the $k \times m$ error matrix. The matching cannot be cast as a problem in linear regression (unless $m = 2$, when the complex representation of the previous section can be used). However, the solution is quite straightforward using a singular value decomposition (Kendall, 1984). In particular, if $\check{Y}$ and $\check{T}$ denote the preshapes of Y and T (centred and scaled to unit size) then the least squares solution (minimizing $s^2(E) = \|E\|^2 = \mathrm{trace}(E^T E)$) is given by $\hat{c} = 0, \hat{\Gamma} = UV^T, \hat{r} = \mathrm{trace}(\Lambda)$ where $\check{Y}^T\check{T} = V\Lambda U^T$ is a singular value decomposition, with $U, V \in SO(m)$. The matrix Λ is diagonal with positive elements given by square roots of the eigenvalues of $\check{T}^T \check{Y}\check{Y}^T\check{T}$, except the smallest element which is negative if $\det(\check{Y}^T\check{T}) < 0$.

A suitable symmetric residual discrepancy measure involves the matching of the preshapes using the full set of similarity transformations:

$$d_{GF}(Y, T) = \inf\|\check{Y} - [1_k, \check{T}][c, r\Gamma^T]^T\| = \left(1 - \{\mathrm{trace}(\Lambda)\}^2\right)^{1/2}$$

An alternative symmetric residual discrepancy measure is obtained by partial matching the preshapes using location and rotation only:

$$d_{GP}(Y, T) = \inf \, \| \check{Y} - [1_k, \check{T}][c, \Gamma^T]^T \| = \sqrt{2}(1 - \text{trace}(\Lambda))^{1/2}$$

As the set of transformations G is a group we can study the shape space here: Kendall's shape space Σ_m^k. Referring to equations (8.1), (8.2), (8.3) we see that the symmetric residual discrepancy measures again correspond to the full and partial Procrustes distances in shape space.

8.2.5 Affine shape matching

Consider $k \geq m + 1$ points in $\mathbb{R}^m$ and let M^k be $\mathbb{R}^{km} \setminus \mathcal{D}$, where $\mathcal{D}$ is the set of configurations of deficient rank. The set of transformations G is the group of affine transformations. Affine matching using regression reduces to linear regression, for any number m dimensions. Matching T onto Y ($k \times m$ matrices), we have a $(m+1) \times m$ affine transformation matrix B and regression equation

$$Y = [1_k, T]B + E \tag{8.7}$$

and $(k \times m)$ error matrix E and we assume that $\text{rank}(T) = m$. Writing $X = [1_k, T]$ as the design matrix, the familiar least squares solution is given by

$$\hat{B} = (X^T X)^{-1} X^T Y,$$

where we assume X is of rank $m + 1$. Note that the match is exact if $k = m + 1$. A residual discrepancy measure for affine matching is

$$D_G(Y, T) = \| Y - X(X^T X)^{-1} X^T Y \|.$$

A symmetric measure can be obtained by initially centering and transforming so that $Y^T Y = I_m$ giving

$$D_A(Y, T) = \| (I_k - X(X^T X)^{-1} X^T) C Y (Y^T C Y)^{-1/2} \|$$

where C is the $k \times k$ centering matrix. The measure has the property that

$$D_A(Y, T) = D_A(Y, TA + 1_k b^T) = D_A(T, Y) = D_A(YA + 1_k b^T, T)$$

assuming T, Y are of rank m, A is a full rank $m \times m$ matrix and b is an m-vector.

As for Kendall's shape spaces symmetric residual discrepancy measures can be considered as distances in the affine shape space.

8.2.6 Further matching

There are many other situations that are of interest. For example, we may wish to match two point configurations using the set of transformations given by the isometry group (translation and rotation only) and the calculation of a symmetric residual discrepancy measure leads to the partial Procrustes distance in the size-and-shape space of equation (8.4). Also, reflections are often not important and for example reflection shape involves the registration group of translation, rotation, scale and reflection [i.e. $O(m)$ rather than $SO(m)$].

More general matching procedures include quadratic matching where the set of transformations G consists of quadratic transformations (e.g. Bookstein and Sampson, 1990). Obviously here G is not a group and so we cannot consider general shape spaces here.

Another very useful matching procedure is the use of thin-plate spline transformations (Bookstein, 1989) and the generalization of Kriging deformations (Mardia *et al.*, 1996b). Again, the transformations do not form a group.

8.2.7 An application in electrophoresis

We now consider an application of affine shape analysis. A technique for the identification of proteins involves the comparison of electrophoretic gel images. Two examples of such images are gel A and gel B shown in Figure 8.1, taken from Horgan *et al.* (1992). Some biological material is placed on the top left corner of the gel and then a process separates the material down the gel according to molecular weight (highest at the top) and across the gel according to isoelectric point (highest on the right of the gel).

The collection of dark spots indicate the composition of the proteins. These particular images were obtained from strains of parasites which carry malaria and the objective is to use a gel image to be able to identify the strain of parasite using the patterns of the spots. The dark spots in each gel can each be one of two types — invariant or variant spots. The invariant spots are present for all parasites and it is the particular arrangement of variant spots which enables identification of the parasite.

A problem with the technique when used in the field is that the gels are prone to deformations (such as translation, scaling, rotation, affine transformations and smooth non-linear bending). The gel images first need to be registered by matching each image using a set of transfor-

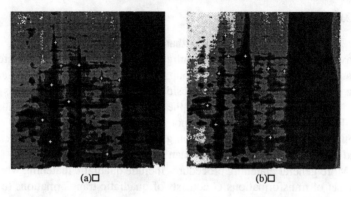

(a)☐   (b)☐

Figure 8.1 *The electrophoretic gels images from a) gel A and b) gel B. Ten invariant spots have been marked on each gel by an expert (+).*

mations (to remove the deformations) so that direct comparisons can be made. In this application ten invariant spots have been picked out by an expert, which are then used to match the configurations. The ultimate aim is to automate the procedure and this project is joint work with Chris Glasbey and Gary Walker.

There are several stages in the application where shape analysis can be helpful:

1. Registration of gel images. After the invariant spots have been expertly or automatically located we wish to register two images in order to compare the variant spots. The procedure should be resistant to poorly located spots.

2. Statistical modelling of the outlines of the invariant spots which can be incorporated into a prior model in an object recognition procedure, such as a Bayesian approach advocated by Grenander and co-workers (e.g. 1991, 1994). It is of interest to describe the average shape and the structure of shape variability of the spots, and in addition to explore the size and rotation of the spots.

3. Statistical modelling of the configurations of invariant spots. From a random sample of 16 images an expert has marked on the approximate centres of ten invariant spots on each image. We wish to provide a suitable model for the shape and registrations of these configurations. The model would also be useful as part of a prior distribution for object recognition.

In Figure 8.2 the fitted gel A image has been superimposed onto the gel B image. New grey levels x_{ij}^N have been calculated from the grey levels from the fitted gel A ($\hat{x}_{ij}^A$) and gel B (x_{ij}^B), using $x_{ij}^N = 2\hat{x}_{ij}^A - x_{ij}^B$. The figures have been thresholded to four levels for display purposes, indicating which spots were in gel A, gel B or both and which region was not dark in either gel. The dark grey regions have been well matched. The black regions are present in gel A only and the white spots are present in gel B only. The affine fitted gel A image matches quite well with the gel B image, as there are many dark grey common spots in both images.

Figure 8.2 *The superimposed gel A onto gel B using an affine match, using the 10 invariant spots chosen by an expert. The four grey levels indicate the correspondence of dark and light pixels in the registered images. The key used is: black — dark pixel in gel A and a light pixel in gel B; dark grey — matched dark pixels in gels A and B; light grey — light pixels matched in gels A and B; white — dark pixel in gel B and a light pixel in gel A.*

8.2.8 Other metrics

The above special cases have considered the Euclidean metric in the manifold M^k. Obviously we could have considered alternative metrics such as the Mahalanobis distance in the space, which leads to weighted least squares procedures (e.g. see Goodall, 1991).

For some choices of manifold M^k a non-Euclidean metric will often be the natural choice. For example if $M = S^p$, the unit sphere, the great circle metric could be used. Le (1989) has considered the shape space for points on a sphere, where the registration group is rotations only. Stoyan and Molchanov (1995) considered the manifold to be the system of non-empty compact sets in $\mathbb{R}^m$, and the registration group consisted of rotations and translations. We shall not consider such generalizations in this paper.

8.3 Estimation by matching

8.3.1 Generalized matching

If we have a random sample of n objects $T_j \in M^k$ ($k \times m$ matrices, $j = 1, \ldots, n$) then it is of interest to obtain an estimate of the average configuration μ ($k \times m$ matrix) and to explore the structure of variability, up to invariances in the set of transformations G.

An estimate of the population mean configuration μ up to invariances in G, denoted by $\hat{\mu}$, can be obtained by simultaneously matching each T_j to μ ($j = 1, \ldots, n$) and choosing μ as a suitable M-estimator (subject to certain constraints on μ). In particular, $\hat{\mu}$ is obtained from the constrained minimization

$$\hat{\mu} = \arg\inf_{\mu} \sum_{j=1}^{n} \inf_{g_j} \phi(s(g_j(T_j) - \mu)), \qquad (8.8)$$

where $\phi(x)$ is a penalty function on $\mathbb{R}^+$, $s(E)$ is the objective function for matching two configurations and in general restrictions need to be imposed on μ to avoid degeneracies. A common choice of estimator has $\phi(x) = x^2$, the least squares choice. More resistant methods (to object outliers) can be achieved with suitable choices of ϕ, as discussed in Section 8.5.

General shape spaces and tangent spaces

If G is a group then μ can be considered to be the population mean general shape, a member of the general shape space. For ease of description we call μ the population mean shape, although it is actually an **icon** (Goodall, 1995) representing the equivalence class of shape.

If the variability in shape is small then it is often convenient to work in the tangent space to shape space at an average shape, rather than the shape space itself. The Euclidean properties of the approximating tangent space are often more statistically convenient than using the non-Euclidean shape space. Consider a configuration $T \in M^k$. Let $\hat{g}(T)$ denote the matched T to the average shape $\hat{\mu}$. Inference can be carried out in the tangent space to the general shape space, by working with the tangent space coordinates

$$v(T) = \text{proj}(\hat{g}(T), \hat{\mu})$$

where $\text{proj}(X, \mu)$ is the projection of X into the tangent space of the general shape space at μ. This approach is only sensible when T is fairly similar to μ in general shape, and this is often the case in practical datasets. Note that the dimension of the general shape space (and the tangent space) is $\dim(M^k) - \dim(G)$.

Since the tangent space is a flat space (unlike shape space which is curved in general) statistical inference is conveniently carried out in the tangent space, for example Hotelling's T^2 test for equal means in independent populations.

Studying the variability in shape in a population is important. Principal component analysis can be carried out on the sample covariance matrix of $v(T_j)$ ($j = 1, \ldots, n$) to explore the structure of variability in general shape. The procedure has been used successfully for prior object modelling in image analysis (Cootes et al., 1992) and for summarizing succinctly the main aspects of shape variability in a random sample (Kent, 1994). Alternatively, the eigenstructure of a more robust estimator of tangent space covariance matrix (an M-estimator) could be used. Robust estimation of the mean μ is considered in Section 8.5.

8.3.2 Generalized shape matching in two dimensions

Consider a random sample of n configurations in the plane $T_1, \ldots, T_n$, ($T_j \in \mathbb{C}^k$, complex k-vectors) and we wish to obtain an average shape, where the set of transformations G is taken to be the Euclidean similarity group. We take $\phi(x) = x^2$ for least squares matching. Let $X_j = [1_k, T_j]$, $j = 1, \ldots, n$ be the $k \times 2$ design matrices. Write

$$\epsilon_j = \mu - X_j B_j$$

where B_j are the 2×1 complex parameters for matching the jth configuration to a complex mean vector μ ($k \times 1$), where μ is restricted to be centred ($\mu^* 1_k = 0$) and to have unit size ($\mu^* \mu = 1$).

Hence, the n configurations can be matched relatively to each other by least squares by minimizing the function $\sum_{j=1}^{n} \epsilon_j^* \epsilon_j$ over $B_1, \ldots, B_n$ and μ. It is clear that, fixing μ, the solution is given by

$$\hat{B}_j = (X_j^* X_j)^{-1} X_j^* \mu \quad (j = 1, \ldots, n),$$

assuming that $X_j^* X_j$ is of rank 2 (i.e. the points in T_j are not all coincident). Hence,

$$\hat{\epsilon}_j = (I_k - H_j)\mu$$

where H_j is the hat matrix for X_j given by

$$H_j = X_j(X_j^T X_j)^{-1} X_j^T \ , \quad (j = 1, \ldots, n).$$

Therefore,

$$\hat{\mu} = \arg \inf_{\|\mu\|=1} \mu^* A \mu$$

where $A = \sum_{j=1}^{n}(I - H_j)$. Hence, $\hat{\mu}$ is given by a complex eigenvector corresponding to the smallest non-zero eigenvalue of A, which is equivalent to finding an eigenvector corresponding to the largest eigenvalue of

$$S = \sum_{j=1}^{n} CT_j(CT_j)^* / ((CT_j)^* CT_j) = \sum_{j=1}^{n} \check{T}_j \check{T}_j^*, \qquad (8.9)$$

where C is the $k \times k$ centering matrix and $\check{T}_j = CT_j/\|CT_j\|$ are the centred preshapes. Hence, the shape corresponding to this eigenvector is the unique least squares estimate for shape providing that there is a single greatest eigenvalue. This result was derived by Kent (1992, 1994). The average shape from this procedure is called the **full** Procrustes mean. An alternative average could be obtained by initially centering and rescaling all the figures, and then matching by rotating and translation only. The resulting average is called the **partial** Procrustes mean (Kent, 1992, 1994), and the solution must be found iteratively.

A choice of shape tangent space coordinates at $\hat{\mu}$ corresponding to a configuration T is given by

$$v(T) = e^{-i\theta}(I_k - \hat{\mu}\hat{\mu}^*)CT/\|CT\|, \qquad (8.10)$$

where $\theta = \arg(\hat{\mu}^* CT)$ and $\hat{\mu}$ is a preshape ($\hat{\mu}^* 1_k = 0$, $\hat{\mu}^* \hat{\mu} = 1$). The coordinates were given by Kent (1994). Note the inverse projection from v to $z = CT/\|CT\|$ is given by,

$$z = e^{i\theta}[(1 - v^* v)^{\frac{1}{2}}\hat{\mu} + v]. \qquad (8.11)$$

For $m > 2$ dimensions generalized matching for shape must proceed with an iterative algorithm. The generalized Procrustes algorithm was given by Gower (1975) and is described in detail by Goodall (1991). Stoyan and Molchanov (1995) also provide a general matching algorithm for the shapes of random sets.

Example 8.1 A random sample of 23 second thoracic (T2) mouse vertebrae outlines was taken from the Large group of mice in the evolu-

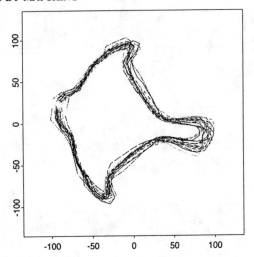

Figure 8.3 *Registered outlines of T2 mouse vertebrae obtained by complex linear regression.*

tionary study described by Johnson *et al.* (1985). Six landmarks were located on each outline according to the semi-automatic method described by Mardia (1989). In between each pair of landmarks 9 equally spaced pseudo-landmarks were placed, giving a total of $k = 60$ landmarks in $m = 2$ dimensions. In Figure 8.3 we see the matched outlines $\hat{T}_j = X_j \hat{B}_j$ obtained by complex linear regression with the mean shape $\hat{\mu}$ given by the shape corresponding to the dominant eigenvector of S in equation (8.9).

The sample covariance matrix in the tangent space (using coordinates of equation (8.10)) is evaluated and in Figure 8.4 we see sequences of shapes evaluated along the first three principal components (PCs). Shapes are evaluated in the tangent space and then projected back using the inverse transformation of equation (8.11) for visualization. The percentages of variability captured by these PCs are 33.0, 22.9 and 14.2%. Interpretation of the principal components is not straight forward, but a possible interpretation is that the first PC includes a measurement of the thickness of the neural spine (the protrusion on the 'right' of the bone) relative to the width of the bone. The effect of the second PC includes the bend in the end of the neural spine with the size of the bump on the far left of the bone. The effect of third

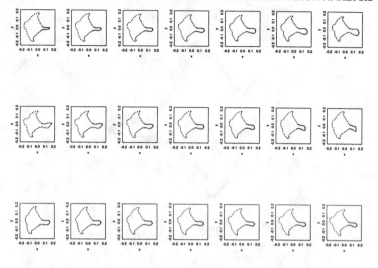

Figure 8.4 *Three rows of series of T2 vertebra shapes evaluated along the first three PCs — the i*th *row shows the shapes at* −3, −2, −1, 0, 1, 2, 3 *standard deviations along the i*th *PC.*

PC includes the size of the bump on the far left of the bone with the curvature of the bone either side of the neural spine.

8.3.3 Generalized affine matching

Consider matching $T_1, \ldots, T_n$ in $\mathbb{R}^{km} \setminus \mathcal{D}$ to estimate the population mean configuration, where the set of transformations G is the affine group of transformations. Let $X_j = [1_k, T_j]$ $(j = 1, \ldots, n)$ be the design matrices. We take $\phi(x) = x^2$ for least squares matching.

Consider matching each configuration B_j to an overall mean μ. The mean configuration takes the role of Y in the above equation (8.7) and so the estimated transformation parameters for matching T_j onto μ are given by

$$\hat{B}_j = (X_j^T X_j)^{-1} X_j^T \mu \quad (j = 1, \ldots, n),$$

assuming X_j is of full rank. Hence, the residual matrix from the match of T_j onto μ is $R_j = (I_k - H_j)\mu$ where H_j is the hat matrix for X_j. Therefore, if we wish to simultaneously estimate B_j and μ to minimize the expression of equation (8.8) we need to solve

$$\hat{\mu} = \arg\inf \ \text{trace} \ \mu^T A \ \mu$$

subject to the constraints $\mu^T \mu = I_m$, where

$$A = \sum_{j=1}^{n} (I_k - H_j). \tag{8.12}$$

Therefore the solution has a basis given by the m eigenvectors of $\sum H_j$ with largest eigenvalues (cf. Hastie and Kishon, 1991; Goodall, 1991). The m vectors are required so that the solution satisfies the constraints $\mu^T \mu = I_m$.

With a general choice of distance in the manifold M^k if $\phi(x) = x^2$ then

$$\hat{\mu} = \arg\inf_{\mu} \text{trace} \sum_{j=1}^{n} \|\mu - X_j \hat{B}_j\|^2.$$

Given $\hat{B}_j$, minimizing the sum of squares with respect to μ leads to

$$\hat{\mu} = \frac{1}{n} \sum_{j=1}^{n} X_j \hat{B}_j$$

which can lead to a simple iterative implementation.

The above generalized affine matching result was termed the linear spectral theorem by Goodall (1991,p335) and was also presented by Rohlf and Slice (1990), Hastie and Kishon (1991) for affine shape in two dimensions and for shapes in two dimensions by Kent (1992).

8.4 Distributions and inference

Statistical inference on population mean shape and population shape covariance using probability distributions in the general shape space is an alternative to generalized matching by minimizing objective functions. Consider a random sample of configurations available $T_1, \ldots, T_n$ $(T_j \in M^k)$. There are two modelling strategies:

1. Propose a distribution for the errors E_j in the regression equation

$$g_j(T_j) = \mu + E_j, \quad (j = 1, \ldots, n)$$

where the parameters in g_j are nuisance parameters and μ is a configuration corresponding to the population mean shape.

2. If the set of transformations G is a group then consider a probability distribution in the general shape space.

8.4.1 Regression models

Regression models involve the specification of the probability distribution in the manifold $M^k \equiv \mathbb{R}^{km} \setminus \mathcal{A}$. Consider a density for E_j given by

$$h(\kappa, g_j) \exp\{-\kappa\{\phi(s(E_j))\} = h(\kappa, g_j) \exp\{-\kappa\{\phi(s(g_j(T_j) - \mu))\}\}, \tag{8.13}$$

where $\phi(x)$ is a penalty function controlling the shape of the density and $s(E)$ is an objective function for matching two configurations. One approach to estimation is to use maximum likelihood and to maximize the likelihood over the unknown parameters μ, κ, g_j $(j = 1, \dots, n)$. The log-likelihood is given by

$$\sum_{j=1}^{n} \log h(\kappa, g_j) - \kappa \sum_{j=1}^{n} \{\phi(s(g_j(T_j) - \mu))\}.$$

It is clear that the maximum likelihood estimator (m.l.e.) of μ is the same as the generalized matching solution from equation (8.8) if and only if the integrating constant $h(\cdot)$ does not depend on g_j (which is not dependent on the value of κ).

In the special case of shape matching using an isotropic Gaussian model for E_j the scale transformations appear in the integrating constant, and hence the m.l.e. solution is different from generalized matching (Goodall, 1995).

8.4.2 General shape distributions

If the set of transformations G is a group then we can consider modelling in the general shape space. An initial task is to derive the uniform measure $d\gamma$ in the general shape space.

A simple density which is useful for inference in the general shape space, with respect to the uniform general shape measure, is the exponential-type density,

$$f([T]) = a(\kappa)e^{-\kappa\phi(d_G(\mu, T))} \tag{8.14}$$

where $[T]$ is the general shape of T, κ is a concentration parameter, d_G is a shape distance measure, $\phi(x)$ is a penalty function and $a(\kappa)$ is the integrating constant, which might be very complicated. Providing the integrating constant does not depend on the mean shape μ, the maximum likelihood estimator of μ from a random sample $T_1, \dots, T_n$

leads to the same mean shape estimate $\hat{\mu}$ from generalized matching using M-estimators in equation (8.8).

The above distribution with density in equation (8.13) could be defined using a distance with radial symmetry (i.e. the function $s(E)$ in calculating d_G has the property that $s(E\Gamma) = s(E)$ for an orthogonal matrix Γ) or more generally a measure of shape distance with non-radial symmetry.

Inference for mean shape μ is straightforward for these densities providing the integrating constant does not depend on μ. Bayesian inference for μ is also fairly straightforward using priors of the above form. Practical implementation using Markov chain Monte Carlo methods (e.g. Gilks, et al., 1996) allows simulation from the joint posterior of μ and κ, although the integrating constant $a(\kappa)$ will need to be known for inference on κ.

Other distributions in the general shape space could be obtained by transforming distributions in the manifold M^k and integrating out the registration parameters G. The resulting offset distributions can be useful for general shape analysis.

We consider some distributions in Kendall's shape space for two dimensional data in particular detail.

8.4.3 Shape distributions for two dimensional data

In order to work with a shape distribution a choice of shape coordinates is required and, for example, for the planar shape of the $k \geq 3$ complex points $z = (z_1, \ldots, z_k)^T$ we obtain the preshape coordinates by premultiplying by the $(k-1) \times k$ Helmert sub-matrix H (e.g. see Kent, 1994) and dividing by the size to give $\check{z} = Hz/\|Hz\|$. Kendall's shape coordinates $w = (w_1, \ldots, w_{k-1})^T$ are given by the ratios of the Helmertized coordinates $w_j = (Hz)_{j+1}/(Hz)_1$, which do not depend on the location, scale or rotation of the original z. For planar shape the uniform measure in the shape space is

$$d\gamma = (k-2)!\pi^{2-k}(1 + w^*w)^{1-k}dw_1 \ldots dw_{k-1}$$

where $dw_1 \ldots dw_{k-1}$ is the Lebesgue measure (Kendall, 1984).

An alternative convenient set of polar coordinates on the preshape sphere were proposed by Kent (1994) for shape analysis. Transform from the preshape $\check{z} = (\check{z}_1, \ldots, \check{z}_{k-1})^T$ to non-standard polar coordinates

$$\text{Re}(\check{z}_j) = s_j^{1/2}\cos\theta_j \,, \quad \text{Im}(\check{z}_j) = s_j^{1/2}\sin\theta_j,$$

for $j = 1, \ldots, k-1$, $s_j \geq 0, 0 \leq \theta_j < 2\pi$. Thus, the coordinates

$s_1, \ldots, s_{k-2}$ are on the $k-2$ dimensional unit simplex, $\mathcal{S}_{k-2}$, and when useful we can use $s_{k-1} = 1 - s_1 - \cdots - s_{k-2}$. By identifying the complex preshape sphere with $\mathcal{S}_{k-2} \times [0, 2\pi)^{k-1}$ we have the volume measure of $\mathbb{C}S^{k-2}$ as

$$2^{2-k} ds_1 \ldots ds_{k-2} d\theta_1 \ldots \theta_{k-1}.$$

Shape coordinates can be obtained by rotating z to a fixed axis. For example, considering the rotation information of the original figure to be in θ_{k-1} then the $2k-4$ shape coordinates are $(s_1, \ldots, s_{k-2}, \phi_1, \ldots, \phi_{k-2})$, where $\phi_i = \theta_i - \theta_{k-1}, i = 1, \ldots, k-2$. So the volume measure on the shape space $\Sigma_m^k = \mathbb{C}P^{k-2}(4)$ is

$$2^{2-k} ds_1 \ldots ds_{k-2} d\phi_1 \ldots d\phi_{k-2}.$$

and the total volume is

$$\frac{\pi^{k-2}}{(k-2)!}$$

since the volume of the j dimensional simplex is $1/j!$. Hence the uniform measure in shape space is

$$d\gamma = \frac{(k-2)!}{(2\pi)^{k-2}} ds_1 \ldots ds_{k-2} d\phi_1 \ldots d\phi_{k-2},$$

where $\int d\gamma = 1$.

For planar shape we choose shape distance $d_G = d_1$ (full Procrustes distance of equation (8.2)) and $\phi(x) = x^2$. From equation (8.14) the exponential-type density of shape with respect to the uniform shape measure $d\gamma$ is

$$f([T]) = a(\kappa) e^{-\kappa d_1^2} = a(\kappa) e^{-\kappa \sin^2 \rho}.$$

This is equivalent to the Dimroth-Watson distribution on the preshape sphere (Dryden, 1991; Prentice and Mardia, 1995) which is a special case of Kent's (1994) complex Bingham distribution (with two distinct eigenvalues). It is simply seen (as in Section 8.3.2) that the m.l.e. of shape under this model leads to the dominant eigenvector of the sums of squares and products of preshapes (if $\kappa > 0$), which is the complex Bingham shape m.l.e..

The offset Gaussian distribution for planar shape was given by Mardia and Dryden (1989a). In particular, the density, with respect to uniform measure, is

$$f([T]) = {}_1F_1\{2 - k; 1; -\kappa \cos^2 \rho\} \exp\{-\kappa \sin^2 \rho\}. \qquad (8.15)$$

where $_1F_1(-a; 1; -x) = \sum_{j=1}^{a} \binom{a}{b}\frac{x^j}{j!}$ is a confluent hypergeometric function (a finite series here). The density is closely related to the non-central beta density. In practice, for large κ there is little difference between the complex Bingham and offset normal approaches.

The generalization to non-radial symmetry was given by Kent (1994) for the complex Bingham distribution and considered by Dryden and Mardia (1991) for the offset Gaussian distribution. The offset complex Gaussian density is given by equation (8.15) but with Euclidean norm in the formula for the Riemannian metric ρ replaced by

$$\|X\|_\Psi = \sqrt{\text{trace}(X^T \Psi^{-1} X)}$$

where Ψ is a $k \times k$ Hermitian matrix. Dryden and Mardia (1991) also considered the general covariance offset Gaussian distribution which has a very complicated density in the non-complex symmetry case. Other offset distributions of interest in shape analysis include the offset central angular complex Gaussian distribution (Kent, 1995).

Inference for random samples is relatively straightforward and some classical inference procedures were given by Kent (1994) and Mardia and Dryden (1989b), for example.

In some applications the shapes of interest are inherently dependent, for example the shapes of Delaunay triangles in a Dirichlet tesselation. Some work taking account of dependencies in such situations has been given by Dryden *et al.* (1997).

8.5 Robust matching

Most of the above discussion has centred on least squares matching, which will obviously perform very poorly in the presence of outlier points or outlier objects. Two choices of objective functions need to be defined: $s(\cdot)$ for matching two point configurations and $\phi(\cdot)$ for obtaining a mean after matching several configurations. Suitable classes to choose for $s(\cdot)$ and $\phi(\cdot)$ include S-estimators (Rousseeuw and Yohai, 1984; Rousseeuw and LeRoy, 1987) and M-estimators, respectively. We consider two situations where resistant methods are required:

1. Resistance to landmark outliers on specimens (i.e. some specimens have particular landmarks that are very unusually located)

2. Resistance to object outliers (i.e. objects that are very different from the rest of the random sample)

8.5.1 Resistance to landmark outliers

Consider the situation where k points are available in m-dimensions. The regression equation for matching T to Y is $Y = g(T) + E$, and denote $(E)_i$ as the ith row of E ($i = 1, \ldots, k$). Least squares involves minimizing the sum of squared norms of the errors at each point,

$$s^2(E) = \|Y - g(T)\|^2 = \|E\|^2 = \sum_{i=1}^{k} \|(E)_i\|^2.$$

Several resistant methods have been proposed including the repeated median technique of Siegel and Benson (1982), also used by Rohlf and Slice (1990). However, residual discrepancy measures from this technique are dependent on the original registration of the two objects, which is clearly undesirable.

Rousseeuw and Yohai (1984) define the S-estimator for a univariate linear regression. When the response is multivariate an isotropic S-estimator could be defined as the solution to the following minimization problem: Minimize $s^2\{(E)_1, \ldots, (E)_k\}$ over the unknown parameters subject to

$$\frac{1}{k} \sum_{i=1}^{k} \xi(\|(E)_i/s\|) = K. \tag{8.16}$$

The positive continuously differentiable function $\xi(x)$, $x \in \mathbb{R}^+$ satisfies $\xi(x) = 0$, is strictly increasing on $[0, c]$, is constant for $x > c$ for a fixed $c > 0$ and $E_\Phi(\xi) = K$ where Φ is standard Gaussian.

An alternative estimator is the least median of squares (LMS) estimator (Rousseeuw, 1984) with objective function

$$s^2(E) = \text{median}(\|(E)_1\|^2, \ldots, \|(E)_k\|^2).$$

The LMS residual discrepancy measure is

$$D_{LMS}^2(Y, T) = \inf_{g \in G} \text{median} \|(Y - g(T))_i\|^2.$$

Note that if we relax some of the conditions for the isotropic S-estimator to allow the choice of indicator function

$$\xi(x) = \begin{cases} 1 & \text{if } x \geq 1 \\ 0 & \text{if } x < 1 \end{cases}$$

and $K = [(k + 1)/2]/k$ then the LMS objective function leads to a solution of equation (8.16).

In our electrophoresis application we consider G to be the affine group. The LMS procedure has a very high breakdown of almost 50% (the breakdown ϵ^* is the minimum percentage of points that can be moved arbitrarily to achieve an infinite discrepancy). Resistant matching using such a distance allows us to detect outliers as those points with very high residuals after a resistant match. The use of the LMS discrepancy measure is new to shape analysis and is joint work with Gary Walker. See Dryden and Walker (1997) for more details.

Note that for points in two dimensions and using the affine transformation group G when $k = 4$ there are 4 solutions which lead to $D_{LMS} = 0$, and when $k = 5$ there are $\binom{5}{3} = 10$ zero solutions. So, the procedure is only sensible for affine matching when $k > 5$. To achieve the maximum possible breakdown from Rousseeuw (1984) one could use the $([k/2] + [(p + 1)/2])$th order statistic rather than the median, where p is the number of explanatory variables ($p = m + 1$ here). This choice avoids zero solutions for $k = 4$ and $k = 5$ in general.

Minimization of the objective function can be difficult as the function is not smooth and there are usually local minima. An approximate procedure based on exact matching of all possible triplets of points and then choosing the triplet which minimizes the objective function leads to an approximate solution. This procedure can be speeded up and made a little more efficient by not considering very thin triplets or very small triplets (see Dryden and Walker, 1997).

If the errors $E \in \mathbb{R}^{km}$ have an isotropic Gaussian distribution then the procedure is quite inefficient. An alternative, more efficient proposal is the least quartile difference (a generalized S-estimator, see Croux *et al.*, 1994) with dissimilarity measure

$$D^2_{LQD}(Y, T) = \inf_{g \in G} Q_1 \left(\| (Y - g(T))_i - (Y - g(T))_j \|^2 \right)$$

where Q_1 is the lower quartile of the differences (for all i, j). Note that translation parameters ('intercepts') cannot be estimated with this approach. To achieve the maximum possible breakdown from Croux *et al.* (1994) one could use the $h_p(h_p - 1)/2$th order statistic rather than the first quartile, where $h_p = ([(k + p + 1)/2])$ and p is the number of explanatory variables ($p = m + 1$ here).

Many other robust regression procedures could be used for matching, including S-estimators, GS-estimators and least absolute deviations. In selecting an estimator one needs to make a compromise between breakdown and efficiency, and any choice will be very much application dependent.

Example 8.2 Consider matching the electrophoretic gel data with an affine transformation. The invariant spots on gel A are located correctly but two of the invariant spots in gel B have been deliberately moved and are thus very poorly identified. Other types of outliers could occur if points were correctly located but wrongly labelled. In Figure 8.5 we see the fitted points in gel A after a least squares affine transformation and after the LMS affine transformation. As expected the least squares fit is dramatically affected by the wrongly identified points, whereas the resistant LMS fit is not affected by the two outliers.

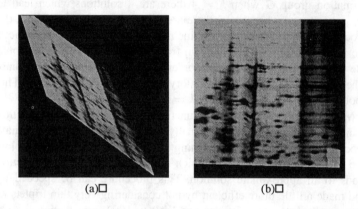

(a)☐   (b)☐

Figure 8.5 *The fitted gel A registered by affine fitting using the invariant points in the gels, with two points poorly located in gel B. a) Least squares affine transformation of gel A, b) LMS affine transformation of gel A.*

8.5.2 Resistance to object outliers

Resistance to object outliers can be achieved by using alternative penalty functions. For planar shape analysis the penalty function $\phi(x)$ could take on a variety of forms in the estimator obtained from equation (8.8). For example, from Kent (1992) if ρ is the Riemannian distance, some choices include

$$\phi(x) = \begin{cases} \rho^2 & \text{(a)} \\ 4\sin^2 \frac{\rho}{2} & \text{(b)} \\ \sin^2 \rho & \text{(c)} \\ (1 - \cos^{2h} \rho)/h. & \text{(d)} \end{cases}$$

The estimator from (b) is the partial Procrustes mean and the estimator from (c) is the full Procrustes mean. As Kent (1992) states, as h

increases in (d) the shape estimator becomes more resistant to outliers. In particular, the full Procrustes estimator is more resistant to outliers than the partial Procrustes estimator (see also Dryden, 1991). Minimization for $m = 2$ dimensions can be carried out as a complex weighted eigenvalue problem (Kent, 1992), whereas for $m \geq 3$ a more general minimization routine could be considered, such as the Newton-Raphson procedure or iteratively re-weighted least squares.

Other choices of $\phi(x)$ can be used. In particular a suitable choice of robust estimator could be using Huber's (1964) function,

$$\phi(\rho) = \begin{cases} \frac{1}{2}\rho^2 & \rho < c \\ c\rho - \frac{1}{2}c^2 & \rho \geq c \end{cases}$$

where $c > 1$. Robust estimators have not been fully exploited in shape analysis but it should be stressed that for small variations different choices of $\phi(\cdot)$ lead to very similar estimates. Kent (1994, 1995) emphasizes this point.

8.6 Smoothed matching

8.6.1 Outline matching

In some applications the data may be recorded with imprecision and there may be quite a large amount of noise. For example, Hastie and Kishon (1991) considered affine matching of handwritten signatures where there was a great deal of added measurement noise. A suitable approach to the problem is to add a roughness penalty to the objective function, in the manner of Rice and Silverman (1991) and Green and Silverman (1994). So for generalized matching of configurations to find an estimate of the population mean shape one could minimize the sum of residual discrepancy measures for each object and a penalty based on the roughness of the solution:

$$\sum_j \phi(s(g_j(T_j) - \mu)) + \lambda R(\mu)$$

where λ is a smoothing parameter and $R(\mu) \geq 0$ is a roughness penalty, which for the least squares case reduces to

$$\sum_j \|g_j(T_j) - \mu\|^2 + \lambda R(\mu).$$

One possibility for sequential series of landmarks (e.g. outlines) could be to use the quadratic penalty $R(\mu) = \text{trace}(\mu^T \Omega \mu)$ where Ω is the $k \times k$ integrated square derivative matrix corresponding to a one-dimensional smoothing cubic spline, as suggested by Hastie and Kishon (1991). Alternatively, the roughness penalty could be based on the bending energy matrix of μ, corresponding to a two-dimensional smoothing thin-plate spline (e.g. see Green and Silverman, 1994).

For the least squares case there are very simple adaptations of the solutions in Section 8.3. In the affine least squares case the solution is simply given by the m eigenvectors of $A+\lambda\Omega$ with smallest eigenvalues (A is the matrix defined in equation (8.12)). In the two-dimensional shape least squares case the solution is the dominant eigenvector of $S - \lambda\Omega$ where S is the complex sum of squares and products matrix.

Consider a series of landmarks on outlines of objects. It would be convenient to label the landmarks sequentially around the outline, so that landmarks i and j are neighbours if $|(i - j) \bmod k| = 1$. We include the modulo k term so that landmarks 1 and k are considered neighbours. In particular, in this periodic case Ω is given by $\Omega = FF$ where

$$(F)_{ij} = \begin{cases} 2 & \text{if } i = j \\ -1 & \text{if } |(i - j) \bmod k| = 1. \end{cases}$$

Smoothing can be introduced in a Bayesian analysis through the prior distribution. For example, a suitable smooth prior for planar shape matching is

$$\pi(\mu) = k(\lambda)e^{-\lambda\mu^*\Omega\mu},$$

which, when used with the exponential-type likelihood of equation (8.13), will have posterior mode of μ given by the above roughness penalty solution. Mardia and Hainsworth (1993) considered a priori using the bending energy matrix (see Bookstein, 1991) for deformable templates in image analysis.

The choice of λ is clearly of vital importance. An automatic choice of λ could be obtained by cross-validation (see Rice and Silverman, 1991). Mardia et al. (1994) also used a related approach in a familiar spinal shape study.

8.6.2 Alternative PCA: Bending energy metrics

An alternative to principal component analysis in the tangent space is Bookstein's (1991) relative warps. The PCA is carried out with respect to the bending energy matrix B_e (see Bookstein, 1989; Mardia, 1995) or the generalized inverse of the bending energy matrix B_e^-. The relative warps are obtained as follows: if S_T is the sample covariance in the tangent space the eigenstructure of $B_e^- S_T$ is examined, which in general emphasizes large scale variability (landmarks far apart). PCA with respect to inverse bending energy B_e^- (i.e. the eigenstructure of $B_e S_T$) emphasizes small scale variability. Other choices of relative PCA may be helpful — i.e. we could consider the eigenstructure of $A S_T$ where A is fixed or dependent on the data. In particular for data with several groups the eigenvectors of $S_W^- S_B$ are the canonical variates, where S_W is the within group sample covariance matrix and S_B is the between group covariance matrix of the tangent space coordinates.

8.7 Discussion

We have reviewed various aspects of shape and registration analysis, and the methodology has been illustrated with an application of electrophoretic gel matching and a biological shape study. The ultimate aim is to fully automate the gel matching procedure and an approach using Bayesian deformable templates looks promising (cf. Grenander *et al.*, 1991). Shape analysis is particularly useful in the prior modelling of the deformable templates and registration analysis is useful in the final automatic matching procedure. Just a few of the many applications of this work include Ripley and Sutherland (1990), Baddeley and van Lieshout (1993), Grenander and Miller (1994), Mardia *et al.* (1995), Phillips and Smith (1994) and Hurn and Rue (1997).

As well as resistance the issue of robustness needs to be considered. In particular, shape and registration analysis should not be sensitive to the particular model choice.

We have concentrated most of our work on point set configurations but the extension to general sets is a promising area for further work. Some initial work has been carried out by Stoyan and Molchanov (1995). Other approaches have been suggested for use on objects which do not have landmarks. For example, there is a large body of work on Fourier analysis of angular functions of outlines (e.g. see Rohlf, 1990). Bookstein (1996) has introduced analysis based on a special type of landmark which is allowed to slip around an outline — this approach appears very promising for the shape analysis of curved outlines. Other 'landmark-free' matching procedures use iterative procedures where a set of points is located on one object and the corresponding closest point on the second object has been used to update the registration (Besl and McKay, 1992). Further matching procedures have been proposed by minimizing pixel based measures (e.g. Glasbey, 1995).

As we have already mentioned there is a great variety of applications of shape and registration analysis in many branches of science, medicine and engineering. We have not attempted to review the broad range of applications, but a flavour of current work can be found in the volumes edited by Mardia and Gill (1995) and Mardia *et al.* (1996a).

Acknowledgements

I wish to thank John Kent for his very helpful comments on this paper. I would also like to thank Wilfrid Kendall and the referees who have also made useful comments. Everyone involved with SEMSTAT deserves special thanks for a highly stimulating conference. Other discussions with Kanti Mardia, Gary Walker, Chris Glasbey and Catherine Thomson have been helpful in presenting this work. The electrophoretic gel data were supplied by Chris Glasbey and Graham Horgan (BioSS) for the EPSRC studentship of Gary Walker. The vertebrae data were kindly provided by Paul O'Higgins (UCL) and David Johnson (Leeds).

References

Ambartzumian, R.V. (1982). Random shapes by factorisation. In Ranneby, B., editor, *Statistics in Theory and Practice*. Swedish University of Agricultural Science, Umea.

Ambartzumian, R.V. (1990). *Factorization, Calculus and Geometric Probability*. Cambridge University Press, Cambridge.

Baddeley, A.J. and van Lieshout, M.N.M. (1993). Stochastic geometry models in high-level vision. In Mardia, K.V. and Kanji, G.K., editors, *Statistics and Images: Vol. 1*, pages 231–256. Carfax, Oxford.

Besl, P.J. and McKay, N.D. (1992). A method for registration of 3D shapes. *IEEE transactions PAMI*, 14:239–256.

Bhavnagri, B. (1995). Connected components of the space of simple, closed non-degenerate polygons. In Mardia, K.V. and Gill, C.A., editors, *Current Issues in Statistical Shape Analysis*, pages 187–188, Leeds. University of Leeds Press.

Bookstein, F.L. (1978). *The Measurement of Biological Shape and Shape change*, volume 24 of *Lecture notes on biomathematics*. Springer-Verlag, New York.

Bookstein, F.L. (1986). Size and shape spaces for landmark data in two dimensions (with discussion). *Statist. Sci.*, 1:181–242.

Bookstein, F.L. (1989). Principal warps: Thin-plate splines and the decomposition of deformations. *IEEE Trans. Pattern Anal. Machine Intell.*, 11:567–585.

Bookstein, F.L. (1991). *Morphometric Tools for Landmark Data: Geometry and Biology*. Cambridge University Press, Cambridge.

Bookstein, F.L. (1996). Applying landmark methods to biological outline data. In Mardia, K.V., Gill, C.A., and Dryden, I.L., editors, *Proceedings in Image*

Fusion and Shape Variability Techniques, pages 59–70. University of Leeds, University of Leeds Press.

Bookstein, F.L. and Sampson, P.D. (1990). Statistical models for geometric components of shape change. *Comm. Statist. -Theory Meth.*, 19:1939–1972.

Carne, T.K. (1990). The geometry of shape spaces. *Proceedings of the London Mathematical Society*, 61:407–432.

Cootes, T.F., Taylor, C.J., Cooper, D.H., and Graham, J. (1992). Training models of shape from sets of examples. In Hogg, D.C. and Boyle, R.D., editors, *British Machine Vision Conference*, pages 9–18. Springer-Verlag.

Croux, C., Rousseeuw, P.J., and Hössjer, O. (1994). Generalized s-estimators. *Journal of the American Statistical Association*, 89:1271–1281.

Dryden, I.L. (1991). Discussion to 'Procrustes methods in the statistical analysis of shape' by C.R. Goodall. *J. Roy. Statist. Soc. B*, 53:327–328.

Dryden, I.L., Faghihi, M.R., and Taylor, C.C. (1997). Procrustes shape analysis of spatial point patterns. *Journal of the Royal Statistical Society, Series B*.

Dryden, I.L. and Mardia, K.V. (1991). General shape distributions in a plane. *Adv. Appl. Prob.*, 23:259–276.

Dryden, I.L. and Mardia, K.V. (1993). Multivariate shape analysis. *Sankhya Series A*, 55:460–480.

Dryden, I.L. and Mardia, K.V. (1997). *The statistical analysis of shape*. Wiley, Chichester.

Dryden, I.L. and Walker, G. (1997). Shape analysis using highly robust regression. In *Proceedings of the ISI, Istanbul, Turkey*. ISI.

Gilks, W.R., Richardson, S., and Spiegelhalter, D.J., editors (1996). *Markov chain Monte Carlo in practice*, London. Chapman and Hall.

Glasbey, C.A. (1995). An overview of image warping algorithms, with applications. Technical report, BioSS.

Goodall, C.R. (1991). Procrustes methods in the statistical analysis of shape (with discussion). *Journal of the Royal Statistical Society B*, 53:285–339.

Goodall, C.R. (1995). Procrustes methods in the statistical analysis of shape revisited. In Mardia, K.V. and Gill, C.A., editors, *Current Issues in Statistical Shape Analysis*, pages 18–33. University of Leeds.

Gower, J.C. (1975). Generalized Procrustes analysis. *Psychometrika*, 40:33–50.

Green, P.J. and Silverman, B.W. (1994). *Non-parametric regression and generalized linear models: A roughness penalty approach*. Chapman and Hall, London.

Grenander, U., Chow, Y., and Keenan, D.M. (1991). *Hands: A pattern theoretic study of biological shapes*, volume 2 of *Research notes in neural computing*. Springer-Verlag, New York.

Grenander, U. and Miller, M.I. (1994). Representations of knowledge in complex systems (with discussion). *Jour. Roy. Statist. Soc. B*, 56:549–603.

Harris, J. (1992). *Algebraic Geometry: A First Course*. Springer-Verlag, Berlin.

Hastie, T. and Kishon, E. (1991). Discussion to Goodall (1991). *Journal of the Royal Statistical Society, Series B*, 53:330–331.

Horgan, G.W., Creasey, A., and Fenton, B. (1992). Superimposing two-dimensional gels to study genetic variation in malaria parasites. *Electrophoresis*, 13:871–875.

Huber, P.J. (1964). Robust estimation of a location parameter. *Annals of Mathematical Statistics*, 35:73–101.

Hurn, M. and Rue, H. (1997). High-level image priors in confocal microscopy applications. Technical Report Statistics No. 2/1997, NTNU Trondheim.

Johnson, D.R., O'Higgins, P., McAndrew, T.J., Adams, L.M., and Flinn, R.M. (1985). Measurement of biological shape: A general method applied to mouse vertebrae. *J.embryol.exp.morph.*, 90:363–377.

Kendall, D.G. (1977). The diffusion of shape. *Advances in Applied Probability*, 9:428–430.

Kendall, D.G. (1984). Shape manifolds, Procrustean metrics and complex projective spaces. *Bulletin of the London Mathematical Society*, 16:81–121.

Kendall, D.G. (1989). A survey of the statistical theory of shape. *Statistical Science*, 4:87–120.

Kent, J.T. (1992). New directions in shape analysis. In Mardia, K. V., editor, *The Art of Statistical Science*, pages 115–127. Wiley, Chichester.

Kent, J.T. (1994). The complex Bingham distribution and shape analysis. *Journal of the Royal Statistical Society B*, 56:285–299.

Kent, J.T. (1995). Current issues for statistical inference in shape analysis. In Mardia, K.V. and Gill, C.A., editors, *Proceedings in Current Issues in Statistical Shape Analysis*, pages 167–175. University of Leeds, University of Leeds Press.

Le, H.-L. (1989). Random spherical triangles I: Geometrical background. *Advances in Applied Probability*, 21:570–580.

Le, H.-L. and Kendall, D.G. (1993). The Riemannian structure of Euclidean shape spaces: a novel environment for statistics. *Annals of Statistics*, 21:1225–1271.

Mardia, K.V. (1989). In discussion to 'A survey of the statistical theory of shape' by D.G. Kendall. *Statistical Science*, 4:108–111.

Mardia, K.V. (1995). Shape advances and future perspectives. In Mardia, K.V. and Gill, C.A., editors, *Proceedings in Current Issues in Statistical Shape Analysis*, pages 57–75. University of Leeds, University of Leeds Press.

Mardia, K.V. and Dryden, I.L. (1989a). Shape distributions for landmark data. *Advances in Applied Probability*, 21:742–755.

Mardia, K.V. and Dryden, I.L. (1989b). The statistical analysis of shape data. *Biometrika*, 76:71–282.

Mardia, K.V., Dryden, I.L., Hurn, M.A., Li, Q., Millner, P.A., and Dickson, R.A. (1994). Familial spinal shape. *Journal of Applied Statistics*, 21:623–641.

Mardia, K.V. and Gill, C.A., editors (1995). *Proceedings in Current Issues in Statistical Shape Analysis*. University of Leeds, University of Leeds Press.

Mardia, K.V., Gill, C.A., and Dryden, I.L., editors (1996a). *Image Fusion and Shape Variability*. University of Leeds, University of Leeds Press.

Mardia, K.V. and Hainsworth, T.J. (1993). Image warping and Bayesian reconstruction with grey-level templates. In Mardia, K.V. and Kanji, G.K., editors, *Statistics and Images: Vol. 1*, pages 257–280. Carfax, Oxford.

Mardia, K.V., Kent, J.T., Goodall, C.R., and Little, J.L. (1996b). Kriging and splines with derivative informations. *Biometrika*, 83:207–221.

Mardia, K.V., Rabe, S., and Kent, J.T. (1995). Statistics, shape and images. In *IMA Proceedings on Complex Stochastic Systems and Engineering Applications*, pages 85–103. Oxford University Press.

Phillips, D.B. and Smith, A.F.M. (1994). Bayesian faces via hierarchical template modeling. *Journal of the American Statistical Association*, 89:1151–1163.

Prentice, M.J. and Mardia, K.V. (1995). Shape changes in the plane for landmark data. *The Annals of Statistics*, 23:1960–1974.

Rice, J.A. and Silverman, B.W. (1991). Estimating the mean and covariance structure non-parametrically when the data are curves. *Journal of the Royal Statistical Society Series B*, 53:233–243.

Ripley, B.D. and Sutherland, A.I. (1990). Finding spiral structures in galaxies. *Phil. Trans. Roy. Soc. London A*, 332:477–485.

Rohlf, F.J. (1990). Fitting curves to outlines. In *Proceedings of the Michigan Morphometrics Workshop*, pages 167–177. University of Michigan Museum of Zoology, Ann Arbor. Special publication No. 2.

Rohlf, F.J. and Slice, D. (1990). Methods for comparison of sets of landmarks. *Systematic Zoology*, 39:40–59.

Rousseeuw, P.J. (1984). Least median of squares regression. *Journal of the American Statistical Association*, 79:871–880.

Rousseeuw, P.J. and LeRoy, A.M. (1987). *Robust regression and outlier detection*. Wiley, New York.

Rousseeuw, P.J. and Yohai, V.J. (1984). Robust regression by means of S-estimators. In Franke, J., Härdle, W., and Martin, R.D., editors, *Robust and non-linear time series analysis*, pages 256–272, New York. Springer-Verlag.

Siegel, A.F. and Benson, R.H. (1982). A robust comparison of biological shapes. *Biometrics*, 38:341–350.

Small, C.G. (1988). Techniques of shape analysis on sets of points. *International Statistical Review*, 56:243–257.

Small, C.G. (1996). *The statistical theory of shape*. Springer, New York.

Sparr, G. (1991). An algebraic-analytic method for affine shapes of point configurations. In *Scandinavian Conference on Image Analysis*, pages 274–281, Aalborg.

Stoyan, D., Kendall, W.S., and Mecke, J. (1995). *Stochastic geometry and its applications, 2nd Edition*. Wiley, Chichester, England.

Stoyan, D. and Molchanov, I.S. (1995). Set-valued means of random particles. Technical Report BS-R9511, CWI, Amsterdam.

Stoyan, D. and Stoyan, H. (1994). *Fractals, Random Shapes and Point Fields: Methods of geometric statistics*. Wiley, Chichester.

CHAPTER 9

Simple examples of the use of Nash inequalities for finite Markov chains

L. Saloff-Coste

CNRS, Université Paul Sabatier, France

9.1 Introduction

This article reports on analytic and geometric techniques that have been used recently by a number of authors to obtain quantitative bounds on the rate of convergence of finite Markov chains to their stationary distribution. It is in part based on joint work with Persi Diaconis and owes much to our numerous conversations. As in most areas of applied mathematics, this is a subject where there are tensions and trade offs between theoretical developments and applications to specific practical examples. This exposition belongs to the theoretical side but is motivated by practical considerations. The aim of this chapter is to describe precise bounds on the time needed by specific Markov chains to approximate their stationary distribution. This task which is of obvious practical interest poses a difficult theoretical challenge. Nash inequalities are one of the theoretical tools that can be used to attack this problem. A good Nash inequality, together with a bound on the spectral gap of the chain under consideration, yields a precise estimate on the time needed to reach equilibrium. This will be illustrated by simple examples.

9.1.1 Finite Markov chains approximate their stationary distribution

Markov chains are used in a number of algorithms to approximate an unknown probability distribution π defined on a state space $\mathcal{X}$. An ergodic chain with stationary distribution π is constructed. It is run for a while and the distribution of the chain at a large time T is then used to approximate π. We refer to this as the Markov Chain Monte Carlo (MCMC) method. Here, we focus on examples where $\mathcal{X}$ is a (typically huge) finite set and we want to draw from $\mathcal{X}$ according to π. This task may be difficult even when π is as simple as the uniform distribution on $\mathcal{X}$ (it is not absurd to consider the uniform distribution as an 'unknown' probability measure since counting how many elements there are in $\mathcal{X}$ may be a hard problem). We start by describing three examples where MCMC might be needed.

An example of statistical interest is given by contingency tables with given margins (row and column sums). Consider the following table

12	6	0	6	24
2	4	5	1	12
5	0	10	4	19
0	5	0	4	9
19	**15**	**15**	**15**	**64**

We want to pick at random another table with the same margins. Here, of course, at random means according to π where π is some specified distribution. For instance, take π to be the uniform distribution on the finite set of all 4 by 4 tables with non-negative integer entries and the given row and column sums. The choice of the uniform distribution is proposed in Diaconis and Efron (1985) as a model that is antagonistic to independence to calibrate the chi-square statistic. See Diaconis and Efron (1985) and Diaconis and Gangolli (1995) for details and further references. For 5 by 5 tables the problem of counting exactly how many tables there are with fixed margins is already close to being intractable.

A second example comes from theoretical computer science. Consider a finite bipartite connected graph with vertex set $V = I \cap O$ (edges go from I to O) with $\#I = \#O = N$. A perfect matching is a set of N edges so that each vertex appears exactly once.

I

O

One problem of interest is to draw uniformly from the set of all perfect matchings of a given graph. In fact, to start with one wants to count how many matchings there are. See Sinclair (1993) for motivations concerning this difficult example (roughly, no deterministic algorithms can do this in a reasonable time).

In a third widespread instance, one wants to draw from a somewhat complicated distribution π on a simple finite set, e.g., a d-dimensional grid. Very simple examples of this type are treated in detail below.

Further examples can be found in Diaconis and Holmes (1995) which also contains pointers to parts of the huge MCMC literature. Volume 55 of the journal of the Royal Statistical Society, Series B, (1993) contains three discussion papers which describe many practical applications and contain numerous references.

A Markov chain on a finite set can be defined through its kernel $K(x, y)$ which describes the basic step of the chain. The kernel $K(x, y)$ must be non-negative with $\sum_y K(x, y) = 1$. Thus, for each x, $K(x, \cdot)$ is a probability distribution. If x is the current position of the chain, the next position is chosen at random according to $K(x, \cdot)$. Starting at x, the probability to be at y after n steps is given by

$$K^n(x, y) = \sum_{z_1, z_2, \ldots, z_{n-1}} K(x, z_1) K(z_1, z_2) \cdots K(z_{n-1}, y)$$

$$= \sum_z K^{n-1}(x, z) K(z, y).$$

There is also a continuous time process associated with K where the discrete jumps occur according to an independent Poisson clock. Thus, at time t, the probability that there have been exactly n jumps is $e^{-t} \frac{t^n}{n!}$ and the probability that starting at x one reaches y at time t is

$$H_t(x, y) = e^{-t} \sum_0^\infty \frac{t^n}{n!} K^n(x, y).$$

Under mild conditions, $K^n(x, \cdot)$ and $H_t(x, \cdot)$ converge to a probability measure $\pi_\infty(\cdot)$ as n or t goes to infinity. Namely,

$$\lim_{t \to \infty} H_t(x, y)$$

exists and is independent of x if and only if K is irreducible. In discrete time, periodicity must further be ruled out. The probability measure π_∞ is called the stationary (or invariant) measure of K. The above convergence results are consequence of the classic Perron–Frobenius Theorem (for instance, see Horn and Johnson (1989), pg. 500) which also identifies π_∞ as the unique right eigenvector of K with eigenvalue 1 and positive entries summing to 1.

Returning to our problem of drawing from $\mathcal{X}$ according to some given distribution π, we are facing the following tasks, in order:

1. Come up with a Markov chain which has the desired limiting distribution $\pi = \pi_\infty$.

2. Prove that indeed our Markov chain converges to π.

3. Find how large n or t should be so that $K^n(x, y)$ or $H_t(x, y)$ is close to π.

None of these three tasks is trivial, in general. Task 1 is made more difficult by the implicit requirement that the chain should be 'easy to run'. Task 2 involves checking irreducibility and aperiodicity (in discrete time).

For instance think again about the example of the 4 by 4 contingency tables with fixed margins. In this case, one possible idea is as follows. Start with a table $x \in \mathcal{X}$. Pick two columns and two rows at random, say $(i, j), i \neq j$, and $(k, \ell), k \neq \ell$. Then change x to y by performing the operations:

$$
\begin{array}{ll}
\text{add } 1 & \text{in position } (i, k) \\
\text{subtract } 1 & \text{in position } (i, \ell) \\
\text{add } 1 & \text{in position } (j, \ell) \\
\text{subtract } 1 & \text{in position } (j, k).
\end{array}
$$

Observe that y has indeed the same row and column sums as x. If all the entries of y are nonnegative, the chain moves to y. If not, the chain stays at x. This describes informally a kernel K on $\mathcal{X}$. This chain can be shown to have the uniform distribution as stationary measure. For more information on this example, see Diaconis and Gangolli (1995).

In the case of perfect matchings the construction of the chain is a serious challenge. In fact, the chain is constructed on the union of the sets of perfect and near-perfect matchings. We refer the reader to Sinclair (1993) for an expository treatment of this interesting example. See also Jerrum and Sinclair (1997).

In situations where one does not know much about π, tasks 1 and 2 are all what one can reasonably expect to do. But, when certain features of π are known, one might expect to quantify the convergence of the Markov chain towards π. For instance, in the case of tables with fixed margins discussed above we want to approximate the uniform distribution, so we know quite a lot about $\pi(x)$. This is typical of a number of examples (e.g., counting problems and Gibbs measures in statistical physics) where π is known up to a normalizing constant.

To turn task 3 into a precise mathematical question, one must choose a distance between probability measures and decide what an acceptable error should be. For instance, we can use the total variation distance

$$\|\mu - \pi\|_{TV} = \sup_{A \subset \mathcal{X}} |\mu(A) - \pi(A)| = \frac{1}{2} \sum_x |\mu(x) - \pi(x)|$$

and ask for n or t such that

$$\|K_x^n - \pi\|_{TV} \leq 1/10 \; ; \quad \|H_t^x - \pi\|_{TV} \leq 1/10.$$

For clarity we will state most of the results in terms of variation distance but we will also work with the chi-square distance $\left(\sum_y |[K_x^n(y)/\pi(y)] - 1|^2 \pi(y)\right)^{1/2}$. Other choices are possible, but this will not be discussed here.

It is worth emphasizing that task 3 can be performed at different level of precision.

From a theoretical point of view, task 3 can roughly be understood as deciding between two different behaviors:

- **slow mixing** which typically corresponds to chains requiring a running time roughly comparable to the size of the state space,

  or

- **rapid mixing** which requires a running time much smaller than the size of the state space.

This point of view appears clearly in the computer science literature where the MCMC method is used to produce randomized algorithms that work in polynomial time for problems where no polynomial time deterministic algorithm is known (e.g., counting the number of perfect matchings). In this context rapid mixing is the key to success.

From a practical point of view, rapid mixing is merely reassuring. Suppose we are running a chain on a state having e^n elements and that we can show that a running time of order n^{10} suffices for convergence. Then the chain is rapidly mixing but, for most practical purposes, this is not very helpful because n^{10} is too big when n is large (there is little difference from this point of view between e^n and n^{10}). So, if our chain really requires a running time of order n^{10}, it is not practical. If it does converge more rapidly, say after a running time of order n^2, but we are not able to prove a better bound than n^{10}, we will probably rely on experimental knowledge for implementation. The point is that quantitative results are directly useful for applications only if they fit closely the behavior of the chain.

We are facing a serious dilemma. In order to bear on applications the theory must produce very precise quantitative results. At the same time these results must be applicable to the complex examples in which the MCMC method is used. The results presented in this paper may serve as good examples of this dilemma: they are quantitative and they describe precisely the behavior of the chains under consideration. Unfortunately, these chains are somewhat too simple to be relevant in real applications of the MCMC method. Still, I believe that these examples are interesting for a number of reasons. First, they show what kind of techniques are needed in order to obtain precise quantitative results. Second, the techniques presented below are useful in certain more complicated situations. Finally, being able to describe the precise behavior of simple chains does improve our understanding of what can happen in more complex examples.

9.1.2 The techniques presented in this paper

This paper focuses on simple examples where task 3 can be tackled reasonably well. These examples do not correspond to real world applications of the MCMC method but they present a number of interesting features beside their simplicity. The techniques presented below are not restricted to these examples and can be used succesfully in more complex problems. See Diaconis and Saloff-Coste (1995a, b, 1996a, b).

The main tools are functional inequalities involving the Dirichlet form $\mathcal{E}_K$ of K which is defined by

$$\mathcal{E}_K(f, f) = \frac{1}{2} \sum_{x,y} |f(x) - f(y)|^2 K(x, y)\pi(x)$$

for any function f.

Poincaré inequalities are inequalities of the type

$$\forall f \in \ell^2(X, \pi), \quad \mathrm{Var}_\pi(f) \leq A \mathcal{E}_K(f, f).$$

Such an inequality implies an eigenvalue estimate that is best stated as

$$\|H_t f - \pi(f)\|_2 \leq e^{-t/A} \|f\|_2$$

where $\|\cdot\|_2$ is the ℓ^2 norm of the Hilbert space $\ell^2(X, \pi)$.

A Nash inequality is an inequality of the type

$$\forall f \in \ell^2(X, \pi), \quad \|f\|_2^{2(1+2/d)} \leq C \left(\mathcal{E}_K(f, f) + \frac{1}{T} \|f\|_2^2 \right) \|f\|_1^{4/d}$$

where d, C, T are constants depending on K. The size of these constants is of course crucial for our applications. This inequality implies

(in fact, is equivalent to)

$$H_t(x, y) \leq B(d)\pi(y)(C/t)^{d/2} \text{ for } 0 < t \leq T$$

where $B(d)$ depends only on d and d, C, T are as above. This implication will be discussed in detail. Intuitively, a Nash inequality captures the smoothing effect of the chain. At time zero, all the mass is concentrated at one point. As time goes by, the distribution of the chain becomes more and more smoothly spread out according to π. A Nash inequality describes this smoothing effect over an infinitesimal time interval.

We will show that these inequalities combine to yield precise results on how large t should be so that $\|H_t - \pi\|_{TV} \leq 1/10$. For instance, this technique can be used to prove the following theorem.

Theorem 9.1 *Consider the chain K described above on the set $\mathcal{X}$ of all $m \times n$ tables with row sums and colum sums $r_i, c_i, i = 1, \ldots, m, j = 1, \ldots, n$. For fixed m, n, there are constants $A = A(m, n), a = a(m, n) > 0$ such that*

$$\|H_t - \pi\|_{TV} \leq A e^{-at/N^2}$$

where $N = \sum r_i = \sum c_i$. Thus, a running time of order N^2 suffices to approximate uniformity.

For fixed m, n, this result is sharp. See Diaconis and Gangolli (1995) and Diaconis and Saloff-Coste (1995a) for details. In the present paper we will discuss some simpler examples of Markov chains which are described in the next section.

9.1.3 The Metropolis algorithm

The Metropolis algorithm introduced in Metropolis *et al.* (1953) is a widely used construction that produces a Markov chain with a given stationary distribution π on a finite state space $\mathcal{X}$. This chain can then be used to approximate π. The construction starts with an arbitrary irreducible Markov kernel $P(x, y)$ that will be used for a 'proposal'. For simplicity, we assume that $P(x, y) = P(y, x)$. The Metropolis chain M is obtained by changing P as follows. Assume the current state is x. Pick y at random according to $P(x, \cdot)$ and compute $A(x, y) = \pi(y)/\pi(x)$. If $A(x, y) \geq 1$, move to y. If $A(x, y) < 1$, flip a coin with probability of heads $A(x, y)$ and move to y only if the coin comes up heads. Otherwise, stay at x. This gives

$$M(x, y) = \begin{cases} P(x, y) & \text{if } A(x, y) \geq 1 \text{ and } x \neq y \\ P(x, y)A(x, y) & \text{if } A(x, y) < 1 \\ P(x, x) + \displaystyle\sum_{z:A(x,z)<1} P(x, z)(1 - A(x, z)) & \text{if } x = y. \end{cases}$$

It is easy to see that

$$\pi(x)M(x, y) = \pi(y)M(y, x).$$

This shows in particular that π is a right eigenvector for M. Since P is irreducible, M is also irreducible. Further M is aperiodic because there exits x such that $M(x, x) > 0$. Thus, the Perron–Frobenius theorem shows that $M^n \to \pi$. Similarly, $H_t = e^{-t(I-M)} \to \pi$. Diaconis and Saloff-Coste (1995b) is a recent survey of precise results concerning the Metropolis algorithm.

To illustrate the theoretical results presented below I will use simple examples of Metropolis chains on

$$\mathcal{X} = \{-n, \ldots, n\} = [-n, n]$$

based on $P(x, y) = P(y, x) = 1/2$ if $x = y - 1 \in \{-n, \ldots, n - 1\}$ $P(-n, -n) = P(n, n) = 1/2$. Given a probability measure π on $[-n, n]$, let $M = M_\pi$ be the corresponding Metropolis chain and set

$$Q((x, y)) = \pi(x)M(x, y).$$

This quantity will appear in future computations. Observe that $Q(x, y) = 0$ if $|x - y| > 1$ whereas for $|x - y| = 1$, $Q((x, y))$ is given by

$$Q((x, y)) = \frac{1}{2} \min\{\pi(x), \pi(y)\}.$$

Example 9.1 Define $\hat{\pi}(i) = \hat{c} \times (n - |i| + 1)$ for $i \in [-n, n]$ with $\hat{c} = (n + 1)^{-2}$.

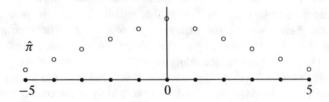

The corresponding Metropolis chain is given by

$$M(x, y) = \begin{cases} \frac{1}{2} & \text{if } \begin{array}{l} x \in [-n, 1], \ y = x + 1 \\ x \in [1, n], \ y = x - 1 \end{array} \\ \frac{n-|x|}{2(n-|x|+1)} & \text{if } |x - y| = 1 \text{ and } |x| < |y| \\ \frac{1}{2(n-|x|+1)} & \text{if } x = y \neq 0 \\ \frac{1}{n+1} & \text{if } x = y = 0 \\ 0 & \text{otherwise.} \end{cases}$$

Example 9.2 Define $\check{\pi}(i) = \check{c} \times (|i| + 1)$, $i \in [-n, n]$, with $\check{c} = [n(n + 3) + 1]^{-1}$.

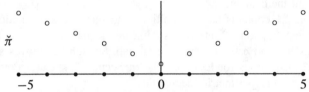

The corresponding Metropolis chain is given by

$$M(x, y) = \begin{cases} \frac{1}{2} & \text{if } \begin{array}{l} x \in [-n + 1, 0], \ y = x - 1 \\ x \in [0, n - 1], \ y = x + 1 \end{array} \\ \frac{|x|}{2(|x|+1)} & \text{if } |x - y| = 1 \text{ and } |x| > |y| \\ \frac{1}{2(|x|+1)} & \text{if } x = y, \ x \neq 0, -n, n \\ \frac{n+2}{2(n+1)} & \text{if } x = y = -n \text{ or } x = y = n \\ 0 & \text{otherwise.} \end{cases}$$

These examples belong to the following two families.

Example 1α For $\alpha > 0$, define $\hat{\pi}_\alpha(i) = \hat{c}(\alpha, n) \times (n - |i| + 1)^\alpha$, $i \in [-n, n]$, with $\hat{c}(\alpha, n)$ defined by $\sum \hat{\pi}_\alpha = 1$.

Example 2α For $\alpha > 0$, define $\check{\pi}_\alpha(i) = \check{c}(\alpha, n) \times (|i| + 1)^\alpha$, $i \in [-n, n]$, with $\sum \check{\pi}_\alpha = 1$.

The constants $\hat{c}(\alpha, n)$, $\check{c}(\alpha, n)$ of Examples 1–2α are distinct but approximately equal. They both satisfy

$$\frac{1}{2} \left(\frac{(n + 1)^{\alpha+1}}{\alpha + 1} + (n + 1)^\alpha \right) \leq c(\alpha, n)^{-1}$$

$$\leq 2 \left(\frac{(n + 1)^{\alpha+1}}{\alpha + 1} + (n + 1)^\alpha \right).$$

The shape of the hat over π will remind us which stationary distribution we are using. These examples are birth and death chains and they can be studied by taking advantage of this special structure. The route that we will take below is very different and can be used, in principle, to study analogous problems in higher dimensions (e.g., on two-dimensional grids).

9.1.4 A guide to the paper

Sections in this chapter are numbered 9.y.z, with 9 standing for the chapter number. In the text, we only use the two last digits to refer to a section. For instance, Section 2.1 refers to the section 2.1 of this chapter which is numbered 9.2.1. Equations are numbered sequentially throughout the chapter.

Section 2.1 contains basic definitions and introduces the notion of spectral gap. Section 2.2 introduces Poincaré inequalities and path techniques as tools to estimate the spectral gap. Section 2.3 proves sharp Poincaré inequalities for Examples 1–2α. Section 2.4 relates Poincaré inequalities to isoperimetric inequalities.

Sections 3.1 and 3.2 describe in detail the equivalence between Nash inequalities and the decay of $H_t(x, y)$ as a function of t and show how Poincaré and Nash inequalities combine and yield bounds on convergence to stationarity. Section 3.3 relates Nash inequalities with isoperimetric inequalities. Section 3.4 applies the previous sections to the Metropolis examples 1–2α.

9.2 Poincaré inequalities

9.2.1 The spectral gap

Throughout this paper we will work with a finite Markov chain K on a finite state space $\mathcal{X}$. We assume that K is irreducible, i.e., $\forall x, y \ \exists n = n(x, y) : K^n(x, y) > 0$. We let π denote the stationary measure of K which is the unique probability measure such that

$$\sum_x \pi(x) K(x, y) = \pi(y).$$

We consider the iterated kernel K^n and set

$$H_t(x, y) = e^{-t} \sum_0^\infty \frac{t^n}{n!} K^n(x, y).$$

This defines a semigroup of operators acting on functions by

$$H_t f(x) = \sum_y H_t(x, y) f(y).$$

The Perron–Frobenius theorem shows that

$$H_t(x, y) \to \pi(y) \quad \text{as } t \to \infty$$

at an exponential rate. One of our aims is to present quantitative forms of this statement. For simplicity we will work with H_t (i.e., in continuous time) instead of K^n (i.e., in discrete time). We define the Dirichlet form $\mathcal{E}_K = \mathcal{E}$ by

$$\mathcal{E}(f, g) = \Re(\langle (I - K)f, g \rangle)$$

where the scalar product is in complex Hilbert space $\ell^2(\pi)$:

$$\langle f, g \rangle = \sum_{x,y} f(x)\overline{g(x)}\pi(x).$$

We also set

$$\text{Var}_\tau(f) = \|f - \pi(f)\|_2^2 = \|f\|_2^2 - \pi(f)^2.$$

Lemma 9.1 *The Dirichlet form satisfies*

$$\mathcal{E}(f, f) = \frac{1}{2} \sum_{x,y} |f(x) - f(y)|^2 K(x, y)\pi(x)$$

and

$$-2\,\mathcal{E}(H_t f, H_t f) = \frac{\partial}{\partial t} \|H_t f\|_2^2. \tag{9.1}$$

Proof. Observe that $\mathcal{E}(f, f) = \|f\|_2^2 - \Re(\langle Kf, f \rangle)$ and

$$\frac{1}{2} \sum_{x,y} |f(x) - f(y)|^2 K(x, y)\pi(x)$$

$$= \frac{1}{2} \sum_{x,y} \left(|f(x)|^2 + |f(y)|^2 - 2\Re(\overline{f(x)}f(y)) \right) K(x, y)\pi(x)$$

$$= \|f\|_2^2 - \Re(\langle Kf, f \rangle).$$

This gives the first equality. The second follows from calculus. □

This simple lemma is crucial. The first identity shows that the Dirichlet form is a manageable quantity because it can be written as a sum of positive terms. The second identity implies the following lemma.

Lemma 9.2 *Define the spectral gap $\lambda = \lambda(K)$ by*

$$\lambda = \inf_{\text{Var}_\pi(f) \neq 0} \frac{\mathcal{E}_K(f, f)}{\text{Var}_\pi(f)}. \tag{9.2}$$

Then,

$$\|H_t f - \pi(f)\|_2^2 \le e^{-2\lambda t} \mathrm{Var}_\pi(f).$$

Proof. Set $g = f - \pi(f)$ and $u(t) = \|H_t g\|_2^2$. Observe that $\|H_t g\|_2^2 = \mathrm{Var}_\pi(H_t g) = \mathrm{Var}_\pi(H_t f)$ and

$$u'(t) = -2\mathcal{E}(H_t g, H_t g) \le -2\lambda u(t).$$

It follows that

$$u(t) \le e^{-2\lambda t} u(0)$$

which is the desired inequality because $u(0) = \mathrm{Var}_\pi(f)$. $\square$

Remarks: 1. It is not hard to see that one can restrict to real functions f only in (9.2) without changing λ.

2. When (K, π) satisfies the extra condition

$$\pi(x)K(x, y) = \pi(y)K(y, x)$$

(detailed balance condition) one says that the chain (K, π) is reversible. In this case K is self-adjoint in $\ell^2(\pi)$ and the operator K is diagonalizable with real eigenvalues. Further, $1 - \lambda$ is the second largest eigenvalue of K. For a non-reversible chain $1 - \lambda$ is not, in general, an eigenvalue of K.

The above lemma and its corollary below are the simplest quantitative results in finite Markov chain theory.

Corollary 9.2 *Set $H_t^x(y) = H_t(x, y)$. Then*

$$\|(H_t^x/\pi) - 1\|_2 \le \sqrt{1/\pi(x)}\, e^{-\lambda t}$$

and

$$2\|H_t^x - \pi\|_{TV} \le \sqrt{1/\pi(x)}\, e^{-\lambda t}.$$

Proof. Let H_t^* be the adjoint of H_t on $\ell^2(\pi)$. By inspection this is a Markov semigroup $H_t^* = e^{-t(I-K^*)}$ associated with the Markov kernel

$$K^*(x, y) = \frac{\pi(y)K(y, x)}{\pi(x)}$$

and given by

$$H_t^*(x, y) = \frac{\pi(y)H_t(y, x)}{\pi(x)}.$$

Observe that the Dirichlet forms of K^* and K satisfy

$$\mathcal{E}_K(f, f) = \mathcal{E}_{K^*}(f, f).$$

It follows that $\lambda(K) = \lambda(K^*)$. Set $\delta_x(y) = 1/\pi(x)$ if $y = x$ and $\delta_x(y) = 0$ otherwise. Then

$$\frac{H_t^x(y)}{\pi(y)} = H_t^* \delta_x(y)$$

and, by Lemma 3 applied to K^*,

$$\|H_t^* \delta_x - 1\|_2 \le e^{-\lambda t} \mathrm{Var}_\pi(\delta_x).$$

Hence

$$\|(H_t^x/\pi) - 1\|_2 \le \sqrt{\frac{1 - \pi(x)}{\pi(x)}} \, e^{-\lambda t}$$

The desired inequality follows by Jensen's inequality since

$$2\|H_t^x - \pi\|_{TV} = \sum_y |H_t(x, y) - \pi(y)|$$

$$= \sum_y |[H_t(x, y)/\pi(y)] - 1|\pi(y)$$

$$\le \left(\sum_y |[H_t(x, y)/\pi(y)] - 1|^2 \pi(y) \right)^{1/2}$$

$$= \|(H_t^x/\pi) - 1\|_2.$$

This ends the proof of Corollary 4. $\square$

Remark. According to Corollary 4 one might think that, in order to reduce the necessary running time for convergence, it is a good idea to start from a state that has $\pi(x)$ as large as possible (so that $\sqrt{1/\pi(x)}$ is small). This is sometimes true and sometimes wrong. The point is that Corollary 4 only gives an upper bound and does not always reflect the real behavior of the chain. To give an example, consider the chains of Examples 1α and 2α with $\alpha = 2$. They converge respectively towards the probability measures $\hat{\pi}_2(i) = \hat{c}(2, n) \times (n - |i| + 1)^2$ and $\check{\pi}_2(i) = \check{c}(2, n) \times (|i| + 1)^2$, $i \in [-n, n]$. Here $\hat{c}(2, n) \simeq \check{c}(2, n) \simeq n^{-3}$.

In the case of $\hat{\pi}_2$, one can show that a running time of order n^2 is necessary and suffices for convergence, independently of the starting point (see Corollary 18 at the end of this paper). Here, the spectral gap is of order $1/n^2$ (Theorem 6 below) and Corollary 4 shows that a running time of order $n^2 \log n$ suffices from any starting point.

In the case of $\check{\pi}_2$, a running time of order n^3 is necessary and suffices for convergence if one starts from n or $-n$ (where $\check{\pi}_2$ attains its maximum) whereas a running time of order n^2 suffices if one starts from 0 (where $\check{\pi}_2$ attains its minimum). So, in this example starting from a point where the stationary measure attains its maximum is the wrong choice; on the contrary, one should start from the point where $\check{\pi}_2$ is minimum. Here, the spectral gap is of order $1/n^3$ (Theorem 7). Corollary 4 shows that a running time of order $n^3 \log n$ suffices for convergence from any starting point. This is roughly the right order of magnitude if one starts from $-n$ or n. It is only a crude upper bound if one starts from 0.

9.2.2 Irreducibility and paths

Let $\mathcal{X}$ be a finite set and K be an irreducible Markov chain on $\mathcal{X}$. Construct a graph with vertex set $\mathcal{X}$ and an edge from x to y if $K(x, y) > 0$. The idea behind this section is to use the geometry and combinatorics of paths in this graph to bound the spectral gap λ defined at (9.2) away from zero.

Recall our basic assumption that K is irreducible. Thus, for each $x, y \in \mathcal{X}$, there exists $n = n(x, y)$ and a path $\gamma(x, y) = (x_0, \ldots, x_n)$ such that $x_0 = x$, $x_n = y$ and $K(x_i, x_{i+1}) > 0$. In what follows we fix, for each $x, y \in \mathcal{X}$, a path $\gamma(x, y)$ as above of length $|\gamma(x, y)| = n(x, y)$ connecting x to y. We do not require these paths to have minimal length but we assume that they do not have repeated vertices. Let Γ be the collection of all the paths $\gamma(x, y)$, $x, y \in \mathcal{X}$. We say that Γ is a 'complete set' of paths for K.

For $\gamma = (x_0, \ldots x_n) \in \Gamma$, set

$$\pi(\gamma) = \pi(x_0)\pi(x_n).$$

We will also view γ as a succession of edges

$$\gamma = ((x_0, x_1), \ldots, (x_{n-1}, x_n)).$$

For $e = (x, y) \in \mathcal{X} \times \mathcal{X}$, set

$$df(e) = f(y) - f(x); \quad Q(e) = \pi(x)K(x, y)$$

so that

$$\mathcal{E}(f, f) = \frac{1}{2} \sum_e |df(e)|^2 Q(e).$$

Finally, fix a weight function

$$w : \mathcal{X} \times \mathcal{X} \to (0, +\infty)$$

on the set $\{(x, y) : K(x, y) > 0\}$ and define the w-length $|\gamma|_w$ of

$\gamma = (x_0, \ldots, x_n) \in \Gamma$ by

$$|\gamma|_w = \sum_{e \in \gamma} \frac{1}{w(e)}.$$

Theorem 9.3 *Let Γ be a complete set of paths and w a weight function. Then $\lambda(K) \geq 1/A$ with*

$$A = \max_{e:\, Q(e)>0} \left\{ \frac{w(e)}{Q(e)} \sum_{\substack{\gamma \in \Gamma: \\ \gamma \ni e}} |\gamma|_w \pi(\gamma) \right\}$$

Proof. Write

$$f(y) - f(x) = \sum_{e \in \gamma(x,y)} df(e).$$

Then, by Cauchy-Schwarz,

$$|f(x) - f(y)|^2 \leq |\gamma(x,y)|_w \sum_{e \in \gamma(x,y)} |df(e)|^2 w(e).$$

Summing over all x, y yields

$$\begin{aligned}
\text{Var}_\pi(f) &= \frac{1}{2} \sum_{x,y} |f(x) - f(y)|^2 \pi(x)\pi(y) \\
&\leq \frac{1}{2} \sum_{\gamma \in \Gamma} \sum_{e \in \gamma} |df(e)|^2 w(e) |\gamma|_w \pi(\gamma) \\
&= \frac{1}{2} \sum_{e \in \gamma} |df(e)|^2 Q(e) \frac{w(e)}{Q(e)} \sum_{\gamma \ni e} |\gamma|_w \pi(\gamma) \\
&\leq \max_{e:\, Q(e)>0} \left\{ \frac{w(e)}{Q(e)} \sum_{\gamma \ni e} |\gamma|_w \pi(\gamma) \right\} \mathcal{E}(f, f).
\end{aligned}$$

This and the definition of λ at (9.2) gives the desired result. □

Remarks: 1. A Poincaré inequality is an inequality of the type

$$\forall\ f,\ \ \text{Var}_\pi(f) \leq A\mathcal{E}(f, f).$$

Clearly, such an inequality is equivalent to $\lambda \geq 1/A$. Theorem 5 proves a Poincaré inequality using paths.

2. Theorem 5 uses paths that respect the oriented graph structure with edge set $\mathcal{A} = \{(x, y) \in \mathcal{X} \times \mathcal{X} : K(x, y) > 0\}$. This set may not be symmetric since it may happen that $K(x, y) > 0$ and $K(y, x) = 0$.

However, the Dirichlet form $\mathcal{E}$ can be writen

$$\mathcal{E}(f, f) = \frac{1}{4} \sum_{x,y} |f(x) - f(y)|^2 [K(x, y)\pi(x) + K(y, x)\pi(y)].$$

It follows that one can always repeat the above argument using the symmetric graph with edge-set $\mathcal{A}' = \{(x, y) \in \mathcal{X} \times \mathcal{X} : Q'(x, y) > 0\}$ where $Q'(x, y) = \frac{1}{2}(K(x, y)\pi(x) + K(y, x)\pi(y))$. This allows us to chose Γ among a larger collection of paths at the cost of using Q' instead of Q in the definition of A in Theorem 5.

3. We will see examples where non-trivial (i.e., $w \not\equiv 1$) weight functions w are useful, for instance $w(e) = Q(e)$ or $w(e) = Q(e)^{1/2}$. The choice of a good weight function is a matter of intuition and experiments. Kahale (1997) contains a useful discussion and interesting examples.

9.2.3 Examples

This section shows how Theorem 5 applies in practice. We only treat the simple examples of section 1.3. Theorem 5 has also been used by different authors in complicated situations. See Sinclair (1992–1993), Jerrum and Sinclair (1993–1997), Diaconis and Stroock (1991), Fill (1991), Diaconis and Saloff-Coste (1995a, b, 1996a), Kahale (1995) among other references.

Example 1 *(The Metropolis chain for $\hat{\pi}$ on $[-n, n]$)* Recall that $\hat{\pi}(i) = \hat{c} \times (n - |i| + 1)$ for $i \in [-n, n]$ with $\hat{c} = (n + 1)^{-2}$. We will use the obvious set of complete paths: $\gamma(x, y) = (x, x+1, \ldots, y)$ if $x < y$ and $\gamma(x, y) = (x, x - 1, \ldots, y)$ is $x > y$ and the constant weight function $w \equiv 1$. Using symmetry, the constant A of Theorem 5 becomes

$$A = \max_{\substack{e=(i,i+1) \\ i \in [0,n]}} \frac{2(n + 1)^2}{(n - i)} \sum_{\substack{(j,k) \\ j \le i < k}} |k - j| \frac{(n - |j| + 1)(n - k + 1)}{(n + 1)^4}$$

$$\le 8n^2$$

where we have used $|k - j| \le 2n, n - |k| + 1 \le n - i, n - |j| + 1 \le n + 1$, $k \in [1, n], j \in [-n, n - 1]$ to obtain the last inequality. Thus, Theorem 5 gives $\lambda \ge 1/[8n^2]$.

Using any function f as a test function in (9.2) gives an upper bound on λ. Here, take $f(x) = x$. Observe that $\hat{\pi}(f) = 0$,

$$\|f\|_2^2 = 2\sum_1^n i^2 \frac{n-i+1}{(n+1)^2}$$

$$= \left(\frac{n^2(n+1)(2n+1)}{3(n+1)^2} - \frac{n^2(n+1)^2}{2(n+1)^2} + \frac{n(n+1)(2n+1)}{3(n+1)^2}\right)$$

$$\geq \frac{n^2}{6}$$

and $\mathcal{E}(f, f) = 2\sum_1^n \pi(x) \leq 1$. Thus $\lambda \leq 6/n^2$.

Example 1α (*The Metropolis chain for $\hat{\pi}_\alpha$*) A similar analysis works for the Metropolis chains corresponding to the measures $\hat{\pi}_\alpha$ with $\alpha > 0$ fixed. The following result which is uniform in α and n is more subtle.

Theorem 9.4 *The spectral gap $\hat{\lambda}(\alpha, n)$ of the Metropolis chain for $\hat{\pi}_\alpha$ on $[-n, n]$ satifies*

$$\frac{1 \vee \alpha^2}{12(n^2 + \alpha^2)} \leq \hat{\lambda}(\alpha, n) \leq \frac{9(\alpha+1)^2}{(n+1)^2} \wedge 2.$$

In particular, for $n \geq \alpha$, $\hat{\lambda}(\alpha, n)$ is of order $(\alpha/n)^2$.

Proof. For the upper bound we use the same argument and the same test function $f(x) = x$ as in Example 1. Assume $(n+1) \geq 3(\alpha+1)$, set $\varepsilon = (\alpha+1)^{-1}$ and write

$$\sum i^2 \hat{\pi}_\alpha(i) \geq 2\hat{c}(\alpha, n)\left(\frac{n+1}{\alpha+1} - 1\right)^2 \sum_{i \geq [\varepsilon(n+1)]} (n-i+1)^\alpha$$

$$\geq \frac{8}{9}\hat{c}(\alpha, n)\left(\frac{n+1}{\alpha+1}\right)^2 \sum_{i \leq n - [\varepsilon(n+1)]} (i+1)^\alpha$$

$$\geq \frac{8(\alpha+1)(n+1)^2}{9(\alpha+1)^2(n+1)^{\alpha+1}} \frac{(\varepsilon(n+1))^{\alpha+1}}{\alpha+1}$$

$$\geq \frac{(n+1)^2}{9(\alpha+1)^2}.$$

the last inequality uses $1 - x \geq e^{-2x}$ for $0 < x \leq 1/2$. This gives a lower bound on $\|f\|_2^2$. The upper bound $\mathcal{E}(f, f) \leq 1$ is obvious and the desired result follows.

For the lower bound, the argument used for $\hat{\pi}$ yields $\lambda \geq 1/(8n^2)$ for any n, α. To obtain a better estimate for large values of α we need to use a nontrivial weight function. Assume $\alpha \geq 4$ and set $w(e) = Q(e)^{1/2}$.

Then

$$|\gamma_{j,k}|_w = \sum_{j \leq \ell < k} Q(\ell, \ell + 1)^{-1/2}$$

$$\leq \sqrt{2}\,\hat{c}(\alpha, n)^{-1/2} \sum_{\ell \geq n - (|j| \vee k) + 1} \ell^{-\alpha/2}$$

$$\leq \sqrt{2}\,\hat{c}(\alpha, n)^{-1/2} \times$$
$$\left(\frac{2(n - (|j| \vee k) + 1)^{1-\alpha/2}}{\alpha - 2} + (n - |j| \vee k) + 1)^{-\alpha/2} \right).$$

For this weight function, Theorem 5 gives $\lambda \geq 1/A$ with

$$A \leq 4 \max_{0 \leq i < n} \left\{ \hat{c}(\alpha, n)^{-1/2} \hat{\pi}_\alpha(i + 1)^{-1/2} \times \right.$$

$$\sum_{\substack{j,k: \\ i < k \leq j}} \left(\frac{2(n - j + 1)^{1-\alpha/2}}{\alpha - 2} + (n - j + 1)^{-\alpha/2} \right) \hat{\pi}_\alpha(j)\hat{\pi}_\alpha(k) \Bigg\}$$

$$\leq 2 \max_{0 \leq i < n} \left\{ \hat{c}(\alpha, n)^{1/2} \hat{\pi}_\alpha(i + 1)^{-1/2} \times \right.$$

$$\sum_{i < j \leq n} \left(\frac{2(n - j + 1)^{1+\alpha/2}}{\alpha - 2} + (n - j + 1)^{\alpha/2} \right) \Bigg\}$$

$$\leq 2 \max_{0 \leq i < n} \left\{ \hat{c}(\alpha, n)^{1/2} \hat{\pi}_\alpha(i + 1)^{-1/2} \times \right.$$

$$6 \left(\frac{(n - i)^{2+\alpha/2}}{\alpha^2} + (n - i)^{\alpha/2} \right) \Bigg\}$$

$$\leq 12 \left(\frac{n^2}{\alpha^2} + 1 \right).$$

This ends the proof of Theorem 5. □

Example 2α (*The Metropolis chain for $\check{\pi}$ on* $[-n, n]$) Recall that $\check{\pi}(i) = \check{c} \times (|i| + 1)$ for $i \in [-n, n]$ with $\check{c} = [n(n + 3) + 1]^{-1}$. We use again the same obvious set of complete paths. This time, we need the non-trivial weight function $w(e) = Q(e)$. Using symmetry, the constant A

of Theorem 5 becomes

$$A = \check{c} \max_{\substack{e=(i,i+1) \\ i\in[0,n]}} \sum_{\substack{(j,k) \\ j \le i < k}} \left(\sum_{\ell=j}^{k-1} \frac{1}{|\ell|+1}\right)(|j|+1)(k+1)$$

$$\le 4\check{c}n^2(n+1)^2 \left(\sum_{\ell=0}^{n-1} \frac{1}{|\ell|+1}\right)$$

$$\le 4n^2(1+\log n).$$

where we have used $\check{c} \le (n+1)^{-2}$ to obtain the last inequality. Hence

$$\lambda \ge \frac{1}{4n^2(1+\log n)}.$$

For an upper bound, use the test function $f(i) = \text{sgn}(i)\log(1+|i|)$. This has $\check{\pi}(f) = 0$,

$$\|f\|_2^2 = 2\check{c}\sum_1^n [\log(1+i)]^2(i+1)$$

$$\ge \frac{3}{4}(\log[(n+1)/2])^2$$

and

$$\mathcal{E}(f,f) = 2\check{c}\sum_1^n |\log(1+i) - \log(i)|^2(i+1)$$

$$\le \frac{4}{(n+1)^2}\sum_1^n \frac{1}{i} \le \frac{4(1+\log n)}{(n+1)^2}.$$

Hence, for $n \ge 4$,

$$\lambda \le \frac{30}{(n+1)^2(1+\log n)}.$$

Example 2α (*The Metropolis chain for $\check{\pi}_\alpha$*) In this case, the behavior of the chain depends very much on α.

Theorem 9.5 *The spectral gap $\check{\lambda}(\alpha, n)$ of the Metropolis chain for $\check{\pi}_\alpha(i)$* $= \check{c}(\alpha, n) \times (i+1)^\alpha$, $\check{c}(\alpha, n) = (\sum_{-n}^n (|i|+1)^\alpha)^{-1}$ *on $[-n, n]$ satisfies:*

$$\frac{\check{c}(\alpha, n)}{4\zeta(\alpha, n)} \le \check{\lambda}(\alpha, n) \le \frac{16\check{c}(\alpha, n)}{\zeta(\alpha, n)}$$

where $\zeta(\alpha, n) = \sum_1^n \ell^{-\alpha}$. Hence

1. For $\varepsilon > 0$ fixed and $0 < \alpha < 1 - \varepsilon$, $\check{\lambda}(\alpha, n) \approx n^{-2}$

2. For $\alpha = 1$, $\frac{1}{4n^2(1+\log n)} \leq \check{\lambda} \leq \frac{30}{(n+1)^2(1+\log n)}$.

3. For $\varepsilon > 0$ *fixed and* $\alpha > 1 + \varepsilon$, $\check{\lambda}(\alpha, n) \approx \check{c}(\alpha, n) \approx n^{-(\alpha+1)}$.

Theorems 6,7 are coherent with the intuition that the needed running time is longer when π has two modes separated by a valley. It is interesting to observe that, in the linear case $\alpha = 1$, $\hat{\lambda}(n)$ and $\check{\lambda}(n)$ differ only by a factor of $\log n$. When α is larger, (e.g. $\alpha = 2$) the 'valley effect' becomes much stronger.

Proof. For the upper bound , we use the test function

$$f(i) = \text{sgn}(i) \sum_1^{|i|} \ell^{-\alpha}$$

This function satisfies $\check{\pi}_\alpha(f) = 0$, $\|f\|_2^2 \geq \frac{1}{8}\varsigma(\alpha, n)^2$ and $\mathcal{E}(f, f) \leq 2\check{c}(\alpha, n)\varsigma(\alpha, n)$. For the lower bound, use Theorem 5 with the weight function $w(e) = Q(e)$. Then the maximal w-length of a path is bounded by

$$\frac{4\varsigma(\alpha, n)}{\check{c}(\alpha, n)}.$$

Using this bound in Theorem 5 gives the announced result.   □

Theorem 7 can be abstracted as follows.

Theorem 9.6 *Assume that* π *is a probability measure on* $[-n, n]$ *such that* $\pi(i) = \pi(-i)$, π *is non-decreasing on* $[0, n]$ *and*

$$\pi([(n + 1)/2, n]) \geq \varepsilon, \quad \sum_0^{(n-1)/2} \pi(i)^{-1} \geq \eta \sum_0^n \pi(i)^{-1}.$$

Then the Metropolis chain of Section 1.2 corresponding to π *satisfies*

$$\frac{1}{4}\left(\sum_0^n \pi(i)^{-1}\right)^{-1} \leq \lambda \leq \frac{2}{\varepsilon\eta}\left(\sum_0^n \pi(i)^{-1}\right)^{-1}$$

Proof. Taking $w(e) = Q(e)$ in Theorem 5 gives the desired lower bound. For the upper bound, use the test function

$$f(i) = \text{sgn}(i) \sum_1^{|i|} \pi(\ell)^{-1}$$

It satisfies $\|f\|_2^2 \geq \varepsilon\eta \left(\sum_0^n \pi(\ell)^{-1}\right)^2$ and $\mathcal{E}(f, f) \leq 2\sum_0^n \pi(\ell)^{-1}$.   □

Remark. Let T be the first time the chain started at n hits 0. It is possible to compute exactly the mean of T:

$$E(T) = \sum_{i=1}^{n} \frac{1}{Q(i, i-1)} \left(\sum_{j=i}^{n} \pi(j) \right)$$

$$= 2 \sum_{1}^{n} \frac{1}{\pi(i-1)} \pi([i, n]).$$

Under the hypothesis of Theorem 8, we have

$$2\varepsilon\eta \sum_{0}^{n} \pi(i)^{-1} \leq E(T) \leq 2 \sum_{0}^{n} \pi(i)^{-1}.$$

Thus, $\lambda \approx 1/E(T)$.

9.2.4 Isoperimetry

It is well known that spectral gap bounds can be obtained through isoperimetric inequalities via the so called Cheeger's inequality introduced in a different context in Cheeger (1970). See Sinclair (1992-1993), Diaconis and Stroock (1991), Kannan (1994), and the earlier references given there. This section gives an account of this technique, emphasizing the fact that isoperimetric inequalities are ℓ^1 version of Poincaré inequalities. A useful reference in this spirit is Rothaus (1985). Diaconis and Stroock compare bounds using Theorem 5 and bounds using Cheeger's inequality. Most of the times bounds using Cheeger's inequality can be tightened by proving directly a Poincaré inequality. This can be seen on Examples 1–2α (see the end of this section).

Our main motivation for writing this section is to prepare for Section 3.3 where a different type of isoperimetric inequality is discussed. .

For simplicity we assume that (K, π) is a finite Markov chain satisfying

$$\pi(x)K(x, y) = \pi(y)K(y, x). \tag{9.3}$$

All the statements given below hold true without this hypothesis (and with only minor changes of notation in the proofs). Recall that, for any $e = (x, y) \in \mathcal{X} \times \mathcal{X}$, we have set

$$df(e) = f(y) - f(x), \quad Q(e) = \pi(x)K(x, y).$$

Here we will view $Q(e)$ as a probability measure on $\mathcal{X} \times \mathcal{X}$.

We need to define the 'boundary' ∂A of $A \subset \mathcal{X}$. We set

$$\partial A = \{e = (x, y) \in \mathcal{X} \times \mathcal{X} : x \in A, y \in A^c \text{ or } x \in A^c, y \in A\}.$$

Thus, the boundary is the set of all pairs connecting A and A^c. This is a rather large boundary, but only the portion that has positive Q-measure will be of interest to us. Define the isoperimetric constant of the chain (K, π) by

$$I = I(K, \pi) = \min_{\substack{A: \\ \pi(A) \leq 1/2}} \left\{ \frac{Q(\partial A)}{\pi(A)} \right\}. \tag{9.4}$$

The next result must be compare to the definition of the spectral gap λ given at (9.2). It shows that the constant $I(K, \pi)$ is nothing else than an ℓ^1 version of the spectral gap. The proof of Cheeger's inequality given below is based on this observation. See Lemma 10.

Lemma 9.3 *The constant I satisfies*

$$I = \min_f \left\{ \frac{\sum_e |df(e)| Q(e)}{\min_\alpha \sum_x |f(x) - \alpha| \pi(x)} \right\}.$$

Here the minimum is over all non-constant functions f.

Remark. It is well known and not to hard to prove that

$$\min_\alpha \sum_x |f(x) - \alpha| \pi(x) = \sum_x |f(x) - \alpha_0| \pi(x)$$

if and only if α_0 satisfies

$$\pi(f > \alpha_0) \leq 1/2 \text{ and } \pi(f < \alpha_0) \leq 1/2$$

i.e., if and only if α_0 is a median.

Proof. Let J be the right hand side in the equality above. To prove that $I \geq J$ it is enough to take $f = 1_A$ in the definition of J. Indeed,

$$\sum_e |d1_A(e)| Q(e) = Q(\partial A), \quad \sum_x 1_A(x) \pi(x) = \pi(A).$$

To prove that $J \geq I$, we start with the following observation. For any non-negative function f, set $F_t = \{f \geq t\}$ and $f_t = 1_{F_t}$. Then $f(x) = \int_0^\infty f_t(x) dt$,

$$\pi(f) = \int_0^\infty \pi(F_t) dt$$

and

$$\sum_e |df(e)| Q(e) = \int_0^\infty Q(\partial F_t) dt. \tag{9.5}$$

This is a discrete version of the so-called co-area formula of geometric measure theory. The proof is simple. Write

$$\sum_e |df(e)| Q(e) = 2 \sum_{\substack{e=(x,y) \\ f(y)>f(x)}} (f(y) - f(x)) Q(e)$$

$$= 2 \sum_{\substack{e=(x,y) \\ f(y)>f(x)}} \int_{f(x)}^{f(y)} Q(e) dt$$

$$= 2 \int_0^\infty \sum_{\substack{e=(x,y) \\ f(y) \geq t > f(x)}} Q(e) dt$$

$$= \int_0^\infty Q(\partial F_t) dt.$$

Armed with (9.5), we return to the proof of $J \geq I$. Given a function f, we find α such that $\pi(f > \alpha) \leq 1/2$, $\pi(f < \alpha) \leq 1/2$ and set $f_+ = (f - \alpha) \vee 0$, $f_- = -[(f - \alpha) \wedge 0]$. Then, $f_+ + f_- = |f - \alpha|$ and $|df(e)| \geq |df_+(e)| + |df_-(e)|$. Setting $F_{\pm,t} = \{x : f_\pm(x) \geq t\}$, using (9.5) and the definition of I, we get

$$\sum_e |df(e)| Q(e) \geq \sum_e |df_+(e)| Q(e) + \sum_e |df_-(e)| Q(e)$$

$$= \int_0^\infty Q(\partial F_{+,t}) dt + \int_0^\infty Q(\partial F_{-,t}) dt$$

$$\geq I \int_0^\infty (\pi(F_{+,t}) + \pi(F_{-,t})) dt$$

$$= I \sum_x (f_+(x) + f_-(x)) \pi(x)$$

$$= I \sum_x |f(x) - \alpha| \pi(x).$$

This proves that $J \geq I$. □

Lemma 9.4 (Cheeger's inequality) *The spectral gap λ and the isoperimetric constant I defined at (9.4) are related by*

$$\frac{I^2}{8} \leq \lambda \leq I.$$

Proof. For the upper bound use the test functions $f = 1_A$ in the definition of λ. For the lower bound, apply

$$\sum_e |df(e)| Q(e) \geq I \min_\alpha \sum_x |f(x) - \alpha| \pi(x)$$

to the function $f = |g - c|^2 \text{sgn}(g - c)$ where g is an arbitrary function and $c = c(g)$ is a median of g so that $\sum_x |f(x) - \alpha|\pi(x)$ is minimum for $\alpha = 0$. Then, for $e = (x, y)$,

$$|df(e)| \leq |dg(e)|(|g(x) - c| + |g(y) - c|)$$

and

$$\sum_e |df(e)|Q(e) = \sum_e |df(e)|Q(e)$$

$$\leq \sum_{e=(x,y)} |dg(e)|(|g(x) - c| + |g(y) - c|)Q(e)$$

$$\leq \left(\sum_e |dg(e)|^2 Q(e)\right)^{1/2} \times$$

$$\left(2\sum_{x,y}(|g(x) - c|^2 + |g(y) - c|^2)\pi(x)K(x, y)\right)^{1/2}$$

$$= (8\mathcal{E}(g, g))^{1/2}\left(\sum_x |g(x) - c|^2\pi(x)\right)^{1/2}$$

Hence

$$I\sum_x |g(x) - c|^2\pi(x) = I\min_\alpha \sum_x |f(x) - \alpha|\pi(x)$$

$$\leq \sum_e |df(e)|Q(e)$$

$$\leq (8\mathcal{E}(g, g))^{1/2}\left(\sum_x |g(x) - c|^2\pi(x)\right)^{1/2}$$

and

$$I^2\text{Var}_\pi(g) \leq I^2\sum_x |g(x) - c|^2\pi(x) \leq 8\mathcal{E}(g, g).$$

for all functions g. This proves the desired lower bound. $\square$

Remarks: 1. Compare with Diaconis and Stroock (1991), Section 3.C. There, the same result is proved by a slightly different argument.

2. It is easy to bound the isoperimetric constants of Examples 1–2α. We leave the details to the reader.

Proposition 9.1 *For the Metropolis chain corresponding to $\hat{\pi}_\alpha$ on $[-n, n]$ with $\alpha > 0$*

$$I \approx (1 + \alpha)/(n + \alpha).$$

For the Metropolis chain corresponding to $\check{\pi}_\alpha$ on $[-n, n]$ with $\alpha > 0$

$$I \approx c(\alpha, n) \approx \frac{(1 + \alpha)}{(n + \alpha)(n + 1)^\alpha}.$$

Here $\approx$ indicates that the ratio of the two quantities is bounded above and below by numerical constants. This family of examples illustrates the fact that bounds on the spectral gap based on Cheeger's inequality are often poor. Compare to Theorems 6-7.

9.3 Nash inequalities

9.3.1 Technical facts

Let us recall some well known facts from elementary functional analysis. Given an operator $K : A \to B$ between two Banach spaces A, B, set

$$\|K\|_{A \to B} = \sup_{f:\, \|f\|_A \leq 1} \|Kf\|_B.$$

When $A = \ell^p(\pi)$ and $B = \ell^q(\pi)$ we replace A, B by p, q in this notation. If $K^* : B^* \to A^*$ is the dual operator, then

$$\|K^*\|_{B^* \to A^*} \leq \|K\|_{A \to B}.$$

In the case of $\ell^p(\pi)$ spaces, there is equality because $(\ell^p)^* = \ell^q$ norms ℓ^p, i.e.,

$$\|f\|_p = \sup_{g:\, \|g\|_q \leq 1} \langle f, g \rangle$$

for $1/q + 1/p = 1$, $1 \leq p, q \leq +\infty$ and $\langle f, g \rangle = \sum f(x)\overline{g(x)}\pi(x)$. Returning to our Markov semigroup $H_t = e^{-t(I-K)}$, we can view it as acting on $\ell^p(\pi)$, $1 \leq p \leq +\infty$. It satisfies

$$\|H_t\|_{p \to p} \leq 1$$

by an application of Jensen's inequality. Further,

$$\|H_t\|_{p \to \infty} = \max_{x \in \mathcal{X}} \left(\sum_y [H_t(x, y)/\pi(y)]^q \pi(y) \right)^{1/q}$$

where $1/p + 1/q = 1$. Similarly,

$$\|H_t - \pi\|_{p \to \infty} = \max_{x \in \mathcal{X}} \left(\sum_y |[H_t(x, y)/\pi(y)] - 1|^q \pi(y) \right)^{1/q}$$

In particular we will use many times the identities

$$\|H_t - \pi\|_{2\to\infty} = \|H_t^* - \pi\|_{1\to2} = \max_x \|[H_t^x/\pi] - 1\|_2$$

and the inequality

$$\max_{x,y} |[H_t(x, y)/\pi(y)] - 1| = \|H_t - \pi\|_{1\to\infty}$$

$$\leq \|H_{t/2} - \pi\|_{1\to2}\|H_{t/2} - \pi\|_{2\to\infty}.$$

This last inequality follows from the fact that

$$(H_t - \pi)f = (H_{t/2} - \pi)(H_{t/2} - \pi)f$$

for all $t > 0$ and all functions f.

9.3.2 Nash's argument

Nash introduced his inequality in Nash (1958) to study the decay of the heat kernel of certain parabolic equations in Euclidean space. His argument is very similar to the proof of Corollary 4 and uses only formula (9.1) for the time derivative of $u(t) = \|H_t f\|_2^2$ which reads $u'(t) = -2\mathcal{E}(H_t f, H_t f)$. This formula shows that any functional inequality between the ℓ^2 norm of g and the Dirichlet form $\mathcal{E}(g, g)$ (for all g, thus $g = H_t f$) can be translated into a differential inequality involving u. For instance, assume that a certain Dirichlet form satisfies the Nash inequality

$$\forall\, g, \quad \|g\|_2^{2(1+2/d)} \leq C\mathcal{E}(g, g)\|g\|_1^{4/d}$$

where the norms are in $\ell^p(\pi)$, $p = 1, 2$. Then, fix a function f satisfying $\|f\|_1 \leq 1$ and set $u(t) = \|H_t f\|_2^2$. In terms of u, Nash's inequality reads

$$\forall\, t, \quad u(t)^{1+2/d} \leq -\frac{C}{2}u'(t),$$

since $\|f\|_1 \leq 1$ implies $\|H_t f\|_1 \leq 1$. Setting $v(t) = \frac{dC}{4}u(t)^{-2/d}$ one gets that $v'(t) \geq 1$, thus $v(t) \geq t$ (because $v(0) \geq 0$). Finally,

$$\forall\, t, \quad u(t) \leq \left(\frac{dC}{4t}\right)^{d/2}$$

This translates into

$$\forall\, t, \quad \|H_t\|_{1\to2} \leq \left(\frac{dC}{4t}\right)^{d/4}$$

The same argument applies to adjoint H_t^* and thus

$$\forall\, t, \quad \|H_t\|_{2\to\infty} \le \left(\frac{dC}{4t}\right)^{d/4}.$$

Finally, using $H_t = H_{t/2}H_{t/2}$, we get

$$\forall\, t, \quad \|H_t\|_{1\to\infty} \le \left(\frac{dC}{2t}\right)^{d/2}$$

which is the same as $H_t(x, y) \le \pi(y)\,(dC/2t)^{d/2}$.

The above argument is not very meaningful for finite Markov chains since the constant function $x \to 1$ shows that the hypothesis considered above can not be satisfied. It is not too hard to cope with this difficulty.

Theorem 9.7 *Assume that the finite Markov chain (K, π) satisfies*

$$\forall\, g, \quad \|g\|_2^{2(1+2/d)} \le C\left\{\mathcal{E}(g, g) + \frac{1}{T}\|g\|_2^2\right\}\|g\|_1^{4/d}. \qquad (9.6)$$

Then

$$\forall\, t \le T, \quad \|[H_t^x/\pi]\|_2 \le e\left(\frac{dC}{4t}\right)^{d/4}$$

and

$$\forall\, t \le T, \quad H_t(x, y) \le \pi(y)\left(\frac{dC}{2t}\right)^{d/2}$$

In practice, one wants to have (9.6) with a small values of d and $C \approx T$. Then, the conclusion of Theorem 12 is that after a time of order T the density $H_t^x(y)/\pi(y)$ is bounded. For instance if (9.6) holds true with $d = 4$ and $C = T$ then $H_T^x(y)/\pi(y) \le 4$.

Proof. Fix f satisfying $\|f\|_1 \le 1$ and set $u(t) = e^{-2t/T}\|H_t f\|_2^2$. Then

$$u'(t) = -2e^{-2t/T}\left(\mathcal{E}(H_t f, H_t f) + \frac{1}{T}\|H_t f\|_2^2\right).$$

Thus (9.6) implies

$$u(t)^{1+2/d} \le -\frac{C}{2}u'(t)$$

and Nash's argument yields

$$u(t) \le \left(\frac{dC}{4t}\right)^{d/2}$$

which implies

$$\|H_t\|_{1\to 2} \le e^{t/T}\left(\frac{dC}{4t}\right)^{d/4}$$

The announced results follow since

$$\max_x \|[H_t^x/\pi]\|_2 = \|H_t^*\|_{1\to 2} \le e^{t/T} \left(\frac{dC}{4t}\right)^{d/4}$$

by the same argument applied to H_t^*. □

One of the ideas behind Theorem 12 is that Nash inequalities are most useful to capture the behavior of the chain for relatively small time, i.e., time smaller than T. The next corollary uses a Nash inequality in the time interval $(0, t_0)$, $t_0 \le T$, and the spectral gap in the interval (t_0, t) to bound the rate of convergence.

Corollary 9.8 *If* (K, π) *satisfies* (9.6) *and has spectral gap* λ *then*

$$2\|H_t^x - \pi\|_{TV} \le \|[H_t^x/\pi] - 1\|_2 \le e^{1-c}$$

and

$$|H_{2t}(x, y) - \pi(y)| \le \pi(y) e^{2-2c}$$

for all $c > 0$, $0 < t_0 \le T$ *and*

$$t = t_0 + \frac{1}{\lambda}\left(\frac{d}{4}\log\left(\frac{dC}{4t_0}\right) + c\right).$$

Proof. Write $t = s + t_0$ with $t_0 \le T$ and

$$\|[H_t^x/\pi] - 1\|_2 \le \|(H_s - \pi)H_{t_0}\|_{2\to\infty}$$
$$\le \|H_s - \pi\|_{2\to 2}\|H_{t_0}\|_{2\to\infty}$$
$$\le e(dC/4t_0)^{d/4} e^{-\lambda s}.$$

The result easily follows. □

Again, in practice, a 'good' Nash inequality is (9.6) with a small value of d and $C \approx T$. Indeed, if (9.6) holds with, say $d = 4$ and $C = T$, then

$$\|[H_t^x/\pi] - 1\|_2 \le e^{1-c} \text{ for } t = T + c/\lambda.$$

Compare with Corollary 4.

Remark. Carlen *et al.* (1987) found that there is a converse to Nash's argument. Namely, if (K, π) satisfies (9.3) and

$$\forall t \le T, \quad \|H_t\|_{1\to 2} \le \left(\frac{C}{t}\right)^{d/4}$$

then

$$\|g\|_2^{2(1+2/d)} \le C'\left(\mathcal{E}(f, f) + \frac{1}{2T}\|f\|_2^2\right)\|f\|_1^{4/d}$$

with $C' = 2^{2(1+2/d)}C$.

Examples 1–2α are considered below in Section 3.5. But, before we can treat any example we have to learn how to prove a Nash inequality. Next section shows how Nash inequalities can be deduced from certain isoperimetric inequalities. This technique will easily apply to our Metropolis examples 1–2α. A more powerful technique based on the notions of moderate growth and local Poincaré inequalities is developed and applied in Diaconis and Saloff-Coste (1995a, 1996a).

9.3.3 Isoperimetric inequalities

For simplicity we assume as in Section 2.4 that (K, π) is a reversible chain (i.e., satisfies (9.3)). None of the results presented below really depends on this assumption. We use the notation introduced in Section 2.4. Our goal is to prove the following results.

Theorem 9.9 *Assume that* (K, π) *satisfies*

$$\pi(A)^{(d-1)/d} \le S\left(Q(\partial A) + \frac{1}{R}\pi(A)\right) \qquad (9.7)$$

for all $A \subset \mathcal{X}$ *and some constants* $d \ge 1$, $S, R > 0$. *Then*

$$\forall g, \quad \|g\|_{d/(d-1)} \le S\left(\sum_e |dg(e)|Q(e) + \frac{1}{R}\|g\|_1\right) \qquad (9.8)$$

and further

$$\forall g, \quad \|g\|_2^{2(1+2/d)} \le 16 S^2\left(\mathcal{E}(g, g) + \frac{1}{8R^2}\|g\|_2^2\right)\|g\|_1^{4/d}. \qquad (9.9)$$

Proof. Since $|d|g|(e)| \le |dg(e)|$ it suffices to prove the result for $g \ge 0$. Write $g = \int_0^\infty g_t dt$ where $g_t = 1_{G_t}$, $G_t = \{g \ge t\}$, and set $q = d/(d-1)$. Then

$$\|g\|_q \le \int_0^\infty \|g_t\|_q dt = \int_0^\infty \pi(G_t)^{1/q} dt$$

$$\le S\int_0^\infty \left(Q(\partial G_t) + \frac{1}{R}\pi(G_t)\right) dt$$

$$= S\left(\sum_e |dg(e)|Q(e) + \frac{1}{R}\|g\|_1\right).$$

The first inequality uses Minkowsky's inequality. The second inequality uses (9.7). The last inequality uses the co-area formula (9.5). This proves (9.8). It is easy to see that (9.8) is in fact equivalent to (9.7) (take $g = 1_A$).

To prove (9.9), we observe that

$$\sum_e |dg^2(e)|Q(e) \le [8\mathcal{E}(g,g)]^{1/2}\|f\|_2.$$

Indeed,

$$\sum_e |dg^2(e)|Q(e) = \sum_{e=(x,y)} |dg(e)|(|g(x)| + |g(y)|)Q(e)$$

$$\le \left(\sum_e |dg(e)|^2 Q(e)\right)^{1/2}$$

$$\times \left(2\sum_{x,y}(|g(x)|^2 + |g(y)|^2)\pi(x)K(x,y)\right)^{1/2}$$

$$= (8\mathcal{E}(g,g))^{1/2}\left(\sum_x |g(x)|^2\pi(x)\right)^{1/2}.$$

Thus, (9.8) applied to g^2 yields

$$\|g\|_{2q}^2 \le S\left([8\mathcal{E}(g,g)]^{1/2}\|g\|_2 + \frac{1}{R}\|g\|_2^2\right) \qquad (9.10)$$

with $q = d/(d-1)$. The Holder inequality

$$\|g\|_2 \le \|g\|_1^{1/(1+d)}\|g\|_{2q}^{d/(1+d)}$$

and (9.10) let us bound $\|g\|_2$ by

$$\left(S\left([8\mathcal{E}(g,g)]^{1/2}\|g\|_2 + \frac{1}{R}\|g\|_2^2\right)\right)^{1/[2(1+d)]}\|g\|_1^{1/(1+d)}.$$

We rise this to the power $2(1+d)/d$ and divide by $\|g\|_2$ to get

$$\|g\|_2^{(1+2/d)} \le S\left([8\mathcal{E}(g,g)]^{1/2} + \frac{1}{R}\|g\|_2\right)\|g\|_1^{2/d}.$$

This yields the desired result. □

9.3.4 Application to Examples 1–2α

In general, the isoperimetric inequality (9.7) is not easier to prove than (9.6) but in the case of Examples 1–2α, (9.7) turns out to be quite simple.

Theorem 9.10 *Assume $n \geq \alpha$. For the Metropolis chain corresponding to either $\pi = \hat{\pi}_\alpha$ or $\pi = \check{\pi}_\alpha$ on $[-n, n]$, we have*

$$\pi(A)^{\alpha/(1+\alpha)} \leq 20(n+1)\left(Q(\partial A) + \frac{1}{n+1}\pi(A)\right) \qquad (9.11)$$

for all $A \subset [-n, n]$.

Proof. First observe that it suffices to prove (9.11) when A is an interval because the right hand side is additive for disjoint sets with disjoint boundaries whereas the left hand side is subadditive. Indeed, $(a+b)^s \leq a^s + b^s$ for $a, b \geq 0$ and $0 \leq s = (d-1)/d \leq 1$.

We assume $n \geq \alpha$. We will use the following estimates:

$$\frac{(\ell+1)^{\alpha+1}}{1+\alpha} \leq \sum_0^\ell (i+1)^\alpha \leq 2(\ell+1)^{\alpha+1}$$

and

$$\frac{(n+1)^{\alpha+1}}{1+\alpha} \leq c(\alpha, n)^{-1} \leq 4\frac{(n+1)^{\alpha+1}}{1+\alpha}$$

where $c(\alpha, n) = \hat{c}(\alpha, n)$ or $\check{c}(\alpha, n)$. We will only gives the details for $\hat{\pi}_\alpha$. The argument for $\check{\pi}_\alpha$ is very similar. Set $A = [a, b]$. By symmetry, we can assume that $b \in [0, n]$.

- If $a \in [1, n]$ then

$$Q(\partial[a, b]) \geq 2\hat{\pi}_\alpha(a) \geq \frac{1+\alpha}{2}\frac{(n-a+1)^\alpha}{(n+1)^{\alpha+1}}$$

and

$$\hat{\pi}_\alpha([a, b]) \leq (1+\alpha)\left(\frac{n-a+1}{n+1}\right)^{\alpha+1}$$

Hence,

$$\hat{\pi}_\alpha([a, b])^{\alpha/(1+\alpha)} \leq 2(n+1)Q(\partial[a, b]).$$

- If $a \in [-n, 0]$, we can assume $a = -b$. Then

$$Q(\partial[-b, b]) = 4\hat{\pi}_\alpha(b+1) \geq (1+\alpha)\frac{(n-b)^\alpha}{(n+1)^{1+\alpha}}$$

and

$$(2c+1)\hat{\pi}_\alpha(c) \leq \hat{\pi}_\alpha([-b, b]) \leq (2b+1)\hat{\pi}_\alpha(0)$$

if $b \geq c$. Hence, for $b \geq \varepsilon(n+1)$,

$$\hat{\pi}_\alpha([-b, b]) \geq \frac{(1+\alpha)\varepsilon(1-\varepsilon)^\alpha}{2}.$$

Setting $\varepsilon = 1/4$ if $\alpha \leq 1$ and $\varepsilon = 1/[2(1+\alpha)]$ otherwise, we obtain

$$\hat{\pi}_\alpha([-b, b]) \geq \frac{1}{20}.$$

If instead $b < \varepsilon(n+1)$, then

$$Q(\partial[-b, b]) \geq \frac{(1+\alpha)(1-2\varepsilon)^\alpha}{2(n+1)} \geq \frac{1+\alpha}{20(n+1)}$$

for the same choice of ε. It follows that

$$(n+1)\left(Q(\partial[-b, b]) + \frac{1}{n+1}\hat{\pi}_\alpha([-b, b])\right) \geq \frac{1}{20}.$$

Finally,

$$\hat{\pi}_\alpha(A)^{\alpha/(1+\alpha)} \leq 20(n+1)\left(Q(\partial A) + \frac{1}{n+1}\hat{\pi}_\alpha(A)\right)$$

for all intervals and thus for all sets $A \subset [-n, n]$.   □

Thanks to Theorems 14, Theorem 15 yields

Corollary 9.11 *Assume $n \geq \alpha$. The Metropolis chains corresponding to $\pi = \hat{\pi}_\alpha$ and $\pi = \check{\pi}_\alpha$ on $[-n, n]$, satisfy*

$$\|f\|_2^{2(1+2/d)} \leq C(n+1)^2\left(\mathcal{E}(f, f) + \frac{1}{8(n+1)^2}\|f\|_2^2\right)\|f\|_1^{4/d}$$

where $d = 1 + \alpha$ and $C = 6400$.

Remark. Theorems 6 and 7 shows that, as expected, the Metropolis chains of Examples 1-α and 2-α have very different spectral gaps (for $\alpha \geq 1$) because of the 'valley' effect. In contrast, Corollary 16 states that the two chains satisfy the **same** Nash inequality (one can show that this Nash inequality is roughly optimal in both cases). This reflects the fact that the Nash inequality (9.6) captures the early behavior of the chain, not the asymptotic behavior. Indeed, from the point of view of a random walker, it takes a long time to distinguished between $\hat{\pi}_\alpha$ and $\check{\pi}_\alpha$ so it is not surprising that the early behaviors of the two chains are similar.

Now, Theorem 12 yields

Corollary 9.12 *Assume $n \geq \alpha$. Let M be the Metropolis chain corresponding to either $\pi = \hat{\pi}_\alpha$ or $\pi = \check{\pi}_\alpha$ on $[-n, n]$ and set $H_t = e^{-t(I-M)}$. Then, $\forall\, t \leq 8(n+1)^2$,*

$$\|[H_t^x/\pi]\|_2 \leq e\left(\frac{1600(1+\alpha)(n+1)^2}{t}\right)^{(1+\alpha)/4}$$

and

$$H_t(x, y) \leq \pi(y) \left(\frac{3200(1 + \alpha)(n + 1)^2}{t} \right)^{(1+\alpha)/2}$$

If we consider α fixed and n large, the content of this Corollary is the following: At time $t = 1$ the largest value of $H_t(x, y)/\pi(y)$ is approximately $n^{1+\alpha}$ but, after a running time t of order n^2, the largest value of $H_t(x, y)/\pi(y)$ becomes $O(1)$. There is no difference here between $\hat{\pi}_\alpha$ and $\check{\pi}_\alpha$ (see the remark after Corollary 16).

To obtain more precise results we use Corollarys 13–16 and Theorems 6–7. Together they yield

Corollary 9.13 *Assume $n \geq \alpha$. Let M be Metropolis chain corresponding to either $\pi = \hat{\pi}_\alpha$ or $\pi = \check{\pi}_\alpha$ on $[-n, n]$ and set $H_t = e^{-t(I-M)}$. Then*

$$2\|H_t^x - \pi\|_{TV} \leq \|[H_t^x/\pi] - 1\|_2 \leq e^{1-c}$$

and

$$|H_{2t}(x, y) - \pi(y)| \leq \pi(y)e^{2-2c}$$

for

$$t = \frac{8(n + 1)^2}{(1 + \alpha)^2} + \frac{1}{\lambda} \left(\frac{(1 + \alpha)}{4} \log \left(200(1 + \alpha)^3 \right) + c \right)$$

where λ is the corresponding spectral gap. Hence,

1. For the chain corresponding to $\hat{\pi}_\alpha$, a running time of order

$$\frac{n^2(1 + \log(1 + \alpha))}{1 + \alpha}$$

suffices to approximate stationarity.

2. For the chain corresponding to $\check{\pi}_\alpha$, a running time of order

$$\frac{n^{1+\alpha} \zeta(\alpha, n)(1 + \log(1 + \alpha))}{(1 + \alpha)}$$

suffices to approximate stationarity where $\zeta(\alpha, n) = \sum_1^n i^{-\alpha}$. Hence, a running time of order

$$\begin{array}{ll} n^2 & \text{if} \quad \alpha \ll 1 \\ n^2 \log n & \text{if} \quad \alpha = 1 \\ \frac{\log(1+\alpha)}{1+\alpha} n^{1+\alpha} & \text{if} \quad \alpha \gg 1 \end{array}$$

suffices to approximate stationarity.

Lower bounds can be obtained as follows. The Metropolis chains considered above satisfies (9.3). Thus λ is an eigenvalue of $(I - K)$. Let ϕ

be an associated eigenfunction normalized so that max $|\phi| = \phi(x_0) = 1$.
Then we have

$$\max_x 2\|H_t^x - \pi\|_{TV} = \max_{\|f\| \leq 1} \|H_t f - \pi(f)\|_\infty \geq H_t \phi(x_0) = e^{-t\lambda}.$$

This and Theorems 6-7 show that the upper bounds on running times
stated in Corollary 18 are sharp up to a factor of order $(1+\alpha)\log(1+\alpha)$.
In particular these bounds are sharp for bounded α.

9.4 Conclusion

This paper illustrates with simple examples how methods from geometry and functional analysis can be used in the quantitative study of
finite Markov chains. Another instance of the use of geometric ideas
in finite Markov chain theory can be found in Chung and Yau (1994).
The path technique of Section 2.2 has become one of the main tools
in this field. It has been used in concrete, complicated examples as, for
instance, the matching problem mentioned in the introduction.

Nash inequalities complement the basic spectral gap technique. The
spectral gap captures accurately the long time (asymptotic) behavior of
the chain. Nash inequalities are useful to describe the early behavior of
the chain, that is, what happens when the chain begins to explore the
state space around its starting point. The spectral gap has little to do
with this early behavior. Nash inequalities can be used to prove sharp
results in situations where the geometry stays under control (what we
call moderate growth in Diaconis and Saloff-Coste (1995a, b, 1996a)).
Roughly, they require the 'dimension' to stay bounded as the state
space grows.

There is an important technique that has not been mentioned in this
paper and which leans on another type of functional inequality called
logarithmic Sobolev inequality. Diaconis and Saloff-Coste (1996b) survey this technique and develop its application to finite Markov chains.
Logarithmic Sobolev inequalities are useful in high dimensional problems.

There is a clear increasing order of difficulty in proving Poincaré,
Nash, logarithmic Sobolev inequalities with good constants. This reflects the increasing difficulty in obtaining sharper and sharper running
time upper bounds. A good Poincaré inequality often suffices to prove
rapid mixing. Nash and log-Sobolev inequalities are useful to prove
sharper bounds that reflect more closely the behavior of the chain under consideration.

Finally, another helpful idea related to these techniques and which has not been discussed here is the comparison of Markov chains. Comparison allows us to use our understanding of a nice auxiliary chain in order to study another chain of interest. This idea seems essential when trying to apply the set of techniques presented here to complex examples. See Diaconis and Saloff-Coste (1993, 1995a, b, 1996a, b).

References

Carlen E., Kusuoka S. and Stroock D. (1987). Upper bounds for symmetric Markov transition functions. *Ann. Inst. H. Poincaré, Proba. Stat.* **23**, 245-287.

Cheeger J. (1970) A lower bound for the smallest eigenvalue of the Laplacian. Problems in Analysis, Symposium in Honor of S. Bochner. Princeton University Press. 195-199.

Chung F. and Yau S.-T. (1994) A Harnack inequality for homogeneous graphs and subgraphs. *Communication in Analysis and Geometry*, **2** 627-640.

Diaconis P. and Efron B. (1985) Testing for independence in a two-way table: new interpretations of the chi-square statistic (with discussion), Ann. Stat. 13, 845-913.

Diaconis P. and Gangolli A. (1995) Rectangular arrays with fixed margins. In Discrete Probability and Algorithms, (Aldous *et al.*, ed.) 15-41. The IMA volumes in Mathematics and its Applications, Vol. 72, Springer-Verlag.

Diaconis P. and Holmes S. (1995) Three Examples of Monte-Carlo Markov Chains: at the Interface between Statistical Computing, Computer Science and Statistical Mechanics. In Discrete Probability and Algorithms, (Aldous *et al.*, ed.) 43-56. The IMA volumes in Mathematics and its Applications, Vol. 72, Springer-Verlag.

Diaconis P. and Saloff-Coste L. (1993) Comparison theorems for reversible Markov chains. *Ann. Appl. Prob.* **3** 696-730.

Diaconis P. and Saloff-Coste L. (1995a) Random walk on contingency tables with fixed row and column sums. Preprint.

Diaconis P. and Saloff-Coste L. (1995b) What do we know about the Metropolis algorithm? J. C. S. S. To appear in 1997.

Diaconis P. and Saloff-Coste L. (1996a) Nash inequalities for finite Markov chains. *J. Th. Prob.* **9**, 459-510.

Diaconis P. and Saloff-Coste L. (1996b) Logarithmic Sobolev inequalities for finite Markov chains. *Ann. Appl. Prob.* To appear.

Diaconis P. and Stroock D. (1991) Geometric bounds for eigenvalues of Markov chains. *Ann. Appl. Prob.*, **1**, 36-61.

Fill J. (1991) Eigenvalue bounds on convergence to stationarity for nonreversible Markov chains, with an application to the exclusion process. *Ann. Appl. Prob.*, **1**, 36-61.

Horn R. and Johnson C. (1989) Topics in matrix analysis. Cambridge University Press.

Jerrum M. and Sinclair A. (1993) Polynomial time approximation algorithms for the Ising model, *SIAM Journal of Computing*, **22** 1087-1116.

Jerrum M. and Sinclair A. (1997) The Markov chain Monte Carlo method: an approach to approximate counting and integration. In *Approximation algorithms for NP-hard problems* D.S. Hochbaum (Ed.), PWS Publishing, Boston.

Kahale N. (1995) A semidefinite bound for mixing rates of Markov chains. To appear in Random Structures and Algorithms, 1997.

Kannan R. (1994) Markov chains and polynomial time algorithms. *Proceedings of the 35th IEEE Symposium on Foundations of Computer Science*, Computer Society Press, 656-671.

Metropolis N., Rosenbluth A., Rosenbluth M., Teller A. and Teller E. (1953) Equations of state calculations by fast computing machines. *J. Chem. Phys.* **21**, 1087-1092.

Nash J. (1958) Continuity of solutions of parabolic and elliptic equations. *Amer. J. Math.* 80, 931-954.

Rothaus O. (1985) Analytic inequalities, Isoperimetric inequalities and logarithmic Sobolev inequalities. *J. Funct. Anal.*, 64, 296-313.

Sinclair A. (1992) Improved bounds for mixing rates of Markov chains and multicommodity flow. *Combinatotics, Probability and Computing* 1, 351-370.

Sinclair A. (1993) Algorithms for random generation and counting: A Markov chain approach. Progr. Theoret. Comp. Sci., Birkhaüser, Basel and Boston.

Index